Pollution Mechanisms and Source Apportionment of Typical Pollutants in Aquatic Environments

Pollution Mechanisms and Source Apportionment of Typical Pollutants in Aquatic Environments

Guest Editor

Qianqian Zhang

MDPI

Basel • Beijing • Wuhan • Barcelona • Belgrade • Novi Sad • Cluj • Manchester

Guest Editor
Qianqian Zhang
Institute of Hydrogeology
and Environmental Geology
Chinese Academy of
Geological Sciences
Shijiazhuang
China

Editorial Office
MDPI AG
Grosspeteranlage 5
4052 Basel, Switzerland

This is a reprint of the Special Issue, published open access by the journal *Water* (ISSN 2073-4441), freely accessible at: https://www.mdpi.com/journal/water/special_issues/X5455QK74R.

For citation purposes, cite each article independently as indicated on the article page online and as indicated below:

Lastname, A.A.; Lastname, B.B. Article Title. *Journal Name* **Year**, *Volume Number*, Page Range.

ISBN 978-3-7258-8241-0 (Hbk)
ISBN 978-3-7258-8242-7 (PDF)
https://doi.org/10.3390/books978-3-7258-8242-7

Contents

About the Editor

Qianqian Zhang

Qianqian Zhang is a researcher and doctoral supervisor at the Institute of Hydrogeology and Environmental Geology, Chinese Academy of Geological Sciences. He earned his Ph.D. in Ecology from the Research Center for Eco-Environmental Sciences, Chinese Academy of Sciences in 2013, and subsequently completed postdoctoral research at the Institute of Geology and Geophysics, Chinese Academy of Sciences from 2020 to 2023. His research focuses on isotope geochemistry, and his particular interests comprise the transport and transformation mechanisms of biogenic elements in aquatic environments, the source apportionment of typical pollutants in soil–water systems, and the characteristics and control technologies of urban non-point source pollution. He has led more than 20 national and provincial research projects, including the General and Youth programs of the National Natural Science Foundation of China, the General projects of the China Postdoctoral Science Foundation, and the General and Youth projects of the Natural Science Foundation of Hebei Province. He has published over 60 peer-reviewed papers in domestic and international core journals, including 30 as the first or corresponding author in journals such as *Landscape and Urban Planning, Environmental Pollution, Journal of Cleaner Production, Journal of Environmental Management, Chemosphere, Journal of Environmental Sciences,* and *Journal of Contaminant Hydrology*. He has also contributed to one monograph, holds four authorized national patents, and has been recognized as a Class C High-Level Talent of Fujian Province and Xiamen City, as well as a member of the Third Level of the "Three-Three-Three Talent Project" of Hebei Province.

Preface

This Reprint, titled "Pollution Mechanisms and Source Apportionment of Typical Pollutants in Aquatic Environments", brings together eleven peer-reviewed papers addressing critical challenges in relation to identifying pollutant sources, understanding transport and transformation processes, and supporting science-based water quality management.

The scope of this Reprint spans multiple aquatic settings—including greenhouse agricultural catchments, industrial areas, overexploited alluvial fans, urbanized regions, the agricultural tributaries of large water diversion projects, arid basins, and river systems with distinct seasonal dynamics. The contributions employ a wide range of advanced methodologies: fluorescence spectroscopy combined with machine learning (EEM-PARAFAC, 2D-COS, and SOM), multi-tracer hydrochemistry and stable isotopes ($\delta^{18}O$, and δD), multivariate statistics (PCA), positive matrix factorization (PMF), GIS-based spatial analysis, high-throughput metagenomic sequencing, and conventional pollutant monitoring (nutrients, sulfate, microplastics, BOD, TSS, and metals). The findings reveal that surface–groundwater interactions, irrigation practices, industrial wastewater, domestic sewage, evaporite dissolution, sulfide oxidation, agricultural fertilizers, and even political/institutional factors jointly shape the "source–pathway–sink" behavior of pollutants. Moreover, several studies provide quantitative natural background levels and contamination thresholds, which are essential for regulatory frameworks.

The motivation for compiling this Reprint stems from the recognition that traditional single-method approaches often fail under complex multi-source and non-stationary pollution scenarios. The integrated paradigms demonstrated here—combining hydrochemical, isotopic, statistical, and machine learning techniques—offer practical solutions for precise source apportionment and risk assessment. These advances are crucial for designing targeted remediation, early warning systems, and sustainable water resource policies.

This Reprint is intended for researchers, environmental professionals, and policymakers in the fields of water environment science, analytical chemistry, ecotoxicology, environmental engineering, and hydrology. We hope it will serve as a valuable reference for advancing the mechanistic understanding of pollutant dynamics and for strengthening science-based water quality management worldwide.

Qianqian Zhang
Guest Editor

water

Editorial

Pollution Mechanisms and Source Apportionment of Typical Pollutants in Aquatic Environments: Current Insights and Future Directions

Qianqian Zhang

Hebei and China Geological Survey Key Laboratory of Groundwater Remediation, Institute of Hydrogeology and Environmental Geology, Chinese Academy of Geological Sciences, Shijiazhuang 050061, China; zhangqianqian@mail.cgs.gov.cn

1. Introduction

As a critical component of the Earth's ecosystem, aquatic environments are under sustained pressure from multiple pollution sources, including industrial and agricultural activities, urban runoff, wastewater treatment plant effluent, and atmospheric deposition [1]. The residues, transport, and transformation of typical pollutants (e.g., heavy metals, persistent organic pollutants, microplastics, nutrients (N, P), and dissolved organic matter (DOM)) in water bodies have posed potential threats to aquatic ecological safety and human health [2–4]. Although substantial efforts have been made over the past decades to prevent and control water pollution, significant knowledge gaps remain regarding the whole-process mechanisms of pollutant "source–pathway–sink", particularly the urgent need for breakthroughs in precise source apportionment under complex pollution scenarios.

Accurate source apportionment of pollutants is the logical starting point for "targeted remediation" and "risk prevention and control" of aquatic environments. Traditional source identification methods often rely on single chemical signatures or statistical models, which struggle to address multi-source superposition, non-stationary emissions, and interactions among coexisting pollutants. In recent years, the rapid development of emerging approaches—such as multivariate statistical techniques, the Positive Matrix Factorization (PMF) model, stable isotope analysis, high-resolution mass spectrometry fingerprinting, and machine learning—has provided unprecedented tools for resolving pollution sources and transformation pathways from the molecular to the watershed scale [5–8]. However, how to standardize these advanced methods, reduce their cost, and apply them effectively in routine regulatory frameworks remains a core challenge.

This Special Issue focuses on the theme "Pollution Mechanisms and Source Apportionment of Typical Pollutants in Aquatic Environments". It brings together a collection of original research articles and reviews. The contributions cover, for example, the sources and transport of DOM in surface water and groundwater, source apportionment of groundwater pollution based on the PMF model in a rapidly urbanized region; and the spatiotemporal distribution and removal efficiency of microplastics in wastewater treatment plants. These findings not only deepen our mechanistic understanding of the environmental behavior of pollutants but also demonstrate the applicability of multi-technique integrated source apportionment paradigms in real-world scenarios.

Looking forward, establishing an interdisciplinary framework that is "mechanism-driven, data-enabled, and regulation-oriented" will be key to solving complex water pollution challenges. This Special Issue aims to provide a platform for colleagues in water environment science, analytical chemistry, ecotoxicology, and environmental engineering to exchange recent advances and to inspire original explorations on early pollution warning, source control, and ecological restoration. We believe that the papers presented here will offer strong scientific support for source-level pollution reduction and aquatic ecosystem health protection and will also contribute to the development of more precise and forward-looking water quality management policies.

2. Main Contribution of This Special Issue

After a rigorous peer-review process, eleven papers have been selected for publication in this Special Issue. The contributions and implications of them are discussed below.

Wang et al. (Contribution 1) investigate the dynamic characteristics of dissolved organic matter (DOM) in surface water and groundwater within an intensive greenhouse agriculture catchment in northern China. The authors employed an integrated approach combining EEM-PARAFAC, two-dimensional correlation spectroscopy (2D-COS), self-organizing map (SOM) analysis, hydrochemical measurements, and stable water isotopes ($\delta^{18}O$ and δD). The study found that irrigation pumping drives surface–groundwater mixing, with microbial degradation of organic fertilizers and domestic wastewater being important DOM sources. 2D-COS revealed that terrestrial humic substances preferentially change parallel to the river direction, while microbial humic substances are more sensitive vertically. SOM analysis validated the relationships among DOM components and their correlations with inorganic ions. The findings suggest that combining DOM parameters with geochemical indicators and multi-technique analyses can effectively elucidate DOM sources and migration pathways, providing scientific support for sustainable water resource management in greenhouse agricultural systems.

Wang et al. (Contribution 2) investigate the pollution mechanisms and driving factors of groundwater quality in typical industrial areas, taking Zibo City, Shandong Province, as the study area. The authors collected phreatic and karst confined groundwater samples during both dry and flood seasons and employed Piper diagrams, Gibbs plots, ion ratios, and principal component analysis (PCA). The results show that pore phreatic water exhibits higher exceedance rates of Na^+, Cl^-, and NO_3^- than karst-confined water. The main hydrochemical type is $HCO_3{\cdot}SO_4{\cdot}Ca$, but pore water shifts to $HCO_3{\cdot}SO_4{\cdot}Cl{\cdot}Ca$ in the dry season. Groundwater quality is primarily controlled by water–rock interactions and industrial activities, with seasonal and aquifer-specific differences. The findings highlight the need for customized strategies to mitigate groundwater quality decline in industrial regions.

Wang et al. (Contribution 3) investigate the spatial distribution and sources of sulfate (SO_4^{2-}) in groundwater of the Hutuo River alluvial fan, a typical overexploited region in northern China. The authors employed hydrochemical analysis, multivariate statistics (PCA), and geostatistical techniques. The results show that sulfate concentrations are significantly higher in the river valley plain (175 mg/L) and upper alluvial fan (169 mg/L) than in the central fan (77.3 mg/L). Groundwater depth is identified as a critical factor controlling sulfate distribution ($p < 0.001$). The main sources of sulfate include industrial wastewater, domestic sewage, evaporite dissolution, and sulfide mineral oxidation, with overexploitation accelerating water–rock interactions. The study provides a spatially explicit framework for sulfate pollution control in overexploited aquifers.

Woods (Contribution 4) investigates why U.S. state regulators allow some facilities to discharge more water pollution than similar facilities elsewhere. The author analyzes effluent limits for biochemical oxygen demand (BOD) and total suspended solids (TSS) in National Pollutant Discharge Elimination System (NPDES) permits, using GIS, demo-

https://doi.org/10.3390/w18101157

graphic data, and state political variables. The results show that facilities discharging into interstate rivers receive significantly higher (more lenient) discharge limits, supporting the environmental free-riding hypothesis. Higher poverty rates are associated with less stringent BOD limits, raising environmental justice concerns. More liberal state governments (for BOD) and stronger environmental interest groups (for TSS) lead to tighter permits. These findings indicate that subnational political factors and free-riding incentives shape water quality stringency permits.

Chen et al. (Contribution 5) investigate the spatiotemporal distribution and removal efficiency of microplastics (MPs) in a wastewater treatment plant (WWTP) in Zhengzhou, China. The authors collected influent, process effluent, final effluent, and sludge samples across summer, autumn, and winter, using density separation, H_2O_2 digestion, and Raman spectroscopy. Results show total MP removal efficiencies of 86% (summer), 81% (autumn), and 73% (winter), with fragments and granules dominating (>80%). Polypropylene (PP, 42.6%) and polyethylene terephthalate (PET, 31.8%) are the main polymers. Secondary treatment removed more MPs than tertiary treatment, and residual sludge retained substantial MP loads (14.2–29.1 n/g). The study highlights that WWTPs remain a significant source of MPs to the environment, urging improved management strategies.

Sun et al. (Contribution 6) investigate river hydrograph separation in a hilly catchment with diverse land uses in Southwest China. The authors propose a novel multi-tracer approach combining the ratio of two conservative fluorescent DOM components (C1/C2), ion ratios (Ca^{2+}/NO_3^-, Ca^{2+}/K^+, and Mg^{2+}/NO_3^-), and oxygen-18 ($\delta^{18}O$), using the MixSIAR model. For a rain event with the longest preceding dry period, a set of four tracers successfully distinguished up to eight water sources at the catchment outlet and five sources in a nested agricultural–forestry drainage area. Drier antecedent soil moisture favored more qualified tracers. The study provides a useful tool for prioritizing water source areas for pollution protection in complex landscapes.

Wang et al. (Contribution 7) investigate natural background levels (NBLs) and contamination mechanisms of shallow groundwater in an overexploited region of Xingtai City, North China Plain. Sixty groundwater samples were analyzed using Piper diagrams, cumulative-probability statistics, contamination indices, and principal component analysis (PCA). The dominant hydrochemical types are $HCO_3{\cdot}Na$ and $SO_4{\cdot}Cl{\cdot}Na$. NBLs for Na^+, Ca^{2+}, Mg^{2+}, Cl^-, SO_4^{2-}, and NO_3^- are 32.3, 34.1, 17.8, 46.2, 66.4, and 0.886 mg/L, with corresponding thresholds of 116, 54.6, 33.9, 248, 258, and 44.7 mg/L. Based on thresholds, 56.7% of sites are anthropogenically contaminated. PCA reveals that groundwater over-extraction, industrial activities, water–rock interaction, agricultural fertilization, and domestic sewage are key contamination drivers. The study provides a scientific basis for pollution control in overexploited aquifers.

Wang et al. (Contribution 8) investigate groundwater pollution sources in Shenzhen, a highly urbanized Chinese city. The authors integrated multivariate statistics, the Positive Matrix Factorization (PMF) model, and GIS spatial interpolation. Results show that groundwater is weakly acidic to neutral, with high exceedance rates for pH, NH_4^+, COD, Mn, and Fe (up to 67.1%). The dominant hydrochemical type is $HCO_3{\cdot}Ca{\cdot}Na$, controlled primarily by silicate weathering. The PMF model identified three pollution sources: domestic/industrial wastewater (43.9%), water–rock interaction (37.0%), and agricultural fertilizers (19.1%). Urban groundwater quality is mainly affected by wastewater discharge, while agricultural areas suffer from excessive fertilizer use. The study provides a scientific basis for groundwater protection in rapidly urbanizing regions.

Yang et al. (Contribution 9) investigate water quality characteristics and solute sources of the Jinqian River, an agricultural tributary of the Danjiangkou Reservoir source area for

 https://doi.org/10.3390/w18101157

China's South-to-North Water Diversion Project. The authors collected river water across dry, normal, and flood seasons, using hydrochemical analysis and multivariate statistics. Results show slightly alkaline pH (7.55–8.30), with total nitrogen exceeding Class III surface water standards in all seasons, indicating significant human impact. Ionic sources are Cl^- and Na^+ from halite dissolution; Ca^{2+}, Mg^{2+}, and HCO_3^- from carbonate-silicate weathering; and SO_4^{2-} from evaporite dissolution. Principal component analysis identifies four controlling factors: agricultural fertilizers, halite weathering, evaporite dissolution, and domestic sewage. The study provides a quantitative basis for managing nutrients and salts in agricultural rivers.

Liu et al. (Contribution 10) investigate groundwater hydrochemical evolution and pollution sources in the Datong River Basin, an arid region of northwestern China. Thirty-seven groundwater samples were analyzed using hydrochemical methods and the Positive Matrix Factorization (PMF) model. The dominant water type is HCO_3-Ca(Mg), controlled by silicate and carbonate weathering. The water quality index classifies 70.3% of samples as excellent. PMF resolves three sources: natural weathering (19.2%, enriched in Mn, Na^+, Cl^-), a mixed source (67.5%, domestic sewage + agriculture + rock weathering), and industrial wastewater (13.3%, dominated by Fe). The mixed source contributes most major ions and COD, while the industrial source accounts for 75.7% of Fe. The study provides a basis for groundwater management in arid basins.

Wu et al. (Contribution 11) investigate microbial driving mechanisms of phosphorus cycling in the Chishui River, southwestern China. Water quality parameters and metagenomic analysis were conducted on samples collected in summer and autumn. Results show higher total phosphorus and soluble reactive phosphorus in summer, while manganese peaks in autumn and correlates positively with SRP, suggesting sediment release. Functional genes for phosphorus cycling differ significantly between seasons: summer enriches high-affinity phosphate transporter (pstB), inorganic phosphate dissolution (pqqC), and polyphosphate decomposition (ppx); autumn enriches phosphonate (phn) and glycerophosphate (ugpQ) genes. Proteobacteria dominate the microbial community. Environmental factors (Fe, DO, DOC, and pH) jointly regulate functional gene composition. The study highlights microbial regulation of phosphorus biogeochemistry in river ecosystems.

3. Conclusions

The papers in this Special Issue highlight key advances in understanding pollution mechanisms and source apportionment in aquatic environments. First, water quality is shaped by the interplay of natural processes and human activities. Intensive greenhouse agriculture drives surface–groundwater mixing and alters dissolved organic matter (Contribution 1). Rock weathering and industrial activities control groundwater quality in industrial areas (Contribution 2); over-exploitation accelerates sulfate release (Contribution 3); and in arid catchments, natural weathering dominates, but sewage, fertilizers, and industry add significant pollution (Contribution 10).

Second, advanced analytical tools enable robust source identification. Combining EEM-PARAFAC, 2D-COS, SOM, and stable isotopes unravels DOM transformation pathways (Contribution 1). A Bayesian mixing model using DOM fluorescence ratios, ion ratios, and $\delta^{18}O$ distinguishes up to eight water sources in hilly catchments (Contribution 6). Metagenomic sequencing reveals seasonal shifts in phosphorus-cycling genes (e.g., pstB and phn) and their environmental drivers (Contribution 11).

Third, long-term monitoring captures temporal variability critical for adaptive management. Seasonal monitoring of a wastewater treatment plant shows microplastic removal peaks in summer (86%) and is lowest in winter (73%), with sludge retaining substantial loads (Contribution 5). In an agricultural river, total nitrogen exceeds standards year-round, with solute dynamics following dilution, flushing, and reservoir-modulated patterns (Contribution 9).

Other important findings include political and regulatory factors: U.S. interstate river discharges receive more lenient permits (free-riding); higher poverty correlates with higher BOD limits (environmental justice); and liberal governments enforce stricter permits (Contribution 4). Determining natural background levels and thresholds in over-exploited aquifers shows 56.7% of sites are anthropogenically contaminated (Contribution 7). PMF quantification in urbanized Shenzhen attributes pollution to domestic/industrial wastewater (43.9%), water–rock interaction (37.0%), and fertilizers (19.1%) (Contribution 8). Collectively, these studies demonstrate that an integrated understanding of natural processes, human pressures, seasonal dynamics, and advanced methodologies is essential for effective water quality management and pollution control.

Funding: This research received no external funding.

Conflicts of Interest: The authors declare no conflicts of interest.

List of Contributions

1. Wang, H.; Song, S.; Xu, W.; Yue, F.-J. Sources and Transport of Dissolved Organic Matter (DOM) in Surface and Groundwater Within a Dominated Greenhouse Agriculture Catchment: Insights from Multi-Tracer. *Water* **2025**, *17*, 2681. https://doi.org/10.3390/w17182681.
2. Wang, L.; Wang, Q.; Zheng, D. Study on the Pollution Mechanism and Driving Factors of Groundwater Quality in Typical Industrial Areas of China. *Water* **2025**, *17*, 1420. https://doi.org/10.3390/w17101420.
3. Wang, L.; Wang, Q.; Li, W.; Liu, Y.; Zhang, Q. A New Insight into Sulfate Contamination inOver-Exploited Groundwater Areas: Integrating Multivariate and Geostatistical Techniques. *Water* **2025**, *17*, 1530. https://doi.org/10.3390/w17101530.
4. Woods, N.D. Spillovers and State Politics: Explaining Variation in U.S. Water Quality Permit Stringency. *Water* **2025**, *17*, 1569. https://doi.org/10.3390/w17111569.
5. Chen, X.; Li, Y.; Lu, K.; Liang, X.; Jin, K.; Ao, T.; Zhang, L.; Lv, J.; Dou, Y.; Duan, X. Spatiotemporal Distribution Characteristics and Removal Effi ciency of Microplastics in a Wastewater Treatment Plant. *Water* **2025**, *17*, 2614. https://doi.org/10.3390/w17172614.
6. Sun, Z.-X.; Ao, Y.-T.; Cui, J.-F.; Li, X.-Y.; Tang, X.-Y.; Cheng, J.-H.; Chen, L. Combining Fluorescent Organic Substances, Ions, and Oxygen-18 to Trace Diverse Water Sources of River Flow in a Hilly Catchment. *Water* **2025**, *17*, 1222. https://doi.org/10.3390/w17081222.
7. Wang, Q.; Wang, M.; Li, Y.; Guo, B.; Li, H.; Liu, Y.; Zhao, L.; Ma, C.; Yuan, Z. Study on Natural Background Levels and Mechanisms of Groundwater Contamination in an Overexploited Aquifer Region: A Case Study of Xingtai City, North China Plain. *Water* **2025**, *17*, 2836. https://doi.org/10.3390/w17192836.
8. Wei, Y.; Li, Y.; Zhang, L.; Liu, C.; Meng, Q.; Yin, J.; Wang, L. A Study on the Hydrochemical Evolution Property and Pollution Source Attribution of Groundwater in Highly Urbanized Areas: A Case Study of Shenzhen City. *Water* **2025**, *17*, 2945. https://doi.org/10.3390/w17202945.
9. Yang, C.; Qu, Z.; Shi, X.; Yang, L.; Yang, N.; Yang, F.; Zhang, Q. Key Controlling Factors and Sources of Water Quality in Agricultural Rivers: A Study on the Water Source Area for the South-to-North Water Transfer Project. *Water* **2025**, *17*, 3111. https://doi.org/10.3390/w17213111.
10. Liu, T.; Kang, J.; Yu, Y.; Qi, Y.; Zhang, Z. Hydrochemical Evolution and Pollution Source Apportionment of Groundwater in Arid Regions: A Case Study of the Datong River Basin, Northwest China. *Water* **2026**, *18*, 105. https://doi.org/10.3390/w18010105.
11. Wu, J.; Xiao, Y.; Wu, Q.; Li, Q.; He, Y.; Tang, Y.; Wang, J. Seasonal and Spatial Variations in Riverine Functional Genes Related to Phosphorus Cycling and Their Responses to Environmental Factors in the Chishui River Basin. *Water* **2026**, *18*, 456. https://doi.org/10.3390/w18040456.

References

1. Zhai, T.L.; Zhang, Q.Q.; Wang, L.; Wang, H.W. Temporal and spatial variations hydrochemical components and driving factors in baiyangdian lake in the northern plain of China. *J. Groundw. Sci. Eng.* **2024**, *12*, 293–308. [CrossRef]
2. Zhu, X.; Yang, F.; Pang, Q.; Peng, F.; Xu, B.; Wang, L.; Xie, L.; Zhang, W.; Tian, L.; Hou, J.; et al. Fluvial dissolved organic matter quality modulates microbial nitrate transformation: Enhanced denitrification under low carbon-to-nitrate ratio. *Environ. Sci. Technol.* **2025**, *59*, 23456–23465. [CrossRef] [PubMed]
3. Wu, Y.; Li, Y.; Men, Y.; Zhu, Z.; Sun, Y.; Song, C. Spatial optimization of Best Management Practices (BMPs) for nonpoint source pollution mitigation in agricultural watersheds. *J. Hydrol.* **2025**, *661*, 133739. [CrossRef]
4. Mishra, S.; Ren, Y.; Sun, X.; Lian, Y.; Singh, A.K.; Sharma, N. Microplastics pollution in the asian water tower: Source, environmental distribution and proposed mitigation strategy. *Environ. Pollut.* **2024**, *356*, 124247. [CrossRef] [PubMed]
5. Wang, Q.; Ou, X.; Wu, C.; Yao, S.; Huang, S.; Zhu, C.; Liao, Z.; Wang, K.; Zhang, S.; Chen, J. Machine learning-driven PMF modeling for accurate and objective source identification of VOCs. *Atmos. Pollut. Res.* **2026**, *17*, 102721. [CrossRef]
6. Li, C.; Chen, Z.; Huang, L.; Wen, W.; Guo, W.; Kong, L.; Chen, L.; Chen, G. Carbon-nitrogen coupling in plateau lake ecosystems: Dual-isotope fingerprints of climate warming and anthropogenic intensification. *Water Res.* **2026**, *289*, 124793. [CrossRef] [PubMed]
7. Zhang, Q.; Liu, W.; Sun, H.; Wang, H.; Wang, L.; Zhang, J.; Xu, Z. Cascade reservoir regulations on nitrogen source and transformation in the Tibetan Plateau river: Constraints from high frequency data of Lancang river. *J. Hydrol.* **2025**, *650*, 132563. [CrossRef]
8. Yu, C.; Zhang, Z.P.; Zhang, Y.L.; Zhang, H.X.; Li, X.H. Assessment of the natural background levels of ammonia nitrogen and chloride in alluvial fan groundwater using an improved pre-selection method: Implications for groundwater management. *J. Groundw. Sci. Eng.* **2026**, *14*, 148–164. [CrossRef]

https://doi.org/10.3390/w18101157

water

MDPI

Article

Combining Fluorescent Organic Substances, Ions, and Oxygen-18 to Trace Diverse Water Sources of River Flow in a Hilly Catchment

Zhi-Xiang Sun [1,2], Yan-Ting Ao [3,4], Jun-Fang Cui [1,*], Xiao-Yu Li [3,4], Xiang-Yu Tang [3,4,*], Jian-Hua Cheng [3,4] and Lu Chen [1,2]

1 Institute of Mountain Hazards and Environment, Chinese Academy of Sciences, Chengdu 6100299, China; sunzhixiang915@163.com (Z.-X.S.); chenluluu0713@163.com (L.C.)
2 University of Chinese Academy of Sciences, Beijing 101408, China
3 National Key Laboratory for Development and Utilization of Forest Food Resources, College of Forestry and Biotechnology, Zhejiang A&F University, Hangzhou 311300, China; aoyanting123@163.com (Y.-T.A.); lixy76@163.com (X.-Y.L.); chengjh@zafu.edu.cn (J.-H.C.)
4 Tianmushan Forest Ecosystem Orientation Observation and Research Station of Zhejiang Province, Zhejiang A&F University, Hangzhou 311300, China
* Correspondence: jfcui@imde.ac.cn (J.-F.C.); xiangyu.tang@foxmail.com (X.-Y.T.)

Abstract: Reliable identification of river hydrograph separation is crucial for prioritizing water source areas to be protected from pollution. A field study was carried out in a hilly catchment with diverse land uses, located in Southwest China. A novel water-tracing method, combining the ratio of two conservative fluorescent components of dissolved organic matter, two ion ratios, and oxygen-18, was proposed for river hydrograph separation with MixSIAR. During a rain event with the longest preceding no-rain period, a set of four tracers were found to be applicable to drainage areas with diverse land uses. Notably, a drier antecedent soil moisture condition could favor the occurrence of more tracers qualified for distinguishing multiple water sources of river flow.

Keywords: dissolved organic matter; fluorescence; ions; oxygen-18; river flow; catchment

1. Introduction

Human activities affect hydrological behaviors in varying manners [1]. Given the complex influence of changes in land use on hydrological processes [2–4], reliable identification of water sources contributing to river flow in a catchment is challenging, especially in rapidly developing regions with diversified land uses. Accurate apportionment of river water sources is essential for prioritizing source areas to be protected from pollution in an intensively utilized catchment [5].

A number of tracers, including water isotopes such as oxygen-18 and deuterium, as well as various solutes such as dissolved organic matter, inorganic ions, and anthropogenic pollutants, have been reported to be conservative during mixing processes and, therefore, used in the water source apportionment of river flow [6–8]. Isotopic signatures of water are commonly used to distinguish rain water from soil water or groundwater [6,9]; however, their specificity across pre-event water sources under different land-use types is often limited, precluding their use to track two or more pre-event water sources of river flow [6]. Dissolved organic matter (DOM), as a mixture of biogenic materials, is pervasive in rain water, soils, surface waters, and groundwater. DOM's composition may vary across soils under different land uses, as a result of differences in the input of plant material and the management of water and fertilization [10–12]. Recently, single or multiple DOM attributes were employed to identify water sources as well as flow paths [7,13]. In particular, the ratio of two conservative, humic-like, fluorescent components of DOM was proposed as a useful tracer to distinguish soil waters under different types of land use [14]. Natural inorganic ions (such as NO_3^-, Cl^-, SO_4^{2-}, Ca^{2+}, Na^+, and K^+) and their ratios are often different between rain water, soil water, and groundwater, and therefore can be used to identify the sources of river water [15]. Moreover, some anthropogenic pollutants, such as artificial sweeteners and pharmaceuticals, have the potential for quantifying the contributions of polluted sources to surface waters and groundwater at limited spatiotemporal scales, where they are ubiquitous and exhibit persistence [8,16].

Low specificity of a single tracer between more than two water sources, along with resultant uncertainties in hydrograph separation, has often been reported [6]. Therefore, a combined use of multiple tracers is necessary to obtain more reliable estimation of water contributions from diverse (>2) sources to river flow [17,18]. Water–soil/rock interactions, vegetation cover, and anthropogenic activities (such as fertilization, irrigation, cropping, and waste discharge) jointly affect water sources and result in their diverse hydrochemical characteristics (such as parameters of major ions and dissolved organic matter) [15,19], which can be potentially used to track the diverse sources of river water in a complex landscape.

It is assumed in this study that the involvement of selected inorganic ion ratios of high specificity in water sources that have similar fluorescent DOM component ratios and water isotopic signatures will lead to more reliable results of river hydrograph separation in a catchment with complex land uses. As far as we know, no multiple-spatial-scale study in a catchment has been documented on the combined use of biogenic and abiogenic solutes and water-stable isotopes to quantify the proportions of waters from different land-use types in river flow.

This study aimed to identify as many tracers as possible for distinguishing multiple (at least five) water sources of river flow in a hilly catchment with complex land uses. This study offers a useful tool to aid in the protection of major water sources of river flow from pollution.

2. Materials and Methods

2.1. Catchment Characteristics

This research was conducted in a small agroforestry-dominated catchment (35 ha; Jieliu catchment) (31°16′ N, 105°28′ E) in the upper reaches of the Yangtze River (Figure 1). This catchment is hilly, at 360−600 m above sea level, and features a subtropical monsoon climate. From 2005 to 2023, the average annual precipitation was 922.26 mm, 79% of which took place during May−September, and the annual mean air temperature was 16.6 °C.

This catchment encompasses various land-use types, including sloped croplands (maize–corn rotation), forestlands (mixed alder–cypress), rice fields, residential areas, plantations (mainly lemon trees), meadows, and others (ponds and roads), which account for 42.2%, 32.2%, 9.9%, 7.6%, 5.2%, 1.7%, and 1.2%, respectively. The catchment's slopes are characterized by purple soil (an Entisol containing 27% sand, 52% silt, and 21% clay), while its lowland areas are dominated by paddy soil. Finely fractured mudrock lies under the thin soil (<60 cm) and overlies an impermeable sandstone [20].

Hydrological processes on the slopes are governed successively by vertical infiltration of rain water through the soil, fast flow through mudrock fractures, lateral flow above the underlying sandstone layer, and eventual seepage into the river. Moreover, in rice fields, lateral flow may also occur above the top of the paddy hardpan (at the 45–50 cm depth) to form bank seepage into the river [21]. The river generally flows southeastward in the catchment [14]. Shallow groundwater is available only in wells located in the lowland areas of the catchment [22]. Irrigation is carried out only for rice fields whenever necessary to ensure the optimal rice production [23].

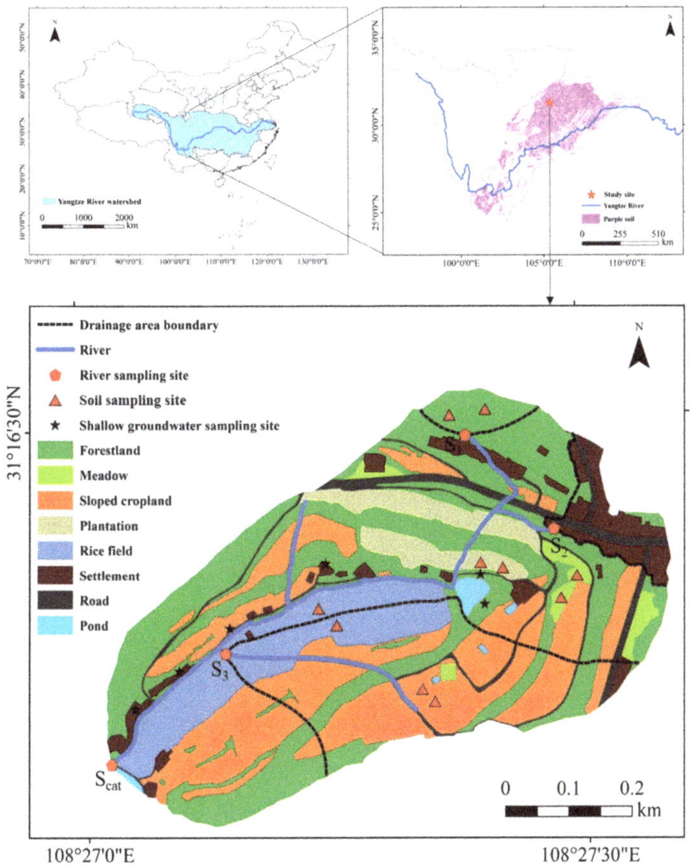

Figure 1. Monitoring locations in the Jieliu catchment, situated in the upper reaches of the Yangtze River.

2.2. Water Sampling and Analyses

Rain water, soil waters under different land uses, groundwater (water in the six wells), and river flow/runoff were sampled for a total of three rain events, i.e., rain events 1, 2, and 3 on 28 June, 27 July, and 25 August 2023, respectively (Figure 1). For each vegetation cover type, two representative areas were selected for sampling. There were a total of ten sampling areas, including two forestlands on the upslope, two meadows on the mid-slope, two sloped croplands on the mid-slope, two plantation lands on the footslope, and two rice fields in the valley bottom. At each area, mobile soil water (if any) was sampled before rainfall with soil water samplers (1900, SEC, Santa Barbara, CA, USA) at four depths (with the porous cup's center at 7.5 cm, 17.5 cm, 27.5 cm, and 47.5 cm, respectively) [24]; on the other hand, disturbed soil cores were taken in triplicate before rainfall to make a composite sample for each depth. Each sample was separated into two subsamples. The fresh soil subsample was extracted for bulk soil water (including both immobile and mobile fractions of soil water) by cryogenic vacuum distillation and measured for oxygen-18 [25]. Another soil subsample was sieved through a 2 mm mesh after air-drying. Notably, mobile soil water was often unavailable at some depths before rainfall, particularly after a long dry period. Therefore, bulk soil water (referred to as "soil water" hereinafter for convenience) was eventually used to represent pre-event soil water in this study. In particular, the extract of the air-dried soil subsample with ultrapure water (at a 1 g/10 mL soil/water ratio) was measured for hydrochemical parameters (major inorganic ions and DOM fluorescence), which were used to represent their pre-event levels in soil water. Moreover, the runoff discharged from the settlement drainage area and water in the six lowland wells were sampled before rainfall and measured for hydrochemical parameters as well as oxygen-18.

During rainfall, river/channel flow samples were taken at varying intervals (depending on rain intensity) from the outlets of the catchment (S_{cat}) and nested forestland (1.6 ha; S_1), settlement (2.02 ha; S_2), and agricultural–forestry (12.1 ha; S_3) drainage areas; meanwhile, rain water samples were also taken at 10, 20, or 30 min intervals. On the other hand, flow discharges at the time of sampling were obtained from water level data recorded by automatic data-logged water level meters using calibrated curves between the water flux and level.

All water (extract) samples were filtered through 0.45 μm cellulose acetate membrane filters and were kept at 4 °C. Rain water, soil water, well water, and river/channel water were measured for $\delta^{18}O$ with an isotope analyzer (L2120-i, Picarro, Santa Clara, CA, USA), concentrations of Ca^{2+}, Mg^{2+}, Na^+, K^+, NO_3^-, Cl^-, and SO_4^{2-} by an ion chromatograph (ICS-900, Dionex, Sunnyvale, CA, USA), and fluorescence excitation–emission matrices (EEMs) with a spectrofluorometer (Aqualog, Horiba, Kyoto, Japan).

The $\delta^{18}O$ content of the water samples was expressed in per mil (‰) and defined relative to the standard (Vienna Standard Mean Ocean Water) in terms of the ^{18}O to ^{16}O ratio (R), as given in Equation (1). The analytical precision was 0.05‰.

$$\delta^{18}O_{sample} = \left(\frac{R_{sample}}{R_{standard}} - 1\right) \times 1000‰ \quad (1)$$

Fluorescence intensity was determined across excitation (Ex) wavelengths of 250–500 nm and emission (Em) wavelengths of 200–600 nm. The blank EEM spectrum of Milli-Q water was employed to remove background noise. Raman scatters and Rayleigh bands were also eliminated. Fluorescent components of DOM were identified and quantified through parallel factor analysis. Then, the maximum fluorescence intensity (F_{max}) was calculated for each identified component of DOM.

2.3. *River Water Sources and Tracers*

Pre-event water sources of river flow were selected according to the land-use composition of the drainage area and hydrological connectivity. Rain water and pre-event forestland soil water were considered to be water sources of river flow discharged from the forestland drainage area (S_1). Rain water, pre-event soil waters of sloped cropland, forestland, and rice fields, and pre-event shallow groundwater were considered to be water sources of river flow discharged from the agricultural–forestry drainage area (S_3). Rain water, pre-event soil waters of plantations, forestland, meadows, sloped cropland, and rice fields, pre-event shallow groundwater, and settlement baseflow runoff were considered to be water sources of river flow discharged from the entire study catchment (S_{cat}).

Given the reported low specificity of the ratio of two conservative, humic-like, fluorescent components of DOM in soil waters under some land uses [14], multiple hydrochemical parameters, including both DOM fluorescence and inorganic ions (Ca^{2+}, Mg^{2+}, Na^{+}, K^{+}, NO_3^{-}, Cl^{-}, and SO_4^{2-}), were considered as potential tracers for identifying multiple water sources of river flow in this study catchment. On the one hand, the ratio of F_{max} of humic-like components C1 (at Ex/Em 255/420 nm) and C2 (at Ex/Em 265 (375)/500 nm) was considered as a biogenic tracer. On the other hand, different ratios of ion concentrations were considered as abiogenic tracers, and their specificity in water sources was evaluated. It should be noted that individual F_{max} and ion concentrations were not considered as tracers, as they were less conservative and of lower specificity as compared to the selected F_{max} and ion concentration ratios.

Using multiple conservative tracers (such as water isotopes and various solutes) with high specificity in potential water sources always leads to more reliable quantification of flow components [26]. The combined use of hydrochemical parameters and $\delta^{18}O$ as tracers is assumedly a better method for hydrograph separation in drainage areas with mixed land uses than the use of any single tracer. Two steps were taken to identify qualified tracers for river flow at each monitoring weir for each rain event. Firstly, hydrochemical parameters significantly correlated with $\delta^{18}O$ were considered to be conservative. Secondly, the largest set of parameters (including the conservative hydrochemical parameters and $\delta^{18}O$), among which at least one parameter showed a significant difference ($p < 0.05$) between any two of the potential water sources, were considered to be qualified tracer(s). The specificity of a tracer set is higher if more parameters exhibiting significant differences between any two potential water sources can be included. The spatial variability of rain water $\delta^{18}O$ was ignored in this small catchment. The means of the $F_{max}(C1)/F_{max}(C2)$ ratio, $\delta^{18}O$, and inorganic ion ratios across rain water samples taken from near the soil sampling locations were employed as tracer values of event water sources. For each vegetation cover type, the mean values of depth-averaged hydrochemical parameters and $\delta^{18}O$ in the water extracts of soil samples collected from two representative areas before rainfall were employed as tracer values of pre-event water sources.

2.4. Hydrograph Separation

The MixSIAR model was employed to estimate the water contributions of potential sources to river flow. This Bayesian mixing model accounts for the uncertainty of multiple sources and integrates prior continuous covariate data [27]. Its mixing equation can be written as shown in Equation (2):

$$Y_j = \sum_k p_k \mu^s_{jk} \tag{2}$$

where Y_j is the value of each tracer after mixing, p_k is the water contribution of each source to the river flow, μ^s_{jk} is the tracer mean of each source, j is the number of tracers, and k is the number of sources.

MixSIAR needs inputs of water source data (observed values of selected tracers in each source) and flow data (observed values of the same tracers in river flow). A default assumption of no tracer fractionation was used. The runtime of the Markov chain Monte Carlo algorithm was initially set to "short", with the process only selected for the error, and subsequently adjusted to "normal", "long", and then "very long" sequentially to achieve the convergence of modeling. Both the Gelman–Rubin and Geweke diagnostics were employed to assess the convergence [28]. The means of the early and late parts of a single chain were compared with the Geweke diagnostic; convergence was achieved when the difference was statistically insignificant. On the other hand, multiple chains were compared with the Gelman–Rubin diagnostic; convergence was achieved when the statistic value was smaller than 1.1. The model's outputs were summarized using the mean posterior values of source contributions in this study.

2.5. Statistical Analyses

Correlations of the $F_{max}(C1)/F_{max}(C2)$ ratio and inorganic ion ratios with $\delta^{18}O$ in different water sources (rain water, various soil waters, shallow groundwater, and settlement runoff) were analyzed to evaluate whether they were as conservative as $\delta^{18}O$. Differences in the selected tracers between water sources were evaluated by one-way analysis of variance.

The performance of each hydrochemical tracer in estimating the water contributions of sources to river flow in the forestland drainage area was evaluated by comparing it with the traditional water isotope tracer ^{18}O. The hydrograph separation results obtained from the data of $\delta^{18}O$ and those obtained from the data of each hydrochemical tracer were taken as true and estimated values, respectively. The accuracy of the results was assessed in terms of Nash–Sutcliffe efficiency (NSE) and percentage bias (PBIAS), which can be calculated using Equations (3) and (4), respectively [29,30]:

$$NSE = 1 - \frac{\sum_{i=1}^{n}(E_i - T_i)^2}{\sum_{i=1}^{n}(T_i - M_i)^2} \quad (3)$$

$$\text{PBIAS} = \sum_{i=1}^{n} \frac{E_i - T_i}{T_i} \times 100 \quad (4)$$

where n represents the number of river flow samples used for validation, M_i represents the mean of the true water contributions of pre-event sources for all river flow samples for each rain event, and T_i and E_i represent the true and estimated water contributions of pre-event sources to each river flow sample, respectively. In this study, the performance of a hydrochemical tracer was considered satisfactory if the NSE was over 0.5 and the absolute value of PBIAS was smaller than 25%.

3. Results

3.1. Differences in Hydrochemical Parameters and $\delta^{18}O$ Between Various Water Sources

The correlations of hydrochemical parameters with $\delta^{18}O$ in rain water and pre-event water sources (various soil waters, shallow groundwater, and settlement runoff) are shown for rain events 1, 2, and 3 (Table 1) in Figure S1. Conservative hydrochemical parameters, which exhibited significant correlations with $\delta^{18}O$ in all potential water sources of river flow at each monitoring weir, are listed with respect to the rain events in Table S1. For all three rain events, the number of conservative hydrochemical parameters decreased with the increase in the number of potential water sources. The number and species of conservative hydrochemical parameters in potential water sources of river flow at all three locations could vary among the three rain events, reflecting that the conservativeness of hydrochemical parameters may be affected by hydro-climatic conditions before rainfall.

The differences in conservative parameters (hydrochemical parameter ratios and $\delta^{18}O$) between potential water sources are shown in Figure S2. Prior to all three rain events, significantly higher values of all 5–6 conservative tracers were observed in soil water from forestland than in rain water. Parameters showing significant differences between soil waters collected from forestland and sloped cropland prior to rainfall differed between rain events. In particular, significantly smaller ratios of $F_{max}(C1)/F_{max}(C2)$ and Ca^{2+}/NO_3^- before rain event 1, $F_{max}(C1)/F_{max}(C2)$, Mg^{2+}/NO_3^-, and Ca^{2+}/NO_3^- before rain event 2, and Mg^{2+}/NO_3^- and Ca^{2+}/NO_3^- before rain event 3 were observed in sloped cropland soil water, whereas significantly smaller values of the Ca^{2+}/NO_3^- ratio in rice field soil water than in forestland soil water were found before rain events 1 and 3. Moreover, significantly smaller values of the Ca^{2+}/NO_3^- ratio were also observed in plantation soil water before rain event 1. Apparently, the application of nitrogen fertilizers was responsible for the observed lower ratios of major divalent cation(s) to NO_3^- in the soil waters of sloped cropland, plantations, and rice fields than in the soil water of forestland. Notably, the Ca^{2+}/NO_3^- ratios in shallow groundwater were not significantly different from those in the soil waters of sloped cropland, indicating that water infiltration through sloped cropland soil probably played a major role in recharging shallow groundwater. The ratios of $F_{max}(C1)/F_{max}(C2)$ were significantly higher in the soil waters of forestland and meadows than in the soil waters of sloped cropland, plantations, and rice fields, which are often amended with crop straw and fertilizers. Moreover, the ratios of $F_{max}(C1)/F_{max}(C2)$ were significantly smaller in rice field soil water than in sloped cropland and plantation soil waters, indicating that water logging can increase the relative abundance of C2, as also observed previously in the same catchment [14]. The patterns of differences in the ratio of $F_{max}(C1)/F_{max}(C2)$ between soil waters under different land uses before rain events 1 and 2 were similar, implying the conservativeness of $F_{max}(C1)/F_{max}(C2)$. The highest $\delta^{18}O$ values in potential water sources were all observed before rain event 1, with the longest preceding no-rain period. Notably, for rain event 2, $\delta^{18}O$ exhibited significant differences among rain water, soil waters, shallow groundwater, and settlement runoff; nevertheless, the differences in $\delta^{18}O$ among soil waters under different vegetation covers were not significant. Moreover, significantly lower values of $\delta^{18}O$ were observed in shallow groundwater than in soil waters under all land-use types before rain event 2, indicating that rain water from previous rainfall (rain event 1) recharged the shallow groundwater preferentially through soil macropores and underlying mudrock fractures [31,32]. For rain events 1 and 2, the number of potential water sources showing significant differences in the ratio of $F_{max}(C1)/F_{max}(C2)$ was higher than that showing significant differences in $\delta^{18}O$.

Table 1. Characteristics of investigated rain events and resultant river flow depths at four monitoring weirs.

	Rain Characteristics *				River Flow Depth §			
Event No. (date)	PNRD	Amount	Time Duration	I_{max}				
	(day)	(mm)	(h)	(mm (30 min)$^{-1}$)	(mm)			
					S_1	S_2	S_3	S_{cat}
1 (28 June 2023)	23	25.95	5.67	6.93	0.29	2.40	0.73	1.64
2 (27 July 2023)	9	54.51	14.67	20.16	0.47	36.29	4.27	17.16
3 (25 August 2023)	2	10.44	6.67	2.17	0.02	5.67	1.69	1.82

Notes: * PNRD represents the number of preceding no-rain days. I_{max} represents the maximum intensity. § River flow depth (mm) was calculated by dividing the cumulative stormflow (excluding baseflow, if present) during each rain event by the corresponding drainage area.

For all three rain events, significant differences in 2–4 conservative tracers were observed between any two of the five potential water sources of river flow at outlet S_3. Significant differences in at least one conservative tracer between any two of the eight potential water sources of river flow at outlet S_{cat} were observed only for rain event 1 (Figure S2). The qualified sets of conservative tracers showing sufficient specificity across all potential water sources of river flow at outlets S_3 and S_{cat} are listed in Table 2. Apparently, the specificity of conservative tracers among different water sources may vary temporally, as a result of changing climatic conditions over time [14,33].

Table 2. Qualified tracers for distinguishing water sources of river flow at three monitoring weirs upon three rain events.

Rain Event No.	Qualified Multiple Tracers for Distinguishing Water Sources of Streamflow *		
	S_1	S_3	S_{cat}
1	$F_{max}(C1)/F_{max}(C2)$, Mg^{2+}/NO_3^-, Ca^{2+}/K^+, Ca^{2+}/NO_3^-, and $\delta^{18}O$	$F_{max}(C1)/F_{max}(C2)$, Ca^{2+}/NO_3^-, Ca^{2+}/K^+, and $\delta^{18}O$	$F_{max}(C1)/F_{max}(C2)$, Ca^{2+}/NO_3^-, Ca^{2+}/K^+, and $\delta^{18}O$
2	$F_{max}(C1)/F_{max}(C2)$, Mg^{2+}/NO_3^-, Ca^{2+}/K^+, Ca^{2+}/NO_3^-, Mg^{2+}/Na^+, and $\delta^{18}O$	$F_{max}(C1)/F_{max}(C2)$, Mg^{2+}/NO_3^-, Ca^{2+}/NO_3^-, and $\delta^{18}O$	
3	$F_{max}(C1)/F_{max}(C2)$, Mg^{2+}/NO_3^-, Ca^{2+}/K^+, Ca^{2+}/NO_3^-, and $\delta^{18}O$	$F_{max}(C1)/F_{max}(C2)$, Mg^{2+}/NO_3^-, Ca^{2+}/NO_3^-, and $\delta^{18}O$	

Notes: * River flow at outlets S_1, S_3, and S_{cat} had 2, 5, and 8 potential water sources, respectively, as described in detail in "2.3 River water sources and tracers". Only one of the listed tracers was needed for hydrograph separation at outlet S_1, while combinations of multiple tracers (if present) were proposed for use in hydrograph separation at outlets S_3 and S_{cat}.

3.2. Temporal Changes in River Discharge, Hydrochemical Parameters, and $\delta^{18}O$ During Rainfall

Dynamics of discharge, conservative hydrochemical parameters, and the $\delta^{18}O$ of river flow at outlets S_1, S_3, and S_{cat} during three rain events in the summer of 2023 are presented in Figure 2, and the depths of rainfall-induced river flow are given in Table 1.

The simultaneity of maximum rain intensity and maximum river discharge and the fastest response of river discharge happened during the light rainfall (rain event 3) at all three sampling outlets. This can be ascribed to the observed highest pre-event soil water contents and shallow groundwater levels resulting from the shortest dry period (2 days) prior to rain event 3. A slight delay in maximum discharge (20 min; relative to I_{max}) occurred during the heavy rainfall (rain event 2) at outlets S_3 and S_{cat}, which can be attributed to the greatest I_{max} and amount of rain event 2, despite the nine preceding dry days. The longest delays in maximum river discharge were observed at outlets S_3 and S_{cat} (3 h and 1.5 h, respectively) upon the medium rainfall (rain event 1), as a result of the lowest detected soil water contents and shallow groundwater levels caused by the longest preceding dry period (23 days).

The lowest ratios of river flow depth to rain amount at outlets S_3 and S_{cat} were observed during the medium rainfall. During all three rain events, the ratios of river flow depth to rain amount across the outlets followed the order $S_1 < S_3 < S_{cat}$. The highest ratio of river flow depth to rain amount (31.48%) was detected at outlet S_{cat} during the heavy rainfall.

The $F_{max}(C1)/F_{max}(C2)$ in the river flow at outlet S_1 exhibited the highest variations, which can be attributed to the more complex composition of DOM in forestland soil due to the presence of dense roots of various plants [34]. The $F_{max}(C1)/F_{max}(C2)$ in the river flow at outlets S_1, S_3, and S_{cat} varied in distinctive ranges among three rain events, implying that hydro-climatic conditions may affect the potential of $F_{max}(C1)/F_{max}(C2)$ for hydrograph separation. At outlet S_1, The Ca^{2+}/NO_3^- ratio increased with increasing rainfall intensity (having greater Ca^{2+}/NO_3^- values than the river's baseflow) during the rising limb of the river flow hydrograph, and then it became stable (Figure 2).

During the medium rainfall, the $\delta^{18}O$ values of all rainfall-induced river flow samples collected at outlets S_1, S_3, and S_{cat} were greater than those of the rain water samples, implying that some proportion(s) of the river flow could be composed of pre-event water(s) with higher $\delta^{18}O$. At the end of the medium rainfall, the greatest dominance of rain water in river flow was found at outlet S_1, as indicated by their similar $\delta^{18}O$ values. At the end of the heavy rainfall, the dominance of rain water in river flow was greatest at outlet S_{cat}, while it was lowest at outlet S_1, as interpreted from the differences in $\delta^{18}O$. At the end of the light rainfall, the river flow at outlet S_{cat} was almost completely composed of rain water.

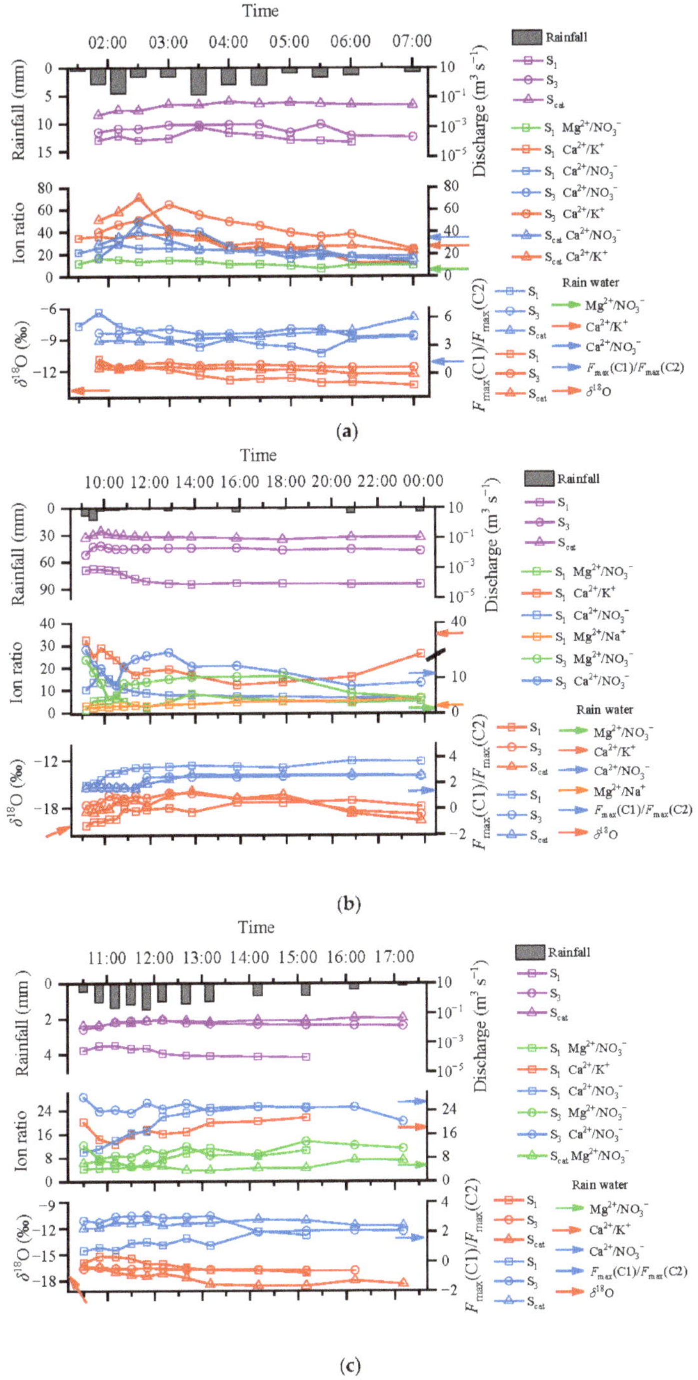

Figure 2. Hydrographs and tracer dynamics of river flow discharged from the study catchment (S_{cat}) and its forestland (S_1) and agricultural–forestry (S_3) drainage areas during rain events on 28 June (**a**), 27 July (**b**), and 25 August (**c**), 2023.

Apparently, the ratios of $F_{max}(C1)/F_{max}(C2)$ and Ca^{2+}/NO_3^- were most responsive hydrochemical parameters to rain events in the river flow at outlet S_1. At outlet S_2, during the medium rain with the longest preceding dry period, the ratios of Ca^{2+}/NO_3^- and Ca^{2+}/K^+ were the most responsive hydrochemical parameters in the river flow; in contrast, during the other two rain events with shorter preceding dry periods, Ca^{2+}/NO_3^- and Mg^{2+}/NO_3^- were more responsive than $F_{max}(C1)/F_{max}(C2)$ and $\delta^{18}O$. At outlet S_{cat}, the most responsive conservative hydrochemical parameter in the river flow to the rain event following the longest no-rain period was the ratio of Ca^{2+}/NO_3^-; notably, the responses of conservative parameters in river flow to the other two rain events following wetter periods were of much smaller magnitudes. Therefore, qualified tracers for distinguishing the water sources of river flow may vary not only with land use but also with hydro-climatic conditions.

3.3. *Performance of Hydrochemical Parameters in the Separation of River Hydrographs in a Drainage Area with a Single Vegetation Cover Type by Comparing with $\delta^{18}O$*

The river hydrograph separation results in the forestland drainage area using each conservative solute tracer are presented in Figure S3. The performance of each conservative hydrochemical parameter in river hydrograph separation is shown in Table S2.

Based on $\delta^{18}O$ data, no delay in the maximum rain water contribution, compared to the maximum river flow discharge, was evidenced for rain events 2 and 3 with short preceding dry periods, implying that piston flow was dominant in the river flow generation under high antecedent soil moistures, as also observed previously in the same forestland drainage area [14]. Nevertheless, a delay of 2 h occurred during rain event 1, following a long no-rain period. This indicates that the infiltration of rain water through large pores was dominant during the rising limb of the river hydrograph, while piston flow through both large and small pores (partially emptied by evaporation before the rainfall and filled gradually by rain water) was dominant during the receding limb [32]. Soil water's contributions to the river flow were 30.8–98.3%, 16.2–89.4%, and 21.5–94.9% during rain events 1, 2, and 3, respectively. According to the PBIAS and NSE values obtained (Table S2), the ratio of $F_{max}(C1)/F_{max}(C2)$ performed best during rain events 1 and 2, while the ratio of Mg^{2+}/NO_3^- showed the best overall performance during rain event 3.

Apparently, the applicability of each potential hydrochemical tracer to hydrograph separation may vary with rain characteristics and pre-event soil moisture (Table S2). For a rain event with a longer dry period, more tracers showing satisfactory performance may be available.

3.4. *Performance of Multiple Tracers in the Separation of River Hydrographs in Drainage Areas with Multiple Land-Use Types*

During rain event 1, a set of four tracers ($F_{max}(C1)/F_{max}(C2)$, Ca^{2+}/NO_3^-, Ca^{2+}/K^+, and $\delta^{18}O$) was applicable to the separation of the river hydrographs at outlets S_{cat} and S_3. During rain events 2 and 3, a different set of four tracers ($F_{max}(C1)/F_{max}(C2)$, Mg^{2+}/NO_3^-, Ca^{2+}/NO_3^-, and $\delta^{18}O$) was applicable at outlet S_3; however, there was no such set of multiple tracers applicable at outlet S_{cat}. The river hydrograph separation results at outlets S_{cat} and S_3, using four tracers, are presented in Figure 3. The statistics and uncertainties of the estimated proportions of water sources in the river flow are given in Table 3. Notably, settlement runoff exhibited very low pre-event discharge, and its inclusion led to unacceptably high uncertainties in the hydrograph separation results at S_{cat}. Therefore, settlement runoff was eventually excluded from the list of river water sources at outlet S_{cat} in the hydrograph separation.

The contributions of all pre-event water sources to river flow at outlet S_3 were 61.3–98.5%, 24.4–84.6%, and 58.8–94.4% during rain events 1, 2, and 3, respectively. Rain water's contribution to the river flow at outlet S_3 was lowest during the medium rainfall with the longest preceding no-rain period (event 1: 18.1%), while it was highest during the heavy rainfall (event 2: 37.3%). During rain event 1, rain water's contribution to the river flow at outlet S_{cat} (9.4%) was markedly lower than that at outlet S_3, which may be ascribed to the involvement of two more water sources (soil waters from plantations and meadows, with the plantation soil having a much greater water-holding capacity than other soils) in the highland areas [35]. Moreover, from rain event 1 through rain event 3, shallow groundwater's contribution to the river flow at outlet S3 increased, which may be partially ascribed to the observed elevations of pre-event shallow groundwater levels (with the mean groundwater level being 7.81, 8.45, and 8.62 m above outlet S_3, respectively). Rain water's maximum contribution to the river flow at outlets S_3 and S_{cat} happened 1 h later than the maximum river discharge during rain event 1, implying that piston flow through small pores with an increased fraction of rain water remained dominant during the early receding limb of the river hydrograph. In contrast, at outlet S_3, no delay in the occurrence time of the maximum contribution of rain water to river flow (relative to the time of maximum river discharge occurrence) took place during rain events 2 and 3, with fewer preceding dry days, which were conducive to piston flow [32].

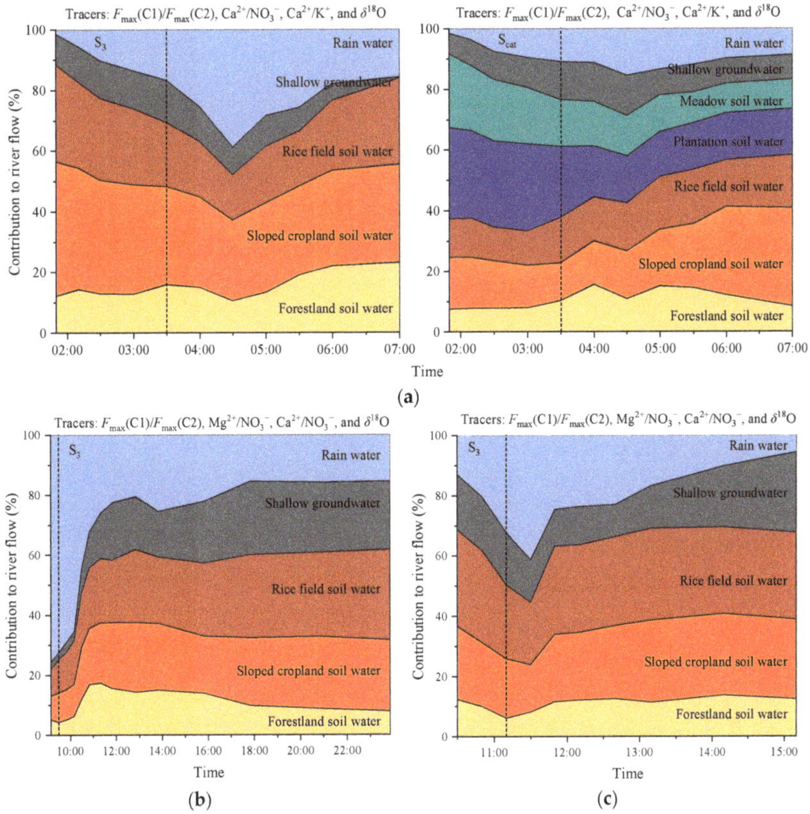

Figure 3. Dynamics of estimated water source contributions to river flow discharged from mixed-land-use drainage areas (S_3 (and S_{cat})) during rain events on 28 June (**a**), 27 July (**b**), and 25 August (**c**), 2023, based on the data of multiple tracers using MixSIAR. Dashed lines represent the times of maximum river discharge occurrence. Water sampling and discharge monitoring were conducted at intervals of 30 min and 15 min, respectively.

Table 3. Separation results and associated river hydrograph uncertainties at monitoring weirs, obtained using single or multiple tracers with MixSIAR.

Rain Event No	Weir	Water Source Proportions in River Flow (%) *							
		Rain Water	Forestland Soil Water	Plantation Soil Water	Sloped Cropland Soil Water	Rice Field Soil Water	Meadow Soil Water	Settlement Runoff	Shallow Groundwater
1	S_1	44.9 ± 21.7 (NU)	55.1 ± 21.7 (NU)	--	--	--	--	--	--
	S_3	18.1 ± 10.8 (7.3–8.9)	15.6 ± 4.1 (5.1–7.9)	--	33.6 ± 5.3 (12.4–14.7)	23.2 ± 5.4 (9.3–14.8)	--	--	9.5 ± 3.9 (3.4–4.2)
	S_{cat}	9.4 ± 4.0 (6.5–7.2)	10.7 ± 3.17 (5.3–5.5)	21.1 ± 6.7 (8.4–14.7)	18.9 ± 6.3 (9.3–12.7)	14.7 ± 2.4 (9.6–11.4)	15.5 ± 5.1 (6.3–8.3)	NS	9.8 ± 2.2 (3.5–6.4)
2	S_1	49.5 ± 25.5 (NU)	50.5 ± 25.5 (NU)	--	--	--	--	--	--
	S_3	37.3 ± 23.2 (12.6–14.3)	10.9 ± 4.7 (4.3–4.6)	--	18.0 ± 5.9 (7.9–10.2)	20.3 ± 6.4 (10.2–11.5)	--	--	13.5 ± 8.4 (5.2–5.6)
3	S_1	30.5 ± 23.8 (NU)	69.5 ± 23.8 (NU)	--	--	--	--	--	--
	S_3	20.9 ± 10.5 (10.8–12.2)	11.0 ± 2.3 (7.6–8.7)	--	23.2 ± 3.6 (10.4–11.2)	28.4 ± 3.3 (9.2–11.8)	--	--	16.4 ± 4.7 (5.7–8.4)

Notes: * Values are means ± standard deviations (uncertainties). NU indicates that no value of uncertainty was obtained; -- indicates that this water source was not included in the river hydrograph separation; NS indicates that no solution was obtained when settlement runoff was included.

During rain event 1, the largest contribution of pre-event water to the river flow at S_3 came from sloped croplands (33.6%). In contrast, during rain events 2 and 3, the largest water source of river flow at S_3 was rice field soil water, accounting for 20.3% and 28.4%, respectively. The largest contribution of pre-event water to river flow at outlet S_{cat} during rain event 1 came from plantations (21.1%), especially at the rising limb of the river hydrograph. Soil waters from rice fields (23.2%) and sloped cropland (18.9%) made the second-greatest contribution to river flow at outlets S_3 and S_{cat} during rain event 1, respectively. Moreover, at the receding limb of the hydrograph during rain event 1, the water contribution of sloped cropland to river flow at outlet S_{cat} gradually increased and eventually became the greatest (32.5%).

4. Discussion

4.1. Selection of Multiple Tracers

Stable isotopes of water have been recognized as the most conservative tracers, which are commonly used to distinguish pre-event and event waters [36–38], but they often exhibit limited specificity in diverse water sources (>2) within a catchment [6]. DOM in soils comes from the leaching of fresh litter, humified organic matter, and microbial biomass [39]. The $F_{max}(C1)/F_{max}(C2)$ ratio has a higher ability than $\delta^{18}O$ to identify water sources of river flow under different vegetation covers or land uses. Nevertheless, it was reported that the differences in $F_{max}(C1)/F_{max}(C2)$ ratio between forestlands and meadows, and between plantations and sloped croplands, were not significant at the 0.05 level, which might partially account for the observed uncertainties in water source apportionment results [14]. Therefore, further involvement of conservative parameters of ions derived from geochemical processes and fertilization is necessary.

The ratio of Ca^{2+}/Mg^{2+} can be employed to trace groundwater flow paths and discern water–rock interactions, and the ratio of Na^+/Cl^- can be used to differentiate groundwater recharge sources and processes [40,41]. Our investigation showed that a total of four ion ratios (Ca^{2+}/NO_3^-, Mg^{2+}/NO_3^-, Ca^{2+}/K^+, and Mg^{2+}/Na^+) could exhibit significant differences between water sources of river flow at one or more outlets (Table 2). For the agricultural–forestry drainage area, the combination of four tracers (including $F_{max}(C1)/F_{max}(C2)$, two ion ratios, and $\delta^{18}O$) applicable to rain event 1, with the longest preceding dry period, was different from that applicable to rain events 2 and 3, with shorter preceding dry periods. In addition to the ratio of Ca^{2+}/NO_3^-, the ratio of Ca^{2+}/K^+ was included for rain event 1, while the ratio of Mg^{2+}/NO_3^- was included instead for rain events 2 and 3. The varying differences in the ratios of hydrochemical parameters can be attributed to the coupled effects of biogeochemical processes (such as rock weathering or carbon and nitrogen transformation) and anthropogenic activities (such as fertilizer application and domestic sewage discharge), as affected by hydro-climatic conditions [6]. For instance, a high soil water content favors a decrease in the NO_3^- concentration in soil water through reduction [41,42], leading to an elevated ratio of Ca^{2+}/NO_3^-. Therefore, caution should be taken in selecting conservative tracers for river hydrograph separation, as their specificity among water sources under different land-use types may change temporally.

Similarly, a comprehensive suite of tracers ($\delta^{18}O$, δD, $\delta^{34}S$-SO_4^{2-}, $\delta^{13}C$-DIC, and $^{87}Sr/^{86}Sr$) was used to analyze the recharge and mixing of groundwater in a karst area [43]. In another study, hydrochemical and isotopic tracers were employed to evaluate the interaction between groundwater and surface water in the Qujiang River, located in Zhejiang Province, China [44]. Apparently, the combined use of isotopic and hydrochemical tracers is a promising approach for identifying water sources of river flow in a catchment with complex land-use patterns.

4.2. Applicability of Multiple Tracers to River Hydrograph Separation

Lands under different uses (such as forestlands, plantations, sloped croplands, meadows, and rice fields) often have different soil hydraulic properties and, therefore, exhibit different patterns of hydrological processes [45]. Distinctive signatures of physically mixed tracers in pre-event water sources enable the quantification of more than two components of river flow. In an intensively utilized catchment, the number of potential water sources may increase with the increasing spatial scale of the drainage area. There is clearly a need to identify as many tracers as possible so as to ensure accurate apportionment of multiple water sources of river flow. However, not all conservative parameters show significant differences between water sources. In this study, there was no such set of tracers among which at least one tracer showed a significant difference between any two of the potential water sources of river flow at outlet S_{cat} during rain events 2 and 3, which had shorter preceding no-rain periods (Tables 1 and 2).

The same set of four tracers was applied successfully to river water source apportionment at outlets S_{cat} and S_3 for rain event 1, with the longest preceding no-rain period, but such successful application of four tracers at both outlets was not achieved for rain events 2 and 3, with shorter preceding no-rain periods (Table 2). This implies that a lower antecedent soil moisture may favor the occurrence of significant differences in more hydrochemical parameters between water sources in a catchment with diverse land uses, probably through affecting various processes such as soil biogeochemical reactions, chemical weathering, and flushing [15].

4.3. Future Perspectives

There are three limitations in the present research: Firstly, a comparison with other hydrologically independent catchments for an extensive validation and applications of the water source tracing method described in our research is lacking. Secondly, potential spatial variations in the tracers were not considered for water sources. Thirdly, the potential effects of the subsoil (below 50 cm depth) and parent rocks were considered implicitly by the measurements of the tracers in well water in the lowland areas only. Despite these limitations, our research offers a useful approach for distinguishing different water sources of river flow using multiple tracers (including both water isotopes and hydrochemical tracers).

Several issues remain to be explored in the future: Firstly, do the multiple selected tracers of different types in water sources remain stable during rainfall? Secondly, are there any other potential tracers representing the distinctive natures of lands under different vegetation, uses, and management? Thirdly, are there significant differences in tracers among waters in soil pores of different sizes? If so, the development of tools for sampling mobile water under low soil moisture is highly desirable. Fourthly, do the tracers in the subsoil and/or parent rocks differ significantly from those in shallow soil water and well water? If so, more subsurface water sources lying between the top 50 cm soil layer and the shallow groundwater table need to be considered to achieve better accuracy of hydrograph separation.

To support the effective control of river contamination in hilly catchments with complex land-use patterns, a protocol for sampling pre-event water sources, particularly source lands with varying soil thicknesses, needs to be developed. Moreover, in the future development of hydrograph separation models, it is necessary to introduce an additional parameter representing the varying connectivities among water sources, as the landscape–stream connectivity may be affected by a number of factors (such as the distance between the landscapes and stream network, topographic characteristics, and antecedent soil moisture) [46,47].

5. Conclusions

A reliable quantification of river water sources in catchments with multiple land-use types still remains challenging. In this study, a novel water source tracing method combining water isotopes, DOM fluorescence, and ion ratio tracers was proposed for hydrograph separation in a hilly catchment with diverse anthropogenic interferences. For the medium rain event with the longest preceding dry period, a combined use of $F_{max}(C1)/F_{max}(C2)$, Ca^{2+}/NO_3^-, Ca^{2+}/K^+, and $\delta^{18}O$ was successful in hydrograph separation of river flow at the outlets of the catchment and nested agricultural–forestry drainage area (with five and seven water sources, respectively). Nevertheless, during the other two rain events, with shorter preceding dry periods, a different set of four tracers was applicable to the nested drainage area with mixed land use but not applicable to the entire catchment. A drier climate condition appeared to favor the occurrence of more qualified tracers for river flow hydrograph separation. A more extensive screening across diverse land uses is needed to build a large collection of conservative tracers, from which as many tracers as possible can be selected to ensure the occurrence of significant differences in at least one tracer between any two potential water sources.

Supplementary Materials: The following supporting information can be downloaded at https://www.mdpi.com/article/10.3390/w17081222/s1: Figure S1: Correlations (r values) of hydrochemical parameters with $\delta^{18}O$ in rain water, soil waters of forestland, plantations, meadows, sloped cropland, and rice fields, shallow groundwater, and settlement runoff for rain events on 28 June (a), 27 July (b), and 25 August (c), 2023; * and ** represent significance at the 0.05 and 0.01 levels, respectively. Figure S2: Differences in selected conservative parameters between water sources for river flow at three monitoring weirs in the study catchment for rain events on 28 June (a), 27 July (b), and 25 August (c), 2023. Different letters indicate significant differences between water sources at the 0.05 level. Figure S3: Dynamics of estimated contributions of rain water and pre-event soil water to river flow at the outlet of the forestland drainage area (S_1) during rain events on 28 June (a), 27 July (b), and 25 August (c), 2023, based on the data of each hydrochemical tracer and $\delta^{18}O$ using MixSIAR. Dashed lines represent the occurrence times of maximum river discharge. Water sampling and discharge monitoring were conducted at intervals of 30 min and 15 min, respectively. Table S1: Conservative hydrochemical parameters in potential water sources of river flow at three monitoring weirs in the study catchment upon three rain events. Table S2: Statistical results of hydrochemical tracers' performance in estimating the water sources of river flow at the outlet of the forestland drainage area (S_1), by comparison with $\delta^{18}O$.

Author Contributions: Conceptualization, X.-Y.T.; methodology, X.-Y.T.; software, Z.-X.S.; validation, Z.-X.S.; formal analysis, Z.-X.S.; investigation, Z.-X.S., Y.-T.A. and L.C.; resources, J.-F.C., X.-Y.T., X.-Y.L. and J.-H.C.; writing—original draft preparation, Z.-X.S.; writing—review and editing, X.-Y.T.; supervision, J.-F.C. and X.-Y.T.; project administration, J.-F.C., X.-Y.L. and X.-Y.T.; funding acquisition, J.-F.C., X.-Y.L. and X.-Y.T. All authors have read and agreed to the published version of the manuscript.

Funding: This research was funded by supported by the National Key Research and Development Program of China (No. 2023YFF0806002) and the National Natural Science Foundation of China (Nos. 42177379, 42371039, 32071581, 42377374, and 32330065).

Data Availability Statement: The data presented in this study are available on request from the corresponding author.

Conflicts of Interest: The authors declare no conflicts of interest.

References

1. Yang, D.W.; Yang, Y.T.; Xia, J. Hydrological cycle and water resources in a changing world: A review. *Geo. Sus.* **2021**, *2*, 115–122. [CrossRef]
2. Musie, M.; Sen, S.; Chaubey, I. Hydrologic responses to climate variability and human activities in Lake Ziway Basin, Ethiopia. *Water* **2020**, *12*, 164. [CrossRef]

3. Osei, M.A.; Amekudzi, L.K.; Wemegah, D.D.; Preko, K.; Gyawu, E.S.; Obiri-Danso, K. The impact of climate and land-use changes on the hydrological processes of Owabi catchment from SWAT analysis. *J. Hydrol. Reg. Stud.* **2019**, *25*, 100620. [CrossRef]
4. Xiao, X.; Zhang, X.P.; Shi, X.N.; Yu, Z.L.; Liu, K.S.; Liu, Z. Identifying river water sources using end-member mixing analysis in a subtropical monsoon basin China. *Hydrol. Process.* **2023**, *37*, e14824. [CrossRef]
5. Lv, Y.J.; Gao, L.; Geris, J.; Verrot, L.; Peng, X.H. Assessment of water sources and their contributions to streamflow by end-member mixing analysis in a subtropical mixed agricultural catchment. *Agric. Water Manag.* **2018**, *203*, 411–422. [CrossRef]
6. Abbott, B.W.; Baranov, V.; Mendoza-Lera, C.; Nikolakopoulou, M.; Harjung, A.; Kolbe, T.; Balasubramanian, M.N.; Vaessen, T.N.; Ciocca, F.; Campeau, A.; et al. Using multi-tracer inference to move beyond single-catchment ecohydrology. *Earth Sci. Rev.* **2016**, *160*, 19–42. [CrossRef]
7. Begum, M.S.; Park, H.Y.; Shin, H.S.; Lee, B.J.; Hur, J. Separately tracking the sources of hydrophobic and hydrophilic dissolved organic matter during a storm event in an agricultural watershed. *Sci. Total Environ.* **2023**, *873*, 162347. [CrossRef] [PubMed]
8. Buerge, I.J.; Buser, H.R.; Kahle, M.; Mueller, M.D.; Poiger, T. Ubiquitous occurrence of the artificial sweetener acesulfame in the aquatic environment: An ideal chemical marker of domestic wastewater in groundwater. *Environ. Sci. Technol.* **2009**, *43*, 4381–4385. [CrossRef]
9. Brenot, A.; Benoît, M.; Carignan, J.; France-Lanord, C. Insights into stable isotope characterization to monitor the signification of soil water sampling for environmental studies dealing with soil water dynamics through the unsaturated zone. *C. R. Geosci.* **2015**, *347*, 317–327. [CrossRef]
10. Cheng, K.; Liu, Z.Q.; Xiong, K.N.; He, Q.F.; Li, Y.; Cai, L.L.; Chen, Y. Migration of dissolved organic matter in the epikarst fissured soil of south China karst. *Land* **2023**, *12*, 887. [CrossRef]
11. Cotrufo, M.F.; Haddix, M.L.; Kroeger, M.E.; Stewart, C.E. The role of plant input physical-chemical properties, and microbial and soil chemical diversity on the formation of particulate and mineral-associated organic matter. *Soil Biol. Biochem.* **2022**, *168*, 108648. [CrossRef]
12. Vogt, R.D.; Porcal, P.; Hejzlar, J.; Paule-Mercado, M.C.; Haaland, S.; Gundersen, C.B.; Orderud, G.I.; Eikebrokk, B. Distinguishing between sources of natural dissolved organic matter (DOM) based on its characteristics. *Water* **2023**, *15*, 3006. [CrossRef]
13. Hu, Y.; Lu, Y.H.; Edmonds, J.W.; Liu, C.; Wang, S.; Das, O.; Liu, J.; Zheng, C.M. Hydrological and land use control of watershed exports of dissolved organic matter in a large arid river basin in northwestern China. *J. Geophys. Res. Biogeosci.* **2016**, *121*, 466–478. [CrossRef]
14. Sun, Z.X.; Cui, J.F.; Cheng, J.H.; Tang, X.Y. A novel tool for tracing water sources of streamflow in a mixed land-use catchment. *Sci. Total Environ.* **2024**, *912*, 168800. [CrossRef] [PubMed]
15. Stewart, B.; Shanley, J.B.; Kirchner, J.W.; Norris, D.; Adler, T.; Bristol, C.; Harpold, A.A.; Perdrial, J.N.; Rizzo, D.M.; Sterle, G.; et al. Streams as mirrors: Reading subsurface water chemistry from stream chemistry. *Water Resour. Res.* **2022**, *58*, e2021WR029931. [CrossRef]
16. Buerge, I.J.; Poiger, T.; Müller, M.D.; Buser, H.R. Caffeine, an anthropogenic marker for wastewater contamination of surface waters. *Environ. Sci. Technol.* **2003**, *37*, 691–700. [CrossRef] [PubMed]
17. Gatesman, T.A.; Liljedahl, A.K.; Douglas, T.A.; Debolskiy, M.V.; Trainor, T.P.; Gaedeke, A. Glacier contribution to lowland streamflow: A multi-year, daily geochemical hydrograph separation study in subarctic *Alaska. Chem. Geol.* **2023**, *621*, 121368. [CrossRef]
18. Ladouche, B.; Probst, A.; Viville, D.; Idir, S.; Baque, D.; Loubet, M.; Probst, J.L.; Bariac, T. Hydrograph separation using isotopic, chemical and hydrological approaches (Strengbach catchment, France). *J. Hydrol.* **2001**, *242*, 255–274. [CrossRef]
19. Ledesma, J.L.J.; Grabs, T.; Futter, M.N.; Bishop, K.H.; Laudon, H.; Koehler, S.J. Riparian zone control on base cation concentration in boreal streams. *Biogeosciences* **2013**, *10*, 3849–3868. [CrossRef]
20. Xian, Q.S.; Li, P.H.; Liu, C.; Cui, J.F.; Guan, Z.; Tang, X.Y. Concentration and spectroscopic characteristics of DOM in surface runoff and fracture flow in a cropland plot of a loamy soil. *Sci. Total Environ.* **2018**, *622*, 385–393. [CrossRef]
21. Liu, H.Y.; Guan, Z.; Cheng, J.H.; Tang, X.Y.; Xian, Q.S. Ambient temperature modulates dissipation and redistribution of chlorpyrifos and 3,5,6-trichloro-2-pyridinol in paddy field. *Soil Sediment Contam.* **2022**, *31*, 533–555. [CrossRef]
22. Zhang, W.; Cheng, J.H.; Xian, Q.S.; Cui, J.F.; Tang, X.Y.; Wang, G.X. Dynamics and sources of colloids in shallow groundwater in lowland wells and fracture flow in sloping farmland. *Water Res.* **2019**, *156*, 252–263. [CrossRef] [PubMed]
23. Zhang, W. Hydrological Processes and Colloid Transport in the Sloping Farmland of Purple Soil. Ph.D. Thesis, Institute of Mountain Hazards and Environment, UCAS, Chengdu, China, 2015.
24. Brooks, J.R.; Barnard, H.R.; Coulombe, R.; McDonnell, J.J. Ecohydrologic separation of water between trees and streams in a Mediterranean climate. *Nat. Geosci.* **2009**, *3*, 100–104. [CrossRef]
25. West, A.G.; Patrickson, S.J.; Ehleringer, J.R. Water extraction times for plant and soil materials used in stable isotope analysis. *Rapid Commun. Mass Spectrom.* **2006**, *20*, 1317–1321. [CrossRef] [PubMed]

26. Bauer, S.; Fulda, C.; Schäfer, W. A multi-tracer study in a shallow aquifer using age dating tracers ^{3}H, ^{85}Kr, CFC-113 and SF6—indication for retarded transport of CFC-113. *J. Hydrol.* **2001**, *248*, 14–34. [CrossRef]
27. Stock, B.C.; Jackson, A.L.; Ward, E.J.; Parnell, A.C.; Phillips, D.L.; Semmens, B.X. Analyzing mixing systems using a new generation of Bayesian tracer mixing models. *PeerJ* **2018**, *6*, e5096. [CrossRef]
28. Hopkins, J.B.; Ferguson, J.M. Estimating the diets of animals using stable isotopes and a comprehensive Bayesian mixing model. *PLoS ONE* **2012**, *7*, e28478. [CrossRef]
29. Nash, J.E.; Sutcliffe, J.V. River flow forecasting through conceptual models: Part I. A discussion of principles. *J. Hydrol.* **1970**, *10*, 282–290. [CrossRef]
30. Moriasi, D.N.; Arnold, J.G.; Van Liew, M.W.; Bingner, R.L.; Harmel, R.D.; Veith, T.L. Model evaluation guidelines for systematic quantification of accuracy in watershed simulations. *Trans. ASABE* **2007**, *50*, 885–900. [CrossRef]
31. Zhao, P.; Tang, X.Y.; Zhao, P.; Wang, C.; Tang, J.L. Tracing water flow from sloping farmland to streams using oxygen-18 isotope to study a small agricultural catchment in southwest China. *Soil Tillage Res.* **2013**, *134*, 180–194. [CrossRef]
32. Zhao, P.; Tang, X.Y.; Zhao, P.; Zhang, W.; Tang, J.L. Mixing of event and pre-event water in a shallow Entisol in sloping farmland based on isotopic and hydrometric measurements. SW China. *Hydrol. Process.* **2016**, *30*, 3478–3493. [CrossRef]
33. Birkel, C.; Correa-Barahona, A.; Martinez-Martinez, M.; Granados-Bolanos, S.; Venegas-Cordero, N.; Gutierrez-Garcia, K.; Blanco-Ramirez, S.; Quesada-Mora, R.; Solano-Rivera, V.; Mussio-Mora, J.; et al. Headwaters drive streamflow and lowland tracer export in a large-scale humid tropical catchment. *Hydrol. Process.* **2020**, *34*, 3824–3841. [CrossRef]
34. Hoffland, E.; Kuyper, T.W.; Comans, R.N.J.; Creamer, R.E. Eco-functionality of organic matter in soils. *Plant Soil* **2020**, *455*, 1–22. [CrossRef]
35. Gbadegesin, L.A. Leaching of Veterinary Antibiotics in Purple Soil Orchard and Farmland. Ph.D. Thesis, Institute of Mountain Hazards and Environment, UCAS, Chengdu, China, 2023.
36. McDonnell, J.J. A rationale for old water discharge through macropores in a steep, humid catchment. *Water Resour. Res.* **1990**, *26*, 2821–2832. [CrossRef]
37. Jasechko, S.; Kirchner, J.W.; Welker, J.M.; McDonnell, J.J. Substantial proportion of global streamflow less than three months old. *Nat. Geosci.* **2016**, *9*, 126–129. [CrossRef]
38. Kirchner, J.W. Quantifying new water fractions and transit time distributions using ensemble hydrograph separation: Theory and benchmark tests. *Hydrol. Earth Syst. Sci.* **2019**, *23*, 303–349. [CrossRef]
39. Kaiser, K.; Kalbitz, K. Cycling downwards—Dissolved organic matter in soils. *Soil Biol. Biochem.* **2012**, *52*, 29–32. [CrossRef]
40. Zhang, X.; Chen, J.S.; Zhan, L.C.; Ma, F.Y.; Yan, J.H.; Wang, T. Study on groundwater recharge based on chloride mass balance and hydrochemistry in the irrigated agricultural area, North China Plain. *Environ. Earth Sci.* **2023**, *82*, 70. [CrossRef]
41. Zhou, Y.H.; Wu, J.H.; Gao, X.Y.; Guo, W.Y.; Chen, W.Q. Hydrochemical background levels and threshold values of phreatic groundwater in the greater Xi'an region, China: Spatiotemporal distribution, influencing factors and implication to water quality management. *Expo. Health.* **2023**, *15*, 757–771. [CrossRef]
42. Dayo-Olagbende, G.O.; Ewulo, B.S. Ionic mobility of cations as affected by redox status of two different soil types. *Bulg. J. Soil Sci.* **2021**, *6*, 18–32.
43. Duan, Y.; Gao, X.B.; Li, C.C.; Wang, H.; Kang, C.Q.; Wang, W.Z.; Zhang, X.; Sun, Z.; Xiong, Y.Z.; Wang, Y.X. Combining hydrodynamics, geochemical and multiple isotopic tracers to understand the hydrogeological functioning of karst groundwater system in Jinci, northern China. *J. Hydrol.* **2025**, *648*, 132375. [CrossRef]
44. Liu, Y.; Zhang, C.Y.; Jiang, J.Y.; Zhang, Y.; Wang, G.H.; Xu, L.L.; Qu, Z.H. Interaction between groundwater and surface water in the Qujiang River basin in China: Evidence from chemical isotope measurements. *Water* **2023**, *15*, 3932. [CrossRef]
45. Wang, H.L.; Tang, X.Y.; Zhang, W.; Song, S.B.; McKenzie, B.M. Within-year changes in hydraulic properties of a shallow entisol in farmland and forestland. *Vadose Zone J.* **2015**, *14*, vzj2014-11. [CrossRef]
46. Jencso, K.G. Hydrologic Connectivity Between Landscapes and Streams: Transferring Reach and Plot Scale Understanding to the Catchment Scale. Ph.D. Thesis, Montana State University, Bozeman, MT, Canada, 2010.
47. Outram, F.N.; Cooper, R.J.; Suennenberg, G.; Hiscock, K.M.; Lovett, A.A. Antecedent conditions, hydrological connectivity and anthropogenic inputs: Factors affecting nitrate and phosphorus transfers to agricultural headwater streams. *Sci. Total Environ.* **2016**, *545*, 184–199. [CrossRef]

water

MDPI

Article

Study on the Pollution Mechanism and Driving Factors of Groundwater Quality in Typical Industrial Areas of China

Li Wang [1], Qi Wang [2] and Dechao Zheng [3,*]

[1] School of Civil Engineering, Nanyang Institute of Technology, Nanyang 473004, China; wangli_0726@nyist.edu.cn
[2] School of Information Engineering, Nanyang Institute of Technology, Nanyang 473004, China; 3022050@nyist.edu.cn
[3] No. 1 Geological Team of Shandong Provincial Bureau of Geology and Mineral Resources, Jinan 250100, China
* Correspondence: zdc2025315@163.com

Abstract: Uncovering the characteristics of groundwater pollution and its driving mechanisms are crucial for maintaining its ecological functions. This study focuses on hydrochemical changes and their driving factors in groundwater from different aquifers in industrial zones, taking Zibo City, Shandong Province, China, as the research area. During the dry and flood seasons of 2022, samples of phreatic water in pore media (17 sites) and karst confined water (23 sites) were collected and monitored. Piper trilinear diagrams, Gibbs diagrams, ion ratio diagrams, and a principal component analysis (PCA) were used for in-depth analyses. Pore phreatic water had higher excess rates of Na^+, Cl^-, and NO_3^- than karst confined water, which indicated a greater degree of human impact compared with karst confined water. The main hydrochemical type was $HCO_3{\cdot}SO_4$-Ca, but in the dry season, pore phreatic water shifted to $HCO_3{\cdot}SO_4{\cdot}$Cl-Ca. The ion ratios and PCA indicated that the groundwater quality was mainly controlled by water–rock interactions and industrial activities. In the flood season, pore phreatic water was influenced by evaporite dissolution, industrial activities, and domestic sewage, while in the dry season, it was affected by halite and carbonate weathering dissolution and domestic sewage. Karst confined water was controlled by water–rock interactions and industrial activities in both seasons. The findings reveal that the key drivers of groundwater quality displayed significant differences depending on the aquifer type and seasonal variations. As such, customized approaches are essential to efficiently address and counteract the decline in groundwater quality.

Keywords: groundwater quality; anthropogenic activities; water–rock interaction; principal component analysis; Zibo city

1. Introduction

Groundwater is crucial for human survival and development, and it is essential for sustainable social development and ecological stability [1]. With the acceleration of industrialization and urbanization, human activities are increasingly influencing groundwater quality, primarily through domestic sewage, industrial activities, and agricultural practices [2]. The concentration of and variation in chemical components in groundwater significantly impact water quality and human health. For instance, excessive nitrate levels can increase cancer risks [3], and high sulfate concentrations may cause diarrhea [4]. Therefore, understanding regional groundwater quality characteristics and their controlling factors is vital for scientific groundwater management and protection [5].

Groundwater quality is influenced by natural factors (e.g., topography, lithology, and recharge–discharge conditions) and anthropogenic factors (e.g., land use and agricultural and industrial activities) [6,7]. Previous studies have shown that in rapidly urbanized areas, groundwater quality is mainly controlled by water–rock interactions and domestic sewage [8], while in agricultural regions, it is influenced by water–rock interactions, fertilization, and sewage [9]. In arid areas, evaporation and human activities are dominant [10]. Although previous studies have explored the driving factors of groundwater quality in different regions, research on the pollution patterns and driving mechanisms of groundwater in large industrial cities remains limited. Compared with other regions, groundwater in industrial areas may have unique pollution mechanisms and driving factors. We hypothesize that industrial activities (such as the discharge of industrial wastewater, leakage of waste residues, and leakage of chemical raw materials) are the main driving forces of groundwater pollution in industrial areas, and its pollution patterns may exhibit characteristics of composite pollution, involving the simultaneous presence of multiple pollutants. In addition, the intensity of human activities in industrial areas is much higher than in other regions, and the rapid changes in land use patterns and urban infrastructure construction may further exacerbate the deterioration of groundwater quality.

Zibo City, a typical old industrial area in Shandong Province, China, with a 100-year industrial history based on coal, ceramics, and glass, has seen reduced groundwater recharge and local quality deterioration due to industrialization and urbanization, threatening the water supply [11]. However, the characteristics of groundwater pollution and its driving factors in this region remain unclear to date, posing a potential threat to the sustainable use of groundwater. This study focused on Zibo City to study groundwater quality and its driving factors across different geological settings. The objectives of this study included (1) analyzing the characteristics of changes in groundwater quality in industrial areas over different periods; (2) studying the evolution patterns of groundwater hydrochemistry and the sources of chemical components in industrial areas based on Piper trilinear diagrams, Gibbs diagrams, and ion ratio methods; and (3) applying a principal component analysis to identify the main controlling factors affecting groundwater quality and ultimately reveal the driving forces of groundwater pollution in industrial areas. This study not only fills the research gap in the pollution mechanisms and driving forces of groundwater in large industrial cities and enriches the theoretical framework of groundwater quality research, but also provides an important scientific basis for the sustainable development of groundwater resources and water ecological protection in Zibo City and similar industrial regions.

2. Materials and Methods

2.1. Overview of the Study Area

Zibo City is located at the junction of the Lu Zhong Mountains and the North China Plain, bordering Yimeng Mountains to the south, the North China Plain to the north, Weifang to the east, and Jinan to the west, with a total area of 5965 km^2, including 1614 km^2 of arable land. The topography includes plains in the north and center and low mountains and hills in the south. The terrain of the study area is generally higher in the south and lower in the north. The southern and central parts are mainly mountainous, hilly, and contain basins, while the northern part mainly comprises plains. The altitude ranges from 67.0 to 72.9 m. The industrial areas are mainly concentrated in the basins and plains in the central and northern parts, and the main agricultural areas are in the southern mountainous areas. The area has a warm temperate continental monsoon climate, with an average annual temperature of 12.5 °C–14.2 °C and precipitation of 650 mm.

2.2. Geology and Hydrogeology

Zibo City is in the northern part of the Lu Zhong Uplift in the Shandong Western Uplift Zone of the North China Plate, adjacent to the Jiyang Depression. The stratigraphy is of the North China type. To the north of Linzi district, Quaternary alluvial, fluvial, and marine deposits are widespread, while to the south, carbonate bedrock predominates. The city can be divided into the following three hydrogeological zones: the northern alluvial plain, the central basin, and a southern hilly area. The primary surface water bodies encompass Zi River, Fu River, and Xiaoqing River, among others. The presence of fault structures has facilitated a close hydrological connection between the surface water and groundwater. The main aquifers are fine sand layers in the north and the permeability is generally between 1×10^{-3} cm/s and 5×10^{-3} cm/s, with carbonate karst fracture and Quaternary porous layers in the central basin and limestone karst fracture layers in the south. Groundwater is primarily recharged by atmospheric precipitation and surface water infiltration, and is discharged through artificial extraction and evaporation. In the northern plain area, the groundwater flows from west to east. The central basin and the southern mountainous area are the main distribution areas of karst aquifers. The overall groundwater flow direction is from south to north. The mountainous area is the recharge area and the central basin is the main catchment area.

2.3. Sample Collection and Analysis

To study groundwater quality characteristics and their driving factors in Zibo City, 40 groundwater samples were collected in June 2022 (dry season) and October 2022 (flood season). These included 17 pore phreatic water samples (G01–G17) from depths of 30 to 90 m in the submountain alluvial flood plain and intermountain river basins and 23 karst confined water samples (G18–G40) from depths of 12 to 400 m in exposed and buried karst areas. All water samples were collected by pumping from wells, and thorough well flushing was conducted prior to pumping until the EC of the water samples stabilized. The sample sites are shown in Figure 1.

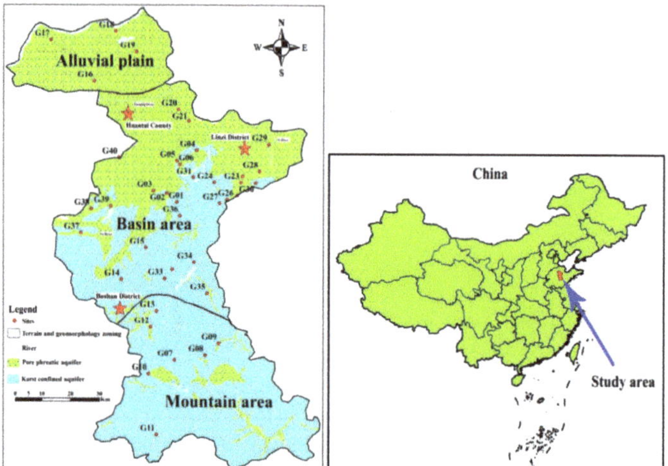

Figure 1. Geographical location map of the study area.

For sampling, 500 mL polyethylene bottles were used. Before sampling, the bottles were cleaned with distilled water and then rinsed three times with the water sample.

Prior to the water sample collection, the wells were purged. Each sample was divided into two parts, one for cation (Na^+, K^+, Ca^{2+}, Mg^{2+}, and Fe^{3+}) analyses, acidified with superior-grade nitric acid to pH < 2 on-site, and the other for anion (HCO_3^-, SO_4^{2-}, Cl^-, and NO_3^-) analyses, with no reagents added. All samples were completely filled; stored without bubbles in a cool, dark place; and analyzed within a week.

The pH was measured on-site using a portable water quality multi-parameter analyzer (Hach-HQ40D, produced by Hach Company Water Quality Analysis Instruments (Shanghai) Co., Ltd., Shanghai, China). Dissolved solids (TDS) were determined using the gravimetric method. Cations were analyzed using inductively coupled plasma mass spectrometry (Agilent 7500ce ICP-MS, Tokyo, Japan). Anions were measured using ion chromatography (Perkin-Elmer Lambda 35, Waltham, MA, USA). Chemical oxygen demand (COD) was measured using alkaline permanganate oxidation and HCO_3^- by hydrochloric acid titration.

2.4. Data Analysis Methods

ArcGIS10.5 was used to create a sampling point distribution map. Excel was used for the statistical analysis of water sample data, including mean and exceedance rate calculations. Origin 2022 was used to create the ion ratio diagrams. Finally, SPSS25.0 was used for the standardization and principal component analysis of groundwater quality data.

3. Results and Discussion

3.1. Groundwater Chemical Composition Characteristics

To analyze the groundwater quality characteristics of pore phreatic water and karst confined water during the two hydrological periods, we statistically analyzed the ranges, means, and exceedance rates of the water quality indicators (Table 1). The results showed that, except for pH, the concentrations of hydrochemical indicators in pore phreatic water were higher than those in karst confined water in both the flood and dry seasons. Specifically, in the flood season, the mean concentrations of K^+, Na^+, Ca^{2+}, Fe^{3+}, HCO_3^-, and NO_3^- in pore phreatic water were higher than in the dry season; for karst confined water, the mean concentrations of K^+, Ca^{2+}, Mg^{2+}, Fe^{3+}, SO_4^{2-}, and COD were higher in the flood season.

Table 1. Statistical water sample index of the study area.

Parameters	Range/(mg·L^{-1})				Average/(mg·L^{-1})				Exceedance Rate/%				Standard
	Flood Season		Dry Season		Flood Season		Dry Season		Flood Season		Dry Season		
	PW	KW	PW	KW	PW	KW	PW	KW	PW	KW	PW	KW	
pH	7.55–8.19	7.74–8.19	7.11–8.03	7.44–8.29	7.86	8.01	7.66	7.85	0.00	0	0	0	6.50–8.50
K^+	0.720–86.2	0.430–4.92	0.266–66.2	0.465–4.61	6.95	1.83	5.78	1.67	–	–	–	–	–
Na^+	12.8–1133	4.40–213	11.6–966	2.64–241	181	40.2	175	40.8	17.7	8.70	11.8	4.36	200
Ca^{2+}	64.5–445	8.75–251	50.4–454	10.4–271	201	120	192	116	–	–	–	–	–
Mg^{2+}	10.3–301	3.63–52.8	10.3–258	4.90–56.0	76.7	23.8	76.7	23.4	–	–	–	–	–
Fe^{3+}	0.040–0.650	0.040–0.280	0.040–0.250	0.040–0.240	0.198	0.088	0.131	0.077	17.7	0	0	0	0.3
Cl^-	18.8–1016	12.9–505	21.11–924	12.6–551	253	72	302	75.2	47.1	4.35	41.2	4.35	250
SO_4^{2-}	28.8–1488	60.9–250	28.5–1215	61.0–275	400	115	408	113	52.9	4.35	52.9	4.35	250
HCO_3^-	146–986	140–426	97.4–1031	103–435	400	265	363	272	–	–	–	–	–
NO_3^-	4.65–346	1.01–128	3.93–431	0.016–139	83.4	43.4	75.5	47.8	29.4	8.70	11.8	17.4	88.6
COD	0.96–1.98	0.95–2.3	0.600–4.30	0.300–3.90	1.54	1.57	1.84	1.09	0	0	5.88	4.35	3.00
TDS	310–5042	362–1408	282–4528	345–1616	1422	565	1431	571	52.9	8.70	47.1	8.70	1000

Note: The units of all water quality parameters are mg·L^{-1}, except for pH, which is dimensionless. Standard is the quality standard for groundwater developed by China (GB/T14848-2017) [12]. PW: Pore phreatic water; KW: karst confined water.

In the flood season, the pH of pore phreatic water ranged from 7.55 to 8.19 (mean value of 7.66) and that of karst confined water ranged from 7.74 to 8.19 (mean value of

7.85), both weakly alkaline. The cation concentration order of both waters was Ca^{2+} > Na^+ > Mg^{2+} > K^+ > Fe^{3+}; Na^+ had the highest exceedance rate (17.7% in pore phreatic water and 8.70% in karst confined water, based on the Grade III standard for groundwater quality in China (GB/T14848-2017)) [12]. For anions, the concentration orders were $SO_4{}^{2-}$ > $HCO_3{}^-$ > Cl^- > $NO_3{}^-$ in pore phreatic water and $HCO_3{}^-$ > $SO_4{}^{2-}$ > Cl^- > $NO_3{}^-$ in karst confined water. $SO_4{}^{2-}$ and $NO_3{}^-$ had the highest exceedance rates (52.9% and 8.70%, respectively). The higher concentration of $HCO_3{}^-$ than $SO_4{}^{2-}$ in karst confined water was due to its occurrence in carbonate aquifers.

In the dry season, the pore phreatic water pH ranged from 7.11 to 8.03 (mean value of 7.66) and that of karst confined water ranged from 7.44 to 8.29 (mean value of 7.85). The cation and anion concentration orders were the same as in the flood season. However, the exceedance rates of $NO_3{}^-$ and COD in karst confined water increased to 17.4% and 4.35% from 8.70% and 0 in the flood season. The new exceedance sample points (G20 and G23) were villages and farmlands in exposed karst areas. Nitrogen from domestic sewage and agricultural fertilizers could infiltrate into deep wells via surface runoff and irrigation return flow, causing $NO_3{}^-$ and COD exceedances.

Overall, during a rainy season, the exceedance rates of groundwater chemical components are significantly higher than those in the dry season. This is mainly due to increased precipitation in the rainy season, which significantly enhances groundwater recharge, allowing more pollutants to enter the groundwater system through surface runoff and infiltration. During flood periods, the rapid recharge of surface water further exacerbates the input of pollutants. In areas with intensive human activities, the input of pollutants such as industrial wastewater discharge, domestic sewage leakage, and agricultural fertilization significantly increases, leading to higher exceedance rates of ions such as Na^+, Cl^-, and $NO_3{}^-$ [13]. Moreover, the exceedance rates of hydrochemical components in pore water are higher than those in karst confined water. This is primarily because pore water is closer to the surface and is more directly affected by human activities. The recharge pathways for pore water are relatively short, allowing pollutants to be more rapidly transported into the aquifer through precipitation and surface runoff [14], thereby exacerbating the exceedance of hydrochemical components. In contrast, karst confined water is mainly distributed in carbonate aquifers, where the longer recharge pathways and greater geological barriers limit the transport and diffusion of pollutants [15], resulting in relatively lower exceedance rates.

3.2. Analysis of Groundwater Chemical Types

A Piper trilinear diagram was used to determine the chemical types of the groundwater in the study area (Figure 2). During the flood season, pore phreatic water and karst confined water had 13 and 8 hydrochemical types, respectively, with $HCO_3{\cdot}SO_4$—Ca being predominant (17.7% and 52.2% of the sampling points). In the dry season, there were 13 and 9 types, with $HCO_3{\cdot}SO_4{\cdot}Cl$—Ca and $HCO_3{\cdot}SO_4$—Ca being the main types (17.7% and 43.5%, respectively). Overall, karst confined water remained predominantly $HCO_3{\cdot}SO_4$—Ca in both seasons, while pore phreatic water shifted from $HCO_3{\cdot}SO_4$—Ca in the flood season to $HCO_3{\cdot}SO_4{\cdot}Cl$—Ca in the dry season. For example, G10 and G11, located in the northern part of the study area in the porous aquifer region, changed from $HCO_3{\cdot}SO_4{\cdot}Cl$—Na·Ca in the flood season to $HCO_3{\cdot}SO_4{\cdot}Cl$—Ca in the dry season, increasing the proportion of this type from 5.88% to 17.7%. Overall, in both seasons, the proportion of sulfate-type water exceeded 50%, indicating a deteriorating trend in groundwater quality in the region, which could be associated with the intense industrial activities in the area [16].

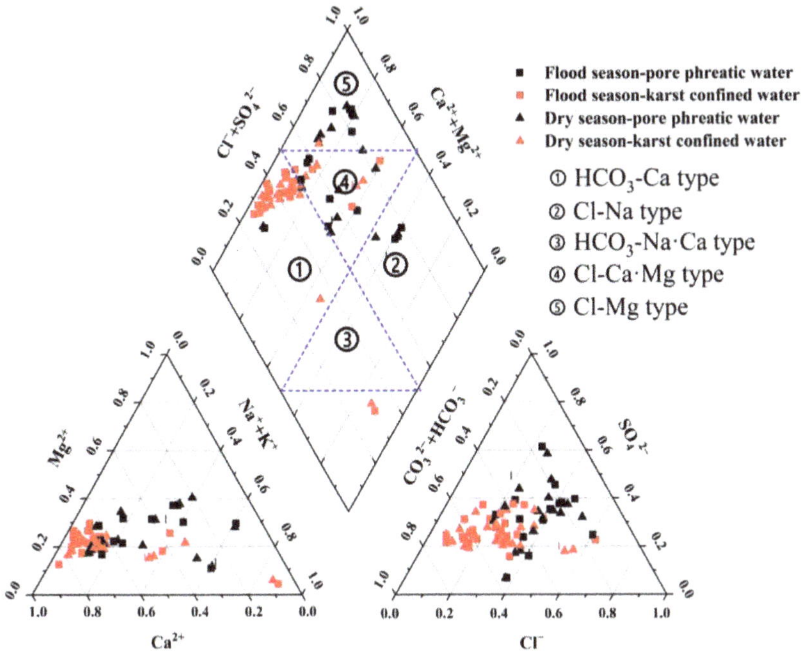

Figure 2. The hydrochemical types' map of groundwater in the study area.

3.3. Control Factors and Sources of Groundwater Ions

3.3.1. Analysis of Control Factors Based on Gibbs Diagrams

Rock weathering, atmospheric precipitation, and evaporation are the main factors affecting the chemical composition of natural waters [17]. In this study, Gibbs diagrams were used to analyze the hydrochemical control factors of pore phreatic water and karst confined water in the study area during both flood and dry seasons. Figure 3a,b show that in both seasons, the sampling points for pore phreatic water and karst confined water exhibited a shifting trend from the rock weathering zone to the evaporation zone, with a more pronounced shift in pore phreatic water. This indicated that rock weathering and evaporation were the primary control factors for the hydrochemistry of both waters, while the influence of atmospheric precipitation on the hydrochemistry was relatively smaller. It is worth noting that evaporation has a more significant impact on pore phreatic water than on karst confined water. This may be related to the shallower burial depth of pore phreatic water, its shorter recharge pathways, and the slower flow velocity and longer residence time of groundwater, which allow evaporation to have more time to influence the chemical composition of groundwater. Although Gibbs diagrams can qualitatively reveal the natural factors controlling groundwater chemistry, they are not effective at distinguishing the impacts of human activities [18], which will be analyzed in subsequent research.

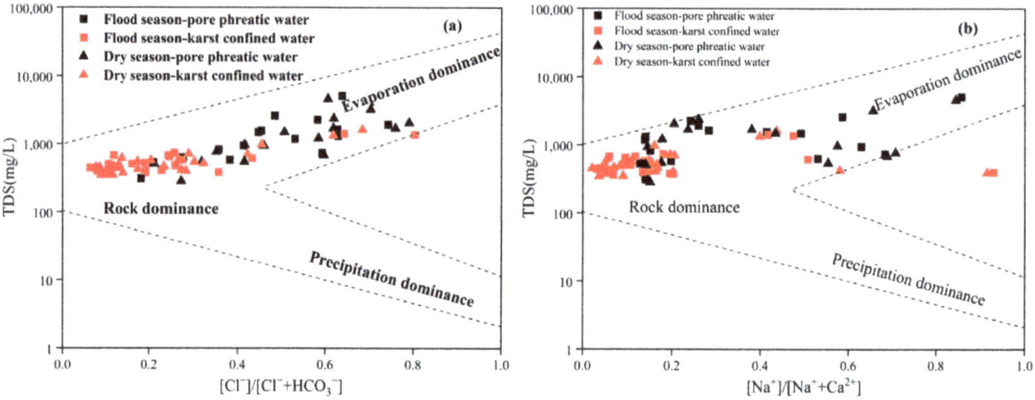

Figure 3. The controlling factors of groundwater hydrochemical variations in the study area. (**a**) $[Cl^-]/[Cl^- + HCO_3^-]$ vs. TDS; (**b**) $[Na^+]/[Na^+ + Ca^{2+}]$ vs. TDS.

3.3.2. Identifying the Sources of Hydrochemical Parameters

Water–rock interaction is a significant factor influencing groundwater hydrochemistry. Research indicates that carbonate rock weathering releases Ca^{2+}, Mg^{2+}, and HCO_3^-; silicate rock weathering releases Ca^{2+}, Mg^{2+}, Na^+, K^+, and HCO_3^-; and evaporite rock weathering releases Ca^{2+}, Mg^{2+}, Na^+, K^+, SO_4^{2-}, and Cl^-. Therefore, ion ratio diagrams can be used to analyze the sources of major ions in groundwater [19].

By examining the relationships between $[Ca^{2+}/Na^+]$ and $[HCO_3^-/Na^+]$ (Figure 4a), the effects of carbonate, silicate, and evaporite weathering on groundwater chemistry could be identified. Xing et al. found that for silicate rock end-members, the $[Ca^{2+}/Na^+]$ and $[HCO_3^-/Na^+]$ ratios are 0.35 ± 0.15 and 2 ± 1, respectively [20]; for carbonate rock end-members, they are 50 and 120; and for evaporite rock end-members, they are 0.25 ± 0.15 and 0.01 ± 0.01. In both flood and dry seasons, the pore phreatic water samples mainly clustered between the silicate and carbonate rock end-member regions, with a few near the evaporite dissolution area. This suggests that the Ca^{2+}, Mg^{2+}, Na^+, and HCO_3^- in the pore phreatic water primarily originated from silicate and carbonate rock weathering, with minor contributions of Na^+ and Ca^{2+} from evaporites like rock salt and gypsum [21]. This was because the phreatic aquifer consisted of sand and gravel layers rich in silicate and carbonate minerals. The karst confined water samples also fell within the carbonate and silicate weathering zones, with 87.0% being closer to the carbonate rock region, indicating a greater influence of carbonate rock weathering on their chemistry that was consistent with previous findings [22]. However, 13% of the karst confined water samples were close to the silicate end-member. This was mainly because these samples were primarily located in areas with a thicker cover of loose sediments. The pore water, acting as one of the recharge sources for the karst confined water in this region, caused a few karst confined water samples to approach the silicate end-member.

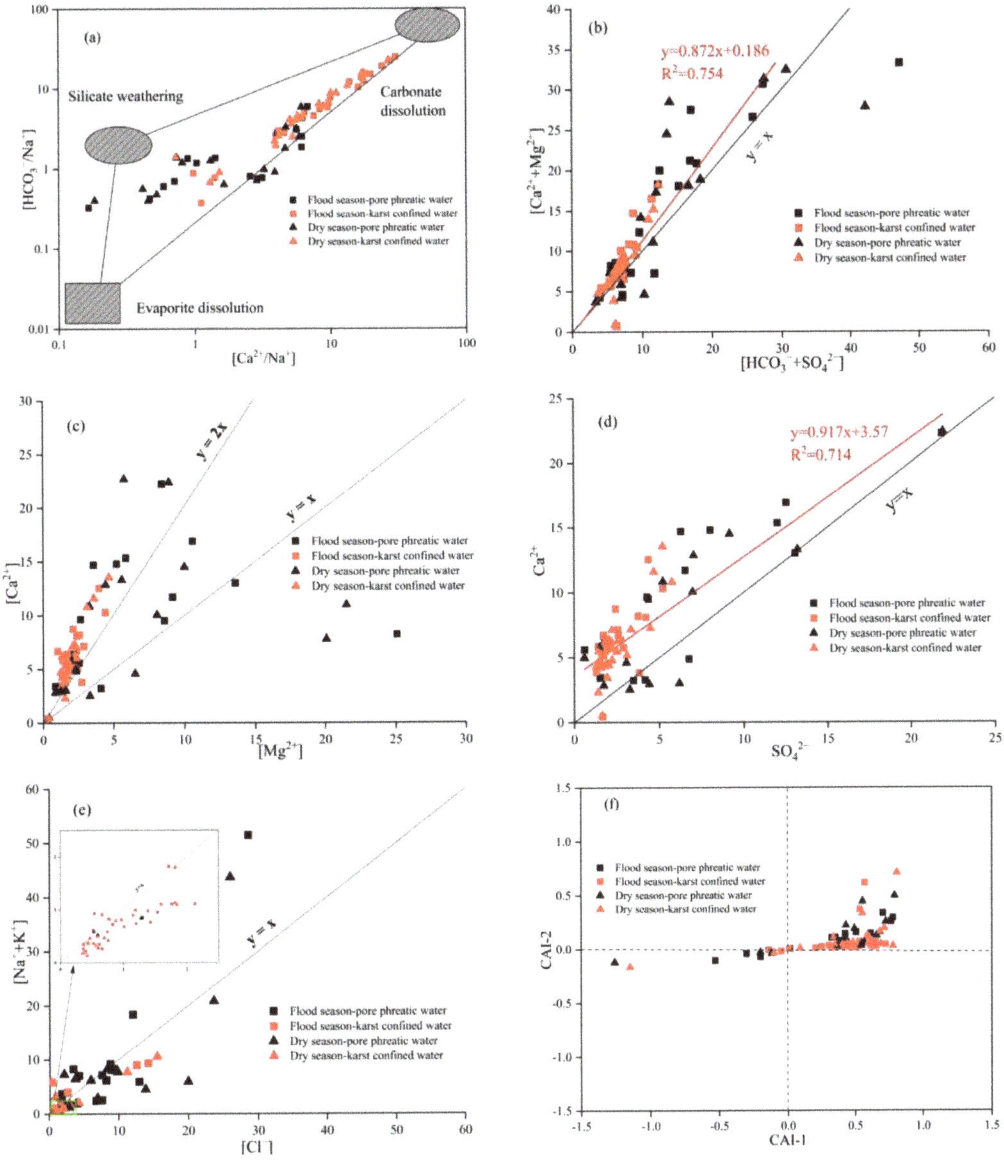

Figure 4. The source identification diagram of groundwater hydrochemical components. (**a**) $[Ca^{2+}/Na^{+}]$ vs. $[HCO_3^{-}/Na^{+}]$; (**b**) $[HCO_3^{-}+SO_4^{2-}]$ vs. $[Ca^{2+} + Mg^{2+}]$; (**c**) $[Mg^{2+}]$ vs. $[Ca^{2+}]$; (**d**) $[SO_4^{2-}]$ vs. $[Ca^{2+}]$; (**e**) $[Cl^{-}]$ vs. $[Na^{+} + K^{+}]$; (**f**) CAI-1 vs. CAI-2. CAI-1 = $(\gamma(Cl^{-}) - [\gamma(Na^{+}) + \gamma(K^{+}))/\gamma(Cl^{-})$. CAI-2 = $(\gamma(Cl^{-}) - [\gamma(Na^{+}) + \gamma(K^{+})]/(\gamma(SO_4^{2-}) + \gamma(HCO_3^{-}) + \gamma(CO_3^{2-}) + \gamma(NO_3^{-}))$.

The molar ratio of $[Ca^{2+} + Mg^{2+}]/[HCO_3^{-} + SO_4^{2-}]$ helps to assess the contribution of carbonate and silicate rock dissolution to ion components (Figure 4b) [23]. In both seasons, 92.3% and 84.6% of the pore phreatic water samples lay above the y = x line, while this proportion was 95.7% for karst confined water. This showed that carbonate rock dissolution dominated in pore phreatic water, with silicate rock dissolution being secondary, and was absolutely dominant in karst confined water. Additionally, [Ca^{2+} + Mg^{2+}] and [HCO_3^- + SO_4^{2-}] exhibited strong correlations (R^2 = 0.754) in both seasons, indicating a common source for some Ca^{2+} and SO_4^{2-}, likely gypsum dissolution [24].

The molar ratio of [Ca^{2+}]/[Mg^{2+}] helps to determine the extent of calcite and dolomite dissolution in carbonate rocks (Figure 4c) [25]. In both seasons, about 17.6% of the pore phreatic water samples fell between y = x and y = 2x and 76.5% fell above y = 2x, indicating that Ca^{2+}, Mg^{2+} and HCO_3^- mainly originated from dolomite dissolution, with calcite being secondary. For karst confined water, 91.3% of the samples were above y = 2x, suggesting that dolomite weathering was the main source of these ions.

The ratio of [Ca^{2+}]/[SO_4^{2-}] indicates the source of SO_4^{2-} (Figure 4d). Most samples from both waters fell above the y = x line in both seasons, with Ca^{2+} and SO_4^{2-} showing a good correlation (R^2 = 0.714). This implied that gypsum dissolution was the primary source of SO_4^{2-} [26], while excess Ca^{2+} originated from carbonate rock weathering.

The molar ratio of (Na^+ + K^+) to Cl^- helps to identify the sources of Na^+ and Cl^-. Most pore phreatic water samples in both seasons clustered around the y = x line with a positive correlation, indicating a common source (rock salt weathering) (Figure 4e). However, 58.8% and 70.6% of the samples fell below the y = x line, suggesting excess Cl^- may have originated from domestic sewage [27]. For karst confined water, 73.9% and 87.0% of the samples were below the y = x line, and most had low (Na^+ + K^+) and Cl^- equivalents (<2 meq/L), indicating a minimal contribution from rock salt weathering. Instead, Na^+, K^+, and Cl^- mainly originated from phreatic water inflow as carbonate rock weathering dominated.

The chlor-alkali index (CAI-1 and CAI-2) assesses cation exchange reactions (Figure 4f). Negative values indicate a forward exchange (Ca^{2+} releases Na^+), while positive values indicate a reverse exchange (Na^+ replaces Ca^{2+}). Most samples in both seasons fell in regions with positive CAI-1 and CAI-2 values, indicating reverse cation exchange. Groundwater Na^+ replaced adsorbed Ca^{2+} in the aquifer medium [28], reducing Na^+ levels and potentially affecting the hydrochemical types, consistent with earlier conclusions.

3.4. *Impact of Human Activities*

Human activities exert a significant influence on groundwater quality. In regions characterized by intense human activities, the concentrations of NO_3^-, Cl^-, SO_4^{2-}, and Na^+ in groundwater are elevated. Specifically, SO_4^{2-} is predominantly influenced by industrial activities, while NO_3^-, Cl^-, and Na^+ are primarily affected by agricultural fertilization, fecal matter, and domestic sewage [29].

Figure 5a shows that in both seasons, 97.5% of the samples had higher [SO_4^{2-}]/[Ca^{2+}] ratios than [NO_3^-]/[Ca^{2+}]. Specifically, in the flood season, [SO_4^{2-}]/[Ca^{2+}] ranged from 0.107 to 3.78, while [NO_3^-]/[Ca^{2+}] ranged from 0.004 to 0.671. In the dry season, [SO_4^{2-}]/[Ca^{2+}] ranged from 0.119 to 3.27 and [NO_3^-]/[Ca^{2+}] ranged from 0.006 to 0.897. This finding indicated that industrial activities were the primary factor influencing groundwater quality, followed by agricultural activities. This was consistent with the study area's long-standing industrial development, particularly its emphasis on heavy chemical industries [30]. Moreover, the pore phreatic water samples were mostly above the karst confined water samples, showing that industrial activities more strongly impacted pore phreatic water. This was because pore phreatic water is shallow and lacks a stable confining layer, allowing SO_4^{2-} from industrial wastewater to more easily infiltrate into groundwater systems through surface runoff and leakage [31].

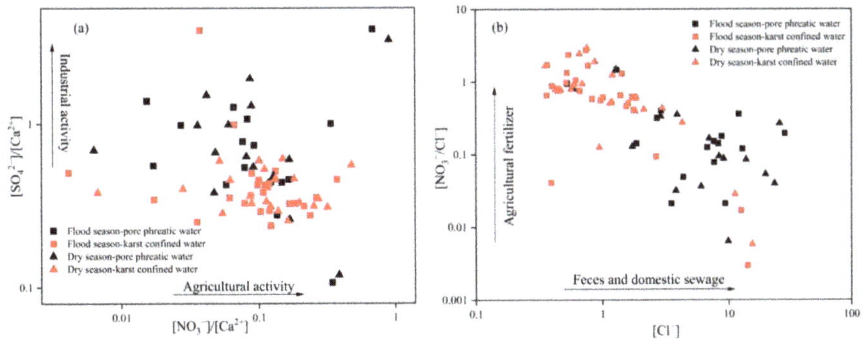

Figure 5. The ratio diagram of groundwater hydrochemical components. (**a**) $[NO_3^-]/[Ca^{2+}]$ vs. $[SO_4^{2-}]/[Ca^{2+}]$ and (**b**) $[Cl^-]$ vs. $[NO_3^-]/[Cl^-]$.

The relationship between $[NO_3^-]/[Cl^-]$ and $[Cl^-]$ helps to identify NO_3^- sources. High $[Cl^-]$ and low $[NO_3^-]/[Cl^-]$ suggest that NO_3^- is from sewage and fecal matter, low $[Cl^-]$ and high $[NO_3^-]/[Cl^-]$ indicate agricultural origins, and low values for both suggest that soil nitrogen is the main source [32]. Figure 5b shows that in both seasons, 88.2% of the pore phreatic water and 52.2% of the karst confined water samples contained higher $[Cl^-]$ than $[NO_3^-]/[Cl^-]$, indicating that NO_3^- in pore phreatic water was mainly from fecal matter and sewage, while in confined water, it originated from fecal matter, sewage, and agricultural fertilizers.

The pore phreatic water samples were mainly from the central and northern plains of the study area, in Linzi District and Huantai County, which are densely populated, with NO_3^- in groundwater mainly from fecal matter and sewage [33]. The karst confined water samples were from the plains and southern Boshan district, where farmland is mostly dry land (96.9%). According to the Zibo Statistical Yearbook, Boshan district's average fertilizer application is 0.328 t/hm^2, below the study area's average of 0.491 t/hm^2, but limited fertilizer absorption by dry-land crops could lead to excess fertilizer entering groundwater via rainwater runoff, increasing NO_3^- in karst confined water.

3.5. Identification of the Main Factors Controlling Groundwater Quality

To gain a deeper understanding of the impact of natural factors and human activities on the groundwater quality of the study area, we selected 12 parameters (K^+, Na^+, Ca^{2+}, Mg^{2+}, Fe^{3+}, Cl^-, SO_4^{2-}, HCO_3^-, NO_3^-, COD, TDS, and pH) and performed a PCA on pore phreatic water and karst confined water during both the flood and dry seasons. Prior to data entry, Kaiser–Meyer–Olkin (KMO) and Bartlett sphericity tests were conducted. The results showed that for flood-season pore phreatic water and karst confined water, the KMO values were 0.643 and 0.611, respectively, and the Bartlett sphericity test values were 589 and 497 ($p < 0.001$). For dry-season pore phreatic water and karst confined water, the KMO values were 0.650 and 0.765, respectively, and the Bartlett sphericity test values were 538 and 449 ($p < 0.001$). These results confirmed the suitability of the hydrochemical data for the PCA.

Based on eigenvalues > 1, during the flood season, two principal factors controlling the hydrochemical variations were identified for both pore phreatic water and karst confined water, explaining 80.8% and 77.0% of the total variance, respectively. During the dry season, two and three principal factors were identified for pore phreatic water and karst confined water, respectively, explaining 80.6% and 85.2% of the total variance.

3.5.1. Primary Factors Influencing Groundwater Quality During the Flood Season

As shown in Table 2, during the flood season, PC1 for pore phreatic water explained 62.3% of the information and was strongly correlated with Na^+, Mg^{2+}, Fe^{3+}, TDS, K^+, Cl^-, NO_3^-, SO_4^{2-}, and HCO_3^-. As confirmed in Section 3.3.2, the Na^+ and Cl^- of the pore phreatic water primary originated from the weathering and dissolution of rock salt. Typically, sulfate in groundwater mainly originates from the oxidation of sulfide minerals, the dissolution of gypsum, industrial wastewater, domestic sewage, and fertilizer application [34]. Nitrate in groundwater primarily stems from domestic sewage, fertilizers, soil nitrogen, and precipitation [35]. In this study, it was demonstrated in Section 3.3.2 that SO_4^{2-} mainly originated from industrial activities and evaporite dissolution, while NO_3^- predominantly originated from fecal matter and domestic sewage. Thus, PC1 represented the combined influence of evaporite dissolution, industrial activities, and domestic sewage on the groundwater quality of pore phreatic water. PC2 accounted for 18.5% of the information and was strongly correlated with Ca^{2+} and pH values. Numerous studies have demonstrated that low-pH environments can significantly enhance the dissolution of carbonate rocks [36]. The Ca^{2+} in the pore phreatic water was mainly from the dissolution of carbonate rocks, which was also demonstrated in Section 3.3.2. Therefore, PC2 reflected the impact of carbonate rock dissolution on the quality of pore phreatic water.

Table 2. Principal component analysis of major ions during the flood season.

Parameters	Pore Phreatic Water		Karst Confined Water	
	PC1	PC2	PC1	PC2
Na^+	0.980	0.055	0.953	0.207
Mg^{2+}	0.922	0.341	0.875	−0.228
Fe^{3+}	0.919	0.281	0.960	−0.081
TDS	0.913	0.388	0.993	−0.013
K^+	0.913	−0.109	0.445	0.696
Cl^-	0.901	0.372	0.926	0.282
NO_3^-	0.865	0.143	−0.023	−0.763
SO_4^{2-}	0.843	0.274	0.690	−0.581
HCO_3^-	0.811	0.449	0.854	0.183
COD	0.442	0.054	−0.217	0.672
Ca^{2+}	0.132	0.896	0.852	−0.186
pH	−0.138	−0.782	−0.681	0.233
Eigenvalue	8.27	1.43	7.09	2.15
Contribution Rate (%)	62.3	18.5	59.0	18.0
Cumulative Contribution Rate (%)	62.3	80.8	59.0	77.0

For karst confined water during the flood season, PC1 explained 59.0% of the information and was mainly correlated with TDS, Na^+, Cl^-, SO_4^{2-}, and HCO_3^-. Na^+ and Cl^- predominantly originated from the weathering and dissolution of rock salt, while SO_4^{2-}, HCO_3^-, and Ca^{2+} were mainly influenced by carbonate rocks, gypsum, and industrial activities. Therefore, PC1 represented the influence of the weathering and dissolution of rock salt, carbonate rocks, and gypsum as well as industrial activities on karst confined water. PC2 accounted for 18.0% of the information and was strongly correlated with NO_3^- and pH. As mentioned earlier, in this study area, the high concentration of NO_3^- was primarily attributed to agricultural fertilizers and domestic sewage. Thus, PC2 mainly reflected the impact of agricultural activities and domestic sewage on the chemical composition of karst confined water.

3.5.2. Primary Factors Influencing Groundwater Quality During the Dry Season

During the dry season, PC1 for pore phreatic water explained 46.0% of the variables and was strongly correlated with K^+, Na^+, NO_3^-, HCO_3^-, TDS, and Mg^{2+}. It also showed a certain positive correlation with Cl^- and SO_4^{2-} (Table 3). Na^+, HCO_3^-, and Mg^{2+} mainly originated from the weathering and dissolution of rock salt and carbonate rocks, while NO_3^- predominantly originated from fecal matter and domestic sewage. Therefore, PC1 represented the influence of rock salt and carbonate rock weathering as well as domestic sewage on pore phreatic water. PC2 accounted for 34.6% of the information and was strongly correlated with Ca^{2+}, pH, Cl^-, and SO_4^{2-}. The common source of Ca^{2+} and SO_4^{2-} is mainly gypsum and industrial activities can also increase the SO_4^{2-} and Cl^- in groundwater. Thus, PC2 could be summarized as the impact of gypsum dissolution and industrial activities on pore phreatic water.

Table 3. Principal component analysis of major ions during the dry season.

Parameters	Pore Phreatic Water		Karst Confined Water		
	PC1	PC2	PC1	PC2	PC3
K^+	0.951	0.021	0.253	0.150	0.909
Na^+	0.918	0.324	0.725	0.530	0.353
NO_3^-	0.892	0.081	0.068	−0.896	−0.112
HCO_3^-	0.882	0.288	0.723	0.553	−0.152
TDS	0.747	0.659	0.912	0.240	0.310
Mg^{2+}	0.690	0.640	0.807	0.466	−0.205
COD	0.305	0.261	0.518	0.568	0.294
Ca^{2+}	−0.068	0.874	0.847	−0.072	0.398
pH	−0.105	−0.801	−0.778	−0.069	−0.192
Fe^{3+}	0.540	0.763	0.761	0.317	0.456
Cl^-	0.601	0.760	0.801	0.385	0.410
SO_4^{2-}	0.631	0.691	0.852	−0.050	0.232
Eigenvalue	7.90	1.77	7.86	1.35	1.01
Contribution Rate (%)	46.0	34.6	51.1	19.5	14.6
Cumulative Contribution Rate (%)	46.0	80.6	51.1	70.6	85.2

For karst confined water during the dry season, PC1 explained 51.1% of the information and was strongly correlated with Na^+, Mg^{2+}, TDS, HCO_3^-, Ca^{2+}, pH, Cl^-, and SO_4^{2-}. As mentioned earlier, PC1 reflected the influence of water–rock interactions and industrial activities on karst confined water. PC2 accounted for 19.5% of the information and was strongly negatively correlated with NO_3^- and strongly positively correlated with COD. NO_3^- in karst confined water mainly originated from agricultural activities and domestic sewage. PC3 accounted for 14.6% of the variance and was strongly correlated with K^+, while it moderately correlated with Fe^{3+}. Previous studies have found that high concentrations of iron in groundwater can indicate industrial pollution as well as a reducing environment in water [37]. As the principal components were independent of each other, and as PC1 represented industrial pollution, it followed that PC3 represented the influence of the aquifer's reducing environment on the groundwater chemistry.

4. Limitations of This Study

This study has comprehensively analyzed the characteristics of groundwater quality and their controlling factors in different aquifers within the industrial zone of Zibo during flood and dry seasons, revealing the significant impact of industrial activities on groundwater quality. However, the source tracing analysis in this study mainly relied on hydrochemical methods, which, although capable of qualitatively identifying the main sources of pollutants in groundwater, had limitations in distinguishing the migration and transformation patterns of different pollutants and in analyzing the contribution rates of each pollutant to groundwater pollution. For instance, this study identified that the high exceedance rates of SO_4^{2-} and Cl^- were mainly due to the impact of industrial activities, but it was difficult to further distinguish whether these pollutants originated from specific industrial discharge points or were the result of cumulative effects from surface runoff and infiltration. Future studies could apply isotope tracing techniques in combination with source apportionment models to conduct more accurate research on the sources of pollutants in the region's groundwater [38,39]. In addition, the sampling time and sample size of this study were limited, with sampling conducted only during the flood and dry seasons, which may not fully reflect the dynamic changes in groundwater quality. Future research could increase the sampling frequency, especially during flood periods and peak industrial activity periods, to gain a more comprehensive understanding of the trends in groundwater quality changes. Applying machine learning methods to predict pollution trends could provide stronger support for the long-term management and protection of groundwater quality.

5. Conclusions

This study comprehensively analyzed the characteristics of groundwater quality and their controlling factors in different aquifers within the industrial zone of Zibo during flood and dry seasons. The results indicated that, in the industrial area, the exceedance rates of SO_4^{2-} and Cl^- in pore water reached as high as 52.9% and 44.5%, respectively, demonstrating a significant impact of industrial activities on groundwater. Rock weathering and evaporation were the primary control factors for the hydrochemistry of both waters, although evaporation had a more significant impact on pore phreatic water than on karst confined water. Water–rock interactions and industrial activities were the two primary factors affecting groundwater quality, with industrial activities having a more pronounced impact on pore phreatic water. These results demonstrate that the dominant factors influencing groundwater quality exhibited substantial variability across different aquifer media and seasons. Drawing from the research conclusions, we propose that preventing the discharge of untreated industrial wastewater is essential to mitigate the decline in groundwater quality within industrial regions. Although our study has shed light on the significant impact of industrial activities on groundwater pollution, it faced limitations in precisely quantifying the contributions of individual factors due to restricted sampling times and sample sizes. Future investigations should consider utilizing advanced source tracing techniques, such as isotope analysis, alongside machine learning algorithms to delve deeper into the factors influencing groundwater quality in industrial areas. To sum up, this study not only enhances the theoretical understanding of groundwater hydrochemical evolution in industrial settings, but also furnishes a scientific foundation for safeguarding groundwater quality in these locales.

Author Contributions: L.W.: Investigation, Methodology, Software, Data Curation, and Writing—Original Draft Preparation. Q.W.: Investigation, Software, and Methodology. D.Z.: Supervision, Methodology, and Writing—Reviewing and Editing. All authors have read and agreed to the published version of the manuscript.

Funding: This work was supported by the Nanyang Institute of Technology Doctoral Research Startup Fund Project (NGBJ-510069).

Data Availability Statement: The data are available from the authors upon reasonable request.

Acknowledgments: The authors gratefully acknowledge the editor and anonymous reviewers for their valuable comments on this manuscript. The authors also appreciate financial support from different organizations.

Conflicts of Interest: The authors declare no conflicts of interest.

References

1. Velis, M.; Conti, K.I.; Biermann, F. Groundwater and human development: Synergies and trade-offs within the context of the sustainable development goals. *Sustain. Sci.* **2017**, *12*, 1007–1017. [CrossRef] [PubMed]
2. Zhang, Q.; Miao, L.; Wang, H.; Hou, J.; Li, Y. How Rapid Urbanization Drives Deteriorating Groundwater Quality in a Provincial Capital of China. *Pol. J. Environ. Stud.* **2020**, *29*, 441–450. [CrossRef] [PubMed]
3. Picetti, R.; Deeney, M.; Pastorino, S.; Miller, M.R.; Shah, A.; Leon, D.A.; Dangour, A.D.; Green, R. Nitrate and nitrite contamination in drinking water and cancer risk: A systematic review with meta-analysis. *Environ. Res.* **2022**, *210*, 112988. [CrossRef] [PubMed]
4. Backer, L.C. Assessing the acute gastrointestinal effects of ingesting naturally occurring, high levels of sulfate in drinking water. *Crit. Rev. Clin. Lab. Sci.* **2000**, *37*, 389–400. [CrossRef]
5. Teklearegay, T.; Atlabachew, A.; Abebe, A.; Jothimani, M. Comprehensive hydrogeochemical and statistical assessment of groundwater quality for drinking and irrigation in the Demie River catchment, Southern Ethiopia. *SN Appl. Sci.* **2025**, *7*, 1–35. [CrossRef]
6. Qu, S.; Duan, L.; Shi, Z.; Liang, X.; Lv, S.; Wang, G.; Yu, R. Hydrochemical assessments and driving forces of groundwater quality and potential health risks of sulfate in a coalfield, northern Ordos Basin, China. *Sci. Total Environ.* **2022**, *835*, 155519. [CrossRef]
7. Ewusi, A.; Sunkari, E.D.; Seidu, J.; Coffie-Anum, E. Hydrogeochemical characteristics, sources and human health risk assessment of heavy metal dispersion in the mine pit water-surface water-groundwater system in the largest manganese mine in ghana. *Environ. Technol. Innov.* **2022**, *26*, 102312. [CrossRef]
8. Zhu, M.; Xie, Y.; Su, J.; Lu, S. Assessment of groundwater quality of unconfined aquifers in an urbanized area using the water quality index method. *Environ. Sci. Pollut. Res.* **2024**, *31*, 64870–64883. [CrossRef]
9. Zhai, T.L.; Zhang, Q.Q.; Wang, L.; Wang, H.W. Temporal and spatial variations hydrochemical components and driving factors in Baiyangdian lake in the northern plain of China. *J. Groundw. Sci. Eng.* **2024**, *12*, 293–308. [CrossRef]
10. Wang, G.S.; Wang, G.; Xia, J.; Chen, J. Quantification of effects of climate variations and human activities on runoff by a monthly water balance model: A case study of the Chaobai River basin in northern China. *Water Resour. Res.* **2009**, *45*, 1–12. [CrossRef]
11. Li, J.; Zhang, Y.K.; Zhao, Y. Evaluation on quality and health risk of groundwater in a highly urbanized watershed, China. *Environ. Sci. Pollut. Res.* **2024**, *31*, 27136–27154. [CrossRef] [PubMed]
12. *GB/T14848-2017*; Standard for Groundwater Quality. General Administration of Quality Supervision, Inspection and Quarantine of the People' s Republic of China. Ministry of Natural Resources of the People's Republic of China (MNRPRC): Beijing, China, 2017.
13. Li, S.L.; Liu, C.Q.; Li, J.; Liu, X.; Chetelat, B.; Wang, B.; Wang, F. Assessment of the sources of nitrate in the Changjiang River, China using a nitrogen and oxygen isotopic approach. *Environ. Sci. Technol.* **2010**, *44*, 1573–1578. [CrossRef] [PubMed]
14. He, H.; Xiong, X.; Wu, T.; Hu, R.; Chen, Y.F.; Yang, Z. Pore-scale study of particle transport and clogging mechanisms in a porous micromodel. *Sep. Purif. Technol.* **2025**, *362*, 131929. [CrossRef]
15. Campanale, C.; Losacco, D.; Triozzi, M.; Massarelli, C.; Uricchio, V.F. An Overall Perspective for the Study of Emerging Contaminants in Karst Aquifers. *Resources* **2022**, *11*, 105. [CrossRef]
16. Miao, Z.; Carroll, K.C.; Brusseau, M.L. Characterization and quantification of groundwater sulfate sources at a mining site in an arid climate: The Monument Valley site in Arizona, USA. *J. Hydrol.* **2013**, *504*, 207–215. [CrossRef]
17. Xing, J.B.; Long, W.; Jie, Z.; Zhai, T.L. Hydrochemical variation characteristics and driving factors of surface water in arid Areas—A case study of Beichuan River in Northwest China. *Front. Environ. Sci.* **2024**, *12*, 1493390. [CrossRef]
18. Andres, M.; Paul, S. Groundwater chemistry and the gibbs diagram. *Appl. Geochem.* **2018**, *97*, 209–212. [CrossRef]
19. Pan, X.; Sun, B.; Zhang, S.; Li, G.; Tian, Z.; Guo, Z. Ion sources and seasonal recharge characteristics of groundwater around Dali lake in semi-arid region of inner Mongolia plateau, China. *J. Geochem. Explor.* **2025**, *269*, 107612. [CrossRef]
20. Wang, Q.; Dong, S.; Wang, H.; Yang, J.; Huang, H.; Dong, X.; Yu, B. Hydrogeochemical processes and groundwater quality assessment for different aquifers in the Caojiatan coal mine of Ordos Basin, northwestern China. *Environ. Earth Sci.* **2020**, *79*, 1–15. [CrossRef]
21. Zhao, Y.; Wang, C.; Xiang, W.; Zhang, S. Evaluation of the hydrochemical evolution characteristics and renewable capacity of deep fresh groundwater in the Hangzhou bay new zone, China. *Environ. Earth Sci.* **2019**, *78*, 1–15. [CrossRef]

22. He, H.; Liu, G.; Zou, Y.; Li, X.; Ji, M.; Li, D. Coupled action of rock weathering and aquatic photosynthesis: Influence of the biological carbon pump effect on the sources and deposition of organic matter in Ngoring Lake, Qinghai–Tibet Plateau, China. *Catena* **2021**, *203*, 105370. [CrossRef]
23. Xiao, Y.; Shao, J.; Cui, Y.; Zhang, G.; Zhang, Q. Groundwater circulation and hydrogeochemical evolution in Nomhon of Qaidam basin, northwest China. *J. Earth Syst. Sci.* **2017**, *126*, 1–16. [CrossRef]
24. Ren, X.; Zhang, Z.; Yu, R.; Li, Y.; Li, Y.; Zhao, Y. Hydrochemical variations and driving mechanisms in a large linked river-irrigation-lake system. *Environ. Res.* **2023**, *225*, 115596. [CrossRef] [PubMed]
25. Billong, P.T.N.; Feumba, R.; Ndjigui, P.D. Hydrogeochemical appraisal of groundwater quality in Ngoua watershed (Douala-Cameroon): Implication for domestic purposes. *Sci. Afr.* **2023**, *22*, e01910. [CrossRef]
26. Wang, L.; Yang, N.; Zhao, Y.; Zhang, Q. Research on the Features and Driving Factors of Shallow Groundwater Quality in Arid Areas, Northwest China. *Water* **2025**, *17*, 934. [CrossRef]
27. Cao, M.; Hu, A.; Gad, M.; Adyari, B.; Qin, D.; Zhang, L. Domestic wastewater causes nitrate pollution in an agricultural watershed, China. *Sci. Total Environ.* **2022**, *823*, 153680. [CrossRef]
28. Gugulothu, S.; Subbarao, N.; Das, R.; Dhakate, R. Geochemical evaluation of groundwater and suitability of groundwater quality for irrigation purpose in an agricultural region of south India. *Appl. Water Sci.* **2022**, *12*, 1–13. [CrossRef]
29. Zhang, Q.; Wang, H. Assessment of sources and transformation of nitrate in the alluvial-pluvial fan region of north China using a multi-isotope approach. *J. Environ. Sci.* **2020**, *89*, 9–22. [CrossRef]
30. Zhang, Y.; Hou, K.; Qian, H.; Gao, Y.; Fang, Y.; Tang, S.; Zhang, Q. Natural-human driving factors of groundwater salinization in a long-term irrigation area. *Environ. Res.* **2023**, *220*, 115178. [CrossRef]
31. Huang, H.; Liang, X.; Xiao, C.; Wang, Z. Analysis and assessment of confined and phreatic water quality using a rough set theory method in Jilin City, China. *Water Sci. Technol. Water Sup.* **2015**, *15*, 773–783. [CrossRef]
32. Zhang, Y. Hydrochemical characteristics, controlling factors and groundwater sources of Zaozigou gold mine. *Sustainability* **2024**, *16*, 7989. [CrossRef]
33. Cui, R.; Fu, B.; Mao, K.; Chen, A.; Zhang, D. Identification of the sources and fate of NO_3^--N in shallow groundwater around a plateau lake in southwest China using NO_3^- isotopes ($\delta^{15}N$ and $\delta^{18}O$) and a Bayesian model. *J. Environ. Manag.* **2020**, *270*, 110897. [CrossRef] [PubMed]
34. Zhang, D.; Li, X.-D.; Zhao, Z.-Q.; Liu, C.-Q. Using dual isotopic data to track the sources and behaviors of dissolved sulfate in the western North China Plain. *Appl. Geochem.* **2015**, *52*, 43–56. [CrossRef]
35. Wang, F.; Liu, L.; Xu, W.; Li, Y.; Ruan, Q.; Cao, W. Multiple stable isotopic approaches for tracing nitrate contamination sources: Implications for nitrogen management in complex watersheds. *Ecotox. Environ. Saf.* **2024**, *269*, 115822. [CrossRef]
36. Naz, M.; Dai, Z.; Hussain, S.; Tariq, M.; Danish, S.; Khan, I.U. The soil pH and heavy metals revealed their impact on soil microbial community. *J. Environ. Manag.* **2022**, *321*, 115770. [CrossRef]
37. Juahir, H.; Zain, S.M.; Yusoff, M.K.; Hanidza, T.T.; Armi, A.M.; Toriman, M.E. Spatial water quality assessment of Langat River basin (Malaysia) using environmetric techniques. *Environ. Monit. Assess.* **2011**, *173*, 625–641. [CrossRef]
38. Zhang, Q.; Wang, H.; Liu, L.; Zhai, T.; Zhang, X. Multiple isotopes reveal the driving mechanism of high NO_3^- level and key processes of nitrogen cycling in the lower reaches of Yellow River. *J. Environ. Sci.* **2024**, *138*, 597–606. [CrossRef]
39. Xie, C.; Zhao, L.; Eastoe, C.J.; Wang, N.; Dong, X. An isotope study of the Shule River Basin, Northwest China: Sources and groundwater residence time, sulfate sources and climate change. *J. Hydrol.* **2022**, *612*, 128043. [CrossRef]

water

MDPI

Article

A New Insight into Sulfate Contamination in Over-Exploited Groundwater Areas: Integrating Multivariate and Geostatistical Techniques

Li Wang [1], Qi Wang [2], Wenchang Li [3], Yifeng Liu [3] and Qianqian Zhang [4,*]

1 School of Civil Engineering, Nanyang Institute of Technology, Nanyang 473004, China; wangli_0726@nyist.edu.cn
2 School of Information Engineering, Nanyang Institute of Technology, Nanyang 473004, China; 3022050@nyist.edu.cn
3 Shandong Coalfield Geological Planning Research Institute, Jinan 250000, China
4 Hebei and China Geological Survey Key Laboratory of Groundwater Remediation, Institute of Hydrogeology and Environmental Geology, Chinese Academy of Geological Sciences, Shijiazhuang 050061, China
* Correspondence: zhangqianqian@mail.cgs.gov.cn

Abstract: The issue of sulfate (SO_4^{2-}) pollution in groundwater has already attracted widespread attention from scientists. However, at the large-scale regional level, especially in areas with groundwater overexploitation, the pollution mechanisms and sources of sulfate remain unclear. This study innovatively investigates the spatial distribution characteristics and sources of SO_4^{2-} in the groundwater of the Hutuo River alluvial fan area, an understudied region facing significant environmental challenges due to overexploitation. Utilizing a combination of hydrochemical analysis, multivariate statistical methods, and geostatistical techniques, we reveal that the mean concentration of SO_4^{2-} is significantly higher (127 mg/L) in overexploited areas, with an exceedance rate of 5.1%. Our findings uncover substantial spatial heterogeneity in SO_4^{2-} concentrations, with particularly high levels in the river valley plain (RVP) (175 mg/L) and the upper area of the alluvial fan (UAF) (169 mg/L), which we attribute to distinct human activities. A novel contribution of our study is the identification of groundwater depth as a critical factor influencing SO_4^{2-} distribution ($p < 0.001$). We also demonstrate that the higher proportion of sulfate-type waters in overexploited areas is primarily due to the accelerated oxidation of sulfide minerals caused by overexploitation. Principal component analysis (PCA) and correlation analysis further identify the main sources of SO_4^{2-} as industrial wastewater, domestic sewage, the dissolution of evaporites, and the oxidation of sulfide minerals. By integrating geostatistical techniques, we present the spatial distribution of sulfate pollution sources at a fine scale, providing a comprehensive and spatially explicit understanding of the pollution dynamics. These results offer a novel scientific basis for developing targeted strategies to control sulfate pollution and protect the sustainable use of regional groundwater resources. Our study thus fills a critical knowledge gap and provides actionable insights for groundwater management in similar regions facing overexploitation challenges.

Keywords: sulfate; pollution sources; groundwater; geostatistical techniques

1. Introduction

Sulfate (SO_4^{2-}) is one of the common soluble anions in groundwater, characterized by its chemical stability, strong solubility, and ease of migration [1,2]. Sulfate exhibits a certain degree of oxidizing properties, which can influence redox reactions in groundwater and alter the migration and transformation behaviors of other pollutants. Due to its diverse sources and complex migration and transformation processes [3], the distribution characteristics and source identification of sulfate in groundwater have long been a research focus and challenging issue in the field of hydrogeochemistry.

Excessive concentrations of sulfate in groundwater can pose multifaceted threats to ecosystems and human health [4]. For example, they can impact the respiration and reproduction of fish and may even lead to the death of aquatic organisms [5]. Regarding human health, the consumption of groundwater with sulfate levels exceeding the standard may cause gastrointestinal dysfunction, such as diarrhea and bloating, and long-term consumption may also damage organs such as the kidneys [6]. In the industrial sector, groundwater with elevated sulfate concentrations can corrode pipelines and equipment, increasing the difficulty and cost of industrial water treatment [7]. Therefore, studying the distribution characteristics and sources of sulfate in groundwater is of great significance for protecting the ecological environment, safeguarding human health, and the rational utilization of groundwater resources.

The accurate identification of sulfate pollution sources is a crucial step in formulating groundwater pollution prevention strategies and sustainable water resource management plans. Previous studies found that the sources of sulfate in water environments are extensive and diverse, including natural weathering, agricultural activities, industrial emissions, and domestic sewage [8], which greatly increases the complexity of pollution source identification. Current research endeavors to identify sulfate sources within aquatic environments have predominantly concentrated on elucidating the pollution sources of sulfate in surface water systems [8,9], while studies on the sources of sulfate pollution in groundwater are mainly concentrated in industrial zones and karst regions [3,10]. However, there is a relative lack of research on the identification of sulfate sources in large-scale overexploited groundwater areas. Especially in groundwater overexploitation areas where the flow and recharge conditions of the groundwater are complex, making the pollution characteristics and source identification of sulfate more challenging.

The Hutuo River alluvial fan is located in the North China Plain. In recent years, with the rapid increase in water demand for industrial and agricultural purposes, the area has experienced intense groundwater extraction and is considered a typical region of groundwater overexploitation [11]. As the intensity of groundwater extraction has increased, the groundwater level has declined annually, severely affecting the groundwater chemical and hydrodynamic fields. This has intensified the water–rock interactions [12]. Additionally, the region is characterized by well-developed coal-bearing strata and human activities have a significant impact on groundwater. As a result, the concentration of sulfate in groundwater has been increasing [13]. However, the characteristics of groundwater sulfate distribution and its pollution sources in this area have not yet received attention, posing a potential threat to the drinking water safety of local residents.

In this study, the Hutuo River alluvial fan is selected as the research site. By integrating hydrochemical analysis, multivariate statistical methods, and advanced geostatistical techniques, we provide a comprehensive and spatially explicit investigation of the sulfate pollution dynamics. Our study not only elucidates the spatial distribution patterns of sulfate and its primary sources but also reveals the underlying pollution mechanisms in this critical region. These findings fill a significant knowledge gap regarding sulfate pollution in overexploited groundwater systems and offer a novel scientific foundation for developing targeted strategies to mitigate and control sulfate contamination in similar regions worldwide.

2. Materials and Methods

2.1. Study Area Overview

The research area is situated in the central and upper reaches of the Hutuo River alluvial fan (Figure 1), stretching from the Gangnan Reservoir (located within the territory of Pingshan County) in the west to the western boundary of Gaocheng County in the east, covering a drainage area of 2442 km^2. The Hutuo River basin is characterized by a scarcity of available surface water resources. Consequently, human activities such as domestic use, industrial production, and agricultural practices predominantly depend on groundwater [14]. The topography of the study area gradually slopes from west to east. The region experiences a temperate continental monsoon climate, with average annual precipitation and temperature varying between 400 and 750 mm and 13.3–15.0 °C, respectively. The majority of precipitation occurs during the months of June to September, accounting for more than 70% of the annual rainfall. The average annual temperature is approximately 13 °C, with the lowest temperature occurring in January, averaging 0.6 °C, and the highest temperature occurring in July, averaging 25 °C. The study area experiences significant seasonal variations in evaporation. The highest evaporation occurs in summer, ranging from 200 to 250 mm, which accounts for approximately 45% of the annual evaporation. In contrast, the lowest evaporation occurs in winter, ranging from 50 to 80 mm, representing about 10% of the annual evaporation.

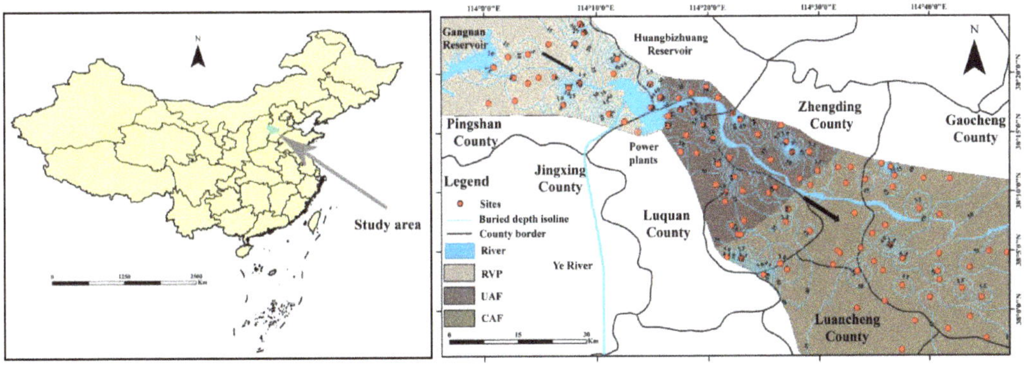

Figure 1. Distribution map of sampling sites in the Hutuo River basin.

2.2. Hydrogeological Conditions

The aquifer within the study area is part of the thick multi-layered Quaternary aquifer system in the North China Plain. This aquifer is relatively water-bearing, with groundwater flowing from the northwest to the southeast. The depth of the water table varies from 12 to 240 m, while the thickness of the aquifer ranges between 20 and 140 m. Both the water table and the aquifer exhibit favorable characteristics in terms of conductivity and water yield. Today, the acceleration of urbanization has led to persistent over-exploitation of groundwater in this area, resulting in a significant decline in groundwater depth. Specifically, the depth of the groundwater table has increased from 36 m in 2002 to 43 m in 2016 [12]. The dominant types of groundwater are fracture water in bedrock and pore water in loose rock layers. The primary sources of recharge are infiltration from atmospheric precipitation, infiltration from surface water, and return flow from agricultural irrigation, while the main discharge mechanism is artificial extraction. The lithology of the Quaternary sediments mainly comprises silty clay, sandy loam, gravelly sandy clay, and gravel and sand layers of various grain sizes (Supplementary Figure S1). The permeability coefficient of the vadose zone in the study area generally ranges from 37 to 145 m/d.

Based on the geological conditions and aquifer characteristics, the study area is divided into three hydrogeological units (Figure 1): (1) the fracture–pore water unit in the river valley plain (RVP) between the Gangnan Reservoir and Huangbizhuang Reservoir; (2) the pore water unit at the upper of the Hutuo River alluvial fan (UAF); and (3) the pore water unit in the central part of the alluvial fan (CAF). In the RVP, the aquifer thickness ranges from 10 to 20 m, with groundwater depths between 2 and 21 m. In the UAF, the aquifer thickness ranges from 20 to 35 m, with groundwater depths between 12 and 20 m. In the CAF, the aquifer thickness ranges from 25 to 60 m, with groundwater depths between 40 and 50 m.

2.3. *Sample Collection and Analysis*

2.3.1. Sample Collection

To eliminate the impact of spatial heterogeneity of groundwater samples, this study established sampling points at an interval of one per 20 km^2, with a total of 118 sampling set ups (Figure 1) and were collected in January 2023. Furthermore, three sewage samples were collected. The groundwater samples were primarily obtained from domestic wells and agricultural irrigation wells, with well depths ranging from 4.0 to 50 m and an average depth of 9.8 m. Prior to sample collection, the wells were purged for 5–10 min until the groundwater pH and electrical conductivity (EC) values stabilized. Subsequently, groundwater samples were obtained through pumping operations from the wells. The sampling bottles were made of polyethylene, with one 1.5 L bottle and one 0.5 L bottle used for analyzing cations and anions, respectively. For cation analysis, the water samples were acidified with 1:1 HNO_3 to adjust the pH to less than 2. Before sampling, the bottles were rinsed three times with distilled water and the original water samples. After collecting, the samples were transported to the laboratory in a portable cooler and tested within 48 h.

2.3.2. Sample Analysis and Testing

The monitoring indicators included the following: (1) on-site measurement indicators: pH, dissolved oxygen (DO); (2) laboratory analysis indicators: total dissolved solids (TDS), K^+, Na^+, Ca^{2+}, Mg^{2+}, NO_3^-, SO_4^{2-}, Cl^-, HCO_3^-, and total hardness (TH, expressed as calcium carbonate). The analytical methods for the hydrochemical indicators are detailed in Supplementary Table S1.

2.4. *Data Analysis*

2.4.1. Data Preprocessing

Due to the requirement of normality in the distribution of each variable for principal component analysis (PCA), an initial assessment of the kurtosis and skewness of all data were conducted prior to PCA. Values of kurtosis and skewness beyond the range of −2 to +2 indicate a significant deviation from normality [15]. The statistical analysis revealed that the distributions of K^+, Na^+, and NO_3^- were skewed and not normally distributed, with standardized kurtosis and skewness values ranging from 10.1 to 30.7 and 2.84 to 7.29, respectively. Following a logarithmic transformation of these parameters, the kurtosis and skewness values were reduced to the range of 0.592 to 1.31 and −0.87 to 0.65, respectively. Subsequently, all water quality parameters underwent further standardization using the z-scale to mitigate the risk of misclassification arising from the substantial variability in data dimensions.

2.4.2. Statistical Analysis and Spatial Interpolation Analysis

In this research, the non-parametric Kruskal–Wallis and Mann–Whitney U tests were utilized to assess the significant differences in SO_4^{2-} concentrations among various hydrogeological units. Spearman's analysis was used to investigate the correlation between hydrochemical components. PCA was conducted to pinpoint the primary sources of sulfate pollution. Pollution sources were spatially visualized by projecting the source identification results from PCA onto the study area map through geostatistical interpolation, primarily using inverse distance weighting. All statistical analyses were performed using the R computing platform (version 4.3.2, Auckland, New Zealand), while geospatial analysis was carried out using ArcGIS v10.5 software (ESRI Inc., Redlands, CA, USA).

3. Results and Discussion

3.1. *Characteristics of Groundwater Quality*

To analyze the characteristics of groundwater quality in the study area, statistical analysis was conducted on the numerical ranges, means, and exceedance rates of various water quality indicators (Table 1). The results show that the pH of groundwater ranges from 6.53 to 8.47, indicating neutral to slightly alkaline conditions, with no exceedance of standards. The concentration of DO ranges from 1.04 to 13.8 mg/L, with a mean value of 4.91 mg/L, indicating an overall oxidizing environment.

Table 1. Statistics of different groundwater parameters in the Hutuo River alluvial fan basin.

Parameters	Min	Max	Mean	SD	RES (%)	Standard
DO	1.04	13.8	4.91	2.66	0	—
pH	6.53	8.47	7.47	0.356	0	6.5–8.5
K^+	0.12	25.9	2.29	2.57	0	—
Na^+	8.78	172	34.7	26.4	0	200
Ca^{2+}	27.2	367	131	63.8	0	—
Mg^{2+}	10.3	102	32.0	13.8	0	—
Cl^-	11.9	379	71.4	59.7	1.7	250
SO_4^{2-}	12.9	395	127	77.2	5.1	250
HCO_3^-	109	565	296	72.6	0	—
NO_3^-	0.24	440	61.8	70.2	22.0	88.6
TH	115	1339	458	198	50.0	450
TDS	174	2091	628	292	8.5	1000

Note: Standard is the grade III standard for groundwater quality in China (GB/T 14848-2017) [16]. The pH value is dimensionless, while the units for all other indices are expressed in mg/L. The sample size for all indicators is 118. RES = rate of exceeded standard; SD = standard deviation. "—" indicates the absence of a corresponding standard.

The average concentrations of cations are ordered as follows: Ca^{2+} (131 mg/L) > Na^+ (34.7 mg/L) > Mg^{2+} (32.0 mg/L) > K^+ (2.29 mg/L). For anions, the concentration order is HCO_3^- (296 mg/L) > SO_4^{2-} (127 mg/L) > Cl^- (71.4 mg/L) > NO_3^- (61.8 mg/L). The highest exceedance rate in the study area is for TH, with a rate of 50.0%, which is mainly related to the geological background of the region. Previous studies have shown that carbonate rocks are widely distributed in the study area, and the dissolution of these rocks leads to high concentrations of Ca^{2+} and Mg^{2+} in groundwater, resulting in significant exceedance of TH levels [17]. Additionally, exceedance of groundwater quality standards for NO_3^-, TDS, SO_4^{2-}, and Cl^- was observed, with exceedance rates of 22.0%, 8.47%, 5.08%, and 1.69%, respectively. It is worth noting that nitrate, with its high exceedance rate and mean value, indicates the significant impact of human activities on groundwater quality [18].

3.2. Spatial Distribution Characteristics of SO_4^{2-} in Groundwater

3.2.1. Variations in SO_4^{2-} Across Distinct Hydrogeological Units

The spatial heterogeneity in sulfate concentrations observed in the Hutuo River alluvial fan groundwater (Figure 2) provides critical insights into the interplay between human activities and natural hydrogeological conditions. The significantly higher sulfate concentrations in the RVP and UAF (175 mg/L and 169 mg/L, respectively) compared to the CAF (77.3 mg/L) ($p < 0.001$) highlight the profound impact of human activity on groundwater quality. In the RVP and UAF, especially around the Huangbizhuang Reservoir, there are large-scale power plants, the wastewater discharged from power plants may be an important cause of the elevated sulfate concentrations in the surrounding groundwater [19]. Additionally, the Ye River basin, a tributary of the Hutuo River (Flow into Huangbizhuang Reservoir), has extensive coal-bearing strata and hosts multiple coal mines and coal-washing plants. The discharge of industrial wastewater from these facilities is the other primary cause of elevated sulfate concentrations in the region. This suggests that even with existing regulations, there remains a critical gap in controlling industrial effluents, particularly in areas where groundwater serves as both a receptor and a resource. In addition, in the RVP, the lack of sewage treatment systems in villages leads to the discharge of sewage directly into the Hutuo River, and ultimately infiltrates into groundwater, which is another significant factor contributing to high sulfate concentrations in groundwater. In conclusion, the spatial differences in sulfate concentrations across the Hutuo River alluvial fan groundwater system represent more than a local pollution issue. They illustrate the complex interplay between development pressures, industrial activities, and natural hydrogeological conditions in an overexploited groundwater region.

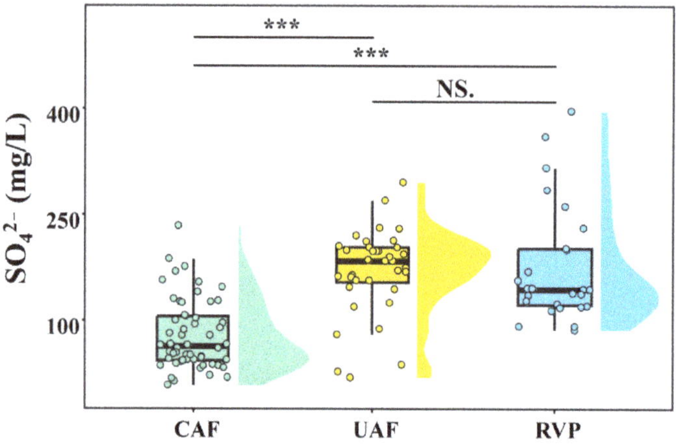

Figure 2. Comparison of spatial sulfate concentrations in the groundwater of the Hutuo River alluvial fan. Note: *** denotes a *p*-value < 0.001; NS: not significant. The shaded plots are violin plots, which are used to show the distribution of data across different groups. The size of each box in the box plot represents the interquartile range (IQR) of the data, which is the range between the first quartile (25th percentile) and the third quartile (75th percentile). The individual points within each region on the *x*-axis represent the actual data points for each group. These points are plotted to show the raw data distribution and to highlight outliers or extreme values that fall outside the whiskers of the box plot.

3.2.2. Variation of SO_4^{2-} with Groundwater Depth

The depth of groundwater burial significantly influences the pathway distance that pollutants must travel to reach the aquifer. In the present study, a significant negative correlation ($p < 0.001$) was observed between groundwater depth and sulfate concentration (Figure 3). Specifically, sulfate concentration decreased with increasing groundwater depth. These findings indicate that groundwater depth is a critical factor influencing the distribution of sulfate. Further analysis suggests that this relationship can be attributed to the fact that shallow groundwater levels facilitate the infiltration of pollutants, such as sulfate, from the surface, while deeper groundwater levels are less susceptible to such direct contamination due to longer flow paths and the filtering effect of the unsaturated zone [20]. Additionally, the redox conditions in deeper aquifers may also play a role in reducing sulfate concentrations, as the reducing environment may promote the precipitation of sulfate or the reduction in sulfate to sulfide [21]. The findings align with previous studies that have highlighted the importance of groundwater depth in determining the vulnerability of aquifers to contamination.

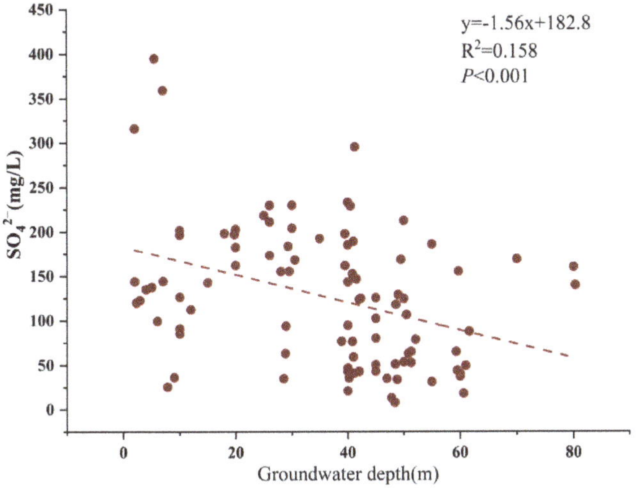

Figure 3. Relation between groundwater depth and sulfate concentration.

3.3. Hydrochemical Types of Groundwater

Using the Piper trilinear diagram, the hydrochemical types of groundwater in the three hydrogeological units of the study area were identified (Figure 4). At the same time, we integrated the Kurlov classification method (which classifies and names the types based on the content of anions and cations in the water, considering ions with a percentage greater than 25% of the milliequivalent as significant) to accurately calculate and distinguish the types and proportions of groundwater chemistry in different hydrogeological units. In the RVP and UAF, the dominant hydrochemical type of groundwater is $HCO_3{\cdot}SO_4$–Ca, accounting for 65.4% and 39.8% of the sampling points, respectively. The secondary type is $HCO_3{\cdot}SO_4$–Ca·Mg, with proportions of 11.5% and 27.8%, respectively. In the CAF, the primary hydrochemical type is HCO_3–Ca·Mg, accounting for 48.2% of the sampling points, followed by $HCO_3{\cdot}SO_4$–Ca·Mg and $HCO_3{\cdot}$Cl–Ca·Mg types, each with a proportion of 14.3%.

Overall, from the RVP to the CAP, the hydrochemical types of groundwater gradually evolve from $HCO_3{\cdot}SO_4$–Ca to HCO_3–Ca·Mg. This transition is mainly related to differences in lithology and human activities [22]. According to the hydrogeological conditions of the study area, coal-bearing strata are widely distributed in the RVP and UAF regions, and there are coal mines and power plants around the Huangbizhuang Reservoir. Both coal-bearing strata and industrial wastewater contribute to elevated SO_4^{2-} concentrations in groundwater, resulting in a higher proportion of SO_4-type water in these regions. In the CAF, the aquifer is primarily composed of quaternary loose sedimentary rocks. This area is mainly a residential zone without large industrial sources, leading to a shift in the hydrochemical type to predominantly HCO_3–Ca·Mg. In the region characterized by overexploitation of the alluvial fan, the prevalence of SO_4-type water significantly surpasses that observed in urban settings (dominated by HCO_3–Ca·Mg-type water) [23], coastal zones (dominated by HCO_3·Cl–Na-type waters) [24], and arid regions (dominated by HCO_3–Ca·Mg-type waters) [25]. These findings imply that the intensification of groundwater extraction may be exacerbating the dissolution of sulfur-bearing minerals within the geological formations.

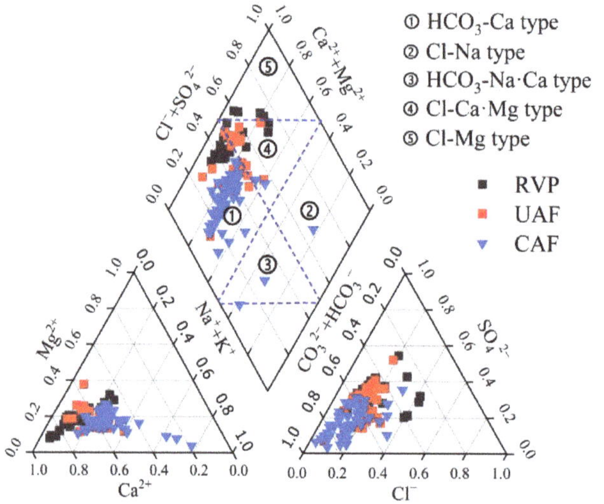

Figure 4. Hydrochemical type of groundwater in the Hutuo River alluvial fan. Note: The blue dashed line represents the boundary of the water chemical type.

3.4. *Identification of SO_4^{2-} Sources in Groundwater*

3.4.1. Identification of SO_4^{2-} Sources Using Hydrochemical Methods

The use of hydrochemical ion ratio diagrams to identify the sources of pollutants in groundwater is a traditional method [26]. In this study, the ratios of $[Ca^{2+}]/[SO_4^{2-}]$ and $[NO_3^-]/[Ca^{2+}]$ to $[SO_4^{2-}]/[Ca^{2+}]$ were used to identify the sources of sulfate in the Hutuo River alluvial fan groundwater.

Based on the $[Ca^{2+}]/[SO_4^{2-}]$ ratio diagram (Figure 5a), a good correlation between Ca^{2+} and SO_4^{2-} was observed in the groundwater of the three hydrogeological units in the study area ($p < 0.001$), indicating a common source for these ions, such as the dissolution of gypsum in coal-bearing strata. This result further corroborates that the spatial variability of sulfate concentrations is also influenced by natural factors, such as the distribution of coal-bearing strata.

To further clarify the sources of sulfate in groundwater, the relationship between $[NO_3^-]/[Ca^{2+}]$ and $[SO_4^{2-}]/[Ca^{2+}]$ was used to characterize the impact of human activities on groundwater. Previous studies have shown that industrial activities are characterized by higher $[SO_4^{2-}]/[Ca^{2+}]$ values, while agricultural activities have higher $[NO_3^-]/[Ca^{2+}]$ values [27]. As shown in Figure 5b, 93.2% of the groundwater samples had higher $[SO_4^{2-}]/[Ca^{2+}]$ values than $[NO_3^-]/[Ca^{2+}]$ in the study area, indicating that industrial activities are the primary human influence on groundwater in the study area. Specifically, the $[SO_4^{2-}]/[Ca^{2+}]$ ratios ranged from 0.167 to 1.15, 0.149 to 1.48, and 0.186 to 2.89 in the RVP, UAF, and CAF, respectively. The $[NO_3^-]/[Ca^{2+}]$ ratios ranged from 0.043 to 0.389, 0.022 to 0.731, and 0.009 to 0.985 in the RVP, UAF, and CAF, respectively. The markedly higher $[SO_4^{2-}]/[Ca^{2+}]$ ratios across all sub-regions provide compelling evidence of the significant role of industrial activities in groundwater sulfate pollution. This conclusion is further supported by the analysis of three sewage samples, which also demonstrated elevated $[SO_4^{2-}]/[Ca^{2+}]$ ratios, aligning with the observed groundwater patterns. These findings collectively underscore the need for targeted industrial pollution control measures to mitigate sulfate contamination in groundwater resources.

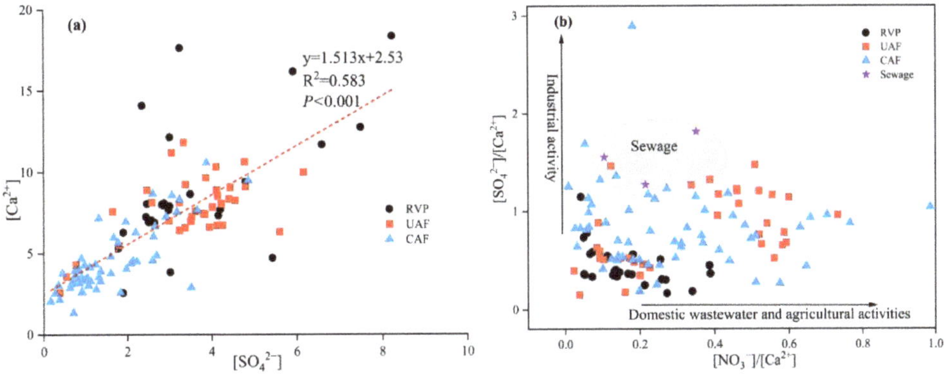

Figure 5. Ion ratio diagrams: (**a**) $[Ca^{2+}]$ vs. $[SO_4^{2-}]$ and (**b**) $[SO_4^{2-}]/[Ca^{2+}]$ vs. $[NO_3^-]/[Ca^{2+}]$.

3.4.2. Identification of SO_4^{2-} Sources Using Correlation Analysis

The relationships between groundwater quality indicators can reflect the sources of these parameters [28]. In this study, Pearson's correlation analysis was conducted to explore the relationships between sulfate and other indicators, which offers crucial insights into the origins of SO_4^{2-} in groundwater. As depicted in Figure 6, indicators with high correlation coefficients to SO_4^{2-} include TDS, TH, Ca^{2+}, Na^+, Mg^{2+}, Cl^-, and NO_3^-, with correlation coefficients of 0.85, 0.80, 0.80, 0.76, 0.56, 0.61, and 0.61, respectively. The strong correlations between SO_4^{2-} and Cl^-, Na^+, and NO_3^- suggest common sources for these ions. Prior research has shown that domestic sewage and industrial activities are common sources of Cl^- and NO_3^- [26], suggesting that SO_4^{2-} in the study area likely originates from industrial activities and domestic sewage. The significant correlation between SO_4^{2-} and Ca^{2+} confirms that gypsum dissolution is one of the sources of sulfate, consistent with the earlier analysis. Additionally, the positive correlation between SO_4^{2-} and TH, Mg^{2+} may be related to enhanced water–rock interactions caused by intensive groundwater extraction [29].

3.4.3. Identification of Sulfate Sources Using PCA

To delve deeper into the sources of sulfate in the groundwater in the area, PCA was conducted on 12 water quality parameters, including DO, pH, K^+, Na^+, Ca^{2+}, Mg^{2+}, Cl^-, SO_4^{2-}, HCO_3^-, NO_3^-, TH, and TDS. Prior to the analysis, the Kaiser–Meyer–Olkin (KMO) and Bartlett's sphericity tests were performed. The results showed that the KMO value was 0.740, and Bartlett's sphericity test value was 2827 ($p < 0.001$), indicating that the water quality data met the requirements for PCA. Based on eigenvalues exceeding 1, three principal components were extracted, which collectively explained 78.5% of the total variance.

	DO	pH	K^+	Na^+	Ca^{2+}	Mg^{2+}	Cl^-	SO_4^{2-}	HCO_3^-	NO_3^-	TH	TDS
DO	1.0	-0.13	0.10	-0.049	0.038	-0.12	-0.089	0.14	-0.10	0.10	0.012	0.029
pH		1.0	-0.027	-0.32	-0.63	-0.40	-0.32	-0.51	-0.60	-0.57	-0.63	-0.59
K^+			1.0	0.23	0.072	0.14	0.057	0.22	0.089	0.066	0.11	0.15
Na^+				1.0	0.59	0.65	0.65	0.76	0.62	0.39	0.64	0.75
Ca^{2+}					1.0	0.61	0.73	0.80	0.74	0.86	0.98	0.95
Mg^{2+}						1.0	0.72	0.56	0.68	0.54	0.73	0.71
Cl^-							1.0	0.61	0.52	0.63	0.78	0.81
SO_4^{2-}								1.0	0.50	0.61	0.80	0.85
HCO_3^-									1.0	0.60	0.78	0.75
NO_3^-										1.0	0.86	0.83
TH											1.0	0.97
TDS												1.0

Figure 6. Spearman's correlation analysis of hydrochemical components in the groundwater of the Hutuo River alluvial fan. Note: n = 118; $p = 0.05$. Very significant correlation—absolute value of the correlation coefficient ≥ 0.8; strongly significant correlation—absolute value of the correlation coefficient is between 0.6 and 0.8; significant correlation—absolute value of the correlation coefficient is between 0.4 and 0.6; not correlated—absolute value of the correlation coefficient < 0.4.

As indicated in Table 2, PC1 accounted for 58.2% of the total variance, with high factor loadings for TH, NO_3^-, Ca^{2+}, pH, HCO_3^-, TDS, SO_4^{2-}, and Mg^{2+}. Common sources of groundwater sulfate include the oxidation of sulfides, the dissolution of evaporites, industrial wastewater, domestic sewage, and precipitation [8]. The positive correlation between NO_3^- and SO_4^{2-} with PC1 suggests a common source for these ions. Previous studies have shown that the NO_3^- of groundwater in the area primarily originates from domestic sewage and fertilizers [14]. Given that nitrogen-based fertilizers are predominantly utilized in local agriculture, coupled with the minimal application of sulfur-containing fertilizers, it is reasonable to posit that domestic sewage constitutes a significant source of sulfate in the groundwater.

Table 2. Results of principal component analysis for groundwater quality data in Hutuo River alluvial fan.

Parameters	PC1	PC2	PC3
TH	0.825	0.529	−0.024
NO_3^-	0.820	0.224	0.178
Ca^{2+}	0.814	0.470	0.064
pH	−0.814	0.157	0.070
HCO_3^-	0.751	0.225	−0.305
TDS	0.739	0.652	0.009
Mg^{2+}	0.589	0.522	−0.264
K^+	−0.112	0.781	0.044
Na^+	0.359	0.769	−0.065
Cl^-	0.523	0.688	−0.138
SO_4^{2-}	0.592	0.595	0.222
DO	−0.009	−0.006	0.940
Eigenvalue	6.98	1.30	1.14
Contribution Rate (%)	58.2	10.8	9.46
Cumulative Contribution Rate (%)	58.2	69.0	78.5

As discussed in Sections 3.3 and 3.4.1, there are large-scale power plants around the Huangbizhuang Reservoir in the UAF of the study area, and the groundwater monitoring sites with exceedances are primarily located around the reservoir. A previous study found that the Ye River exhibits a relatively high average SO_4^{2-} concentration of 216 mg/L [30]. Moreover, the mean SO_4^{2-} concentration in the three sewage samples we collected reached 274.5 mg/L. These findings collectively suggest that industrial wastewater discharge from power plants and other industrial facilities, along with domestic sewage, are the predominant contributors to the elevated sulfate concentrations in the region. In summary, PC1 represents that the main sources of sulfate in groundwater are domestic sewage and industrial wastewater.

PC2 explained 10.8% of the total variance, with high factor loadings for K^+, Na^+, Cl^-, TDS, and SO_4^{2-}. Moderate correlations were also observed between TH, Mg^{2+}, Ca^{2+}, and PC2. Since PC1, PC2, and PC3 are independent of each other, the strong correlation between Na^+ and Cl^- is indicative of rock salt dissolution [31]. The good correlation between SO_4^{2-} and Ca^{2+} with PC2 indicated that the evaporites (gypsum) are an important source of SO_4^{2-}. Previous studies have shown that the increase in TH in groundwater in this region is caused by intensive groundwater extraction [19]. Indeed, in recent years, the Hutuo River alluvial fan area has experienced significant groundwater extraction, leading to a substantial decline in groundwater levels and enhanced water–rock interactions, which in turn accelerate the dissolution of rock salt and evaporites in the strata [32]. It is important to note that while rock salt dissolution contributes to the increase in Na^+ and Cl^-, it does not directly provide sulfate ions. Therefore, the correlation between SO_4^{2-} and Ca^{2+} with PC2, coupled with the presence of evaporites, strongly indicates that the SO_4^{2-} in groundwater originates from the dissolution of evaporites such as gypsum. This conclusion is further supported by the hydrogeochemical processes prevalent in areas with significant groundwater extraction, where enhanced water–rock interactions play a crucial role in shaping groundwater chemistry. Thus, PC2 demonstrates that SO_4^{2-} in groundwater originates from the dissolution of evaporites (gypsum).

PC3 explained 9.46% of the total variance, with high factor loadings for DO, and slight correlations were also observed between SO_4^{2-} and PC3. The mean concentration of DO in groundwater in the study area is 4.91 mg/L, indicating an oxidizing environment. Previous studies have shown that sulfide minerals in the region mainly originate from pyrite in Carboniferous and Permian coal seams [33]. Moreover, large-scale coal mining in the study area has exposed reducing inorganic sulfur minerals, accelerating the processes of weathering, oxidation, and dissolution. These minerals then enter into the groundwater through rainfall runoff, thereby causing an increase in the sulfate concentration of groundwater [34]. The positive correlation between DO and SO_4^{2-}, as reflected in PC3, supports the conclusion that the oxidation of sulfide minerals, particularly pyrite, is a significant source of sulfate in the groundwater. This finding aligns with the hydrogeochemical processes typical of mining-impacted regions, where the disruption of coal seams and associated sulfide minerals introduces additional pathways for oxidation and subsequent sulfate enrichment in groundwater. Therefore, PC3 demonstrates that SO_4^{2-} in groundwater originates from the oxidation of sulfide minerals.

3.5. Spatial Distribution of Sulfate Sources in Groundwater

To analyze the spatial distribution characteristics of SO_4^{2-} sources in the Hutuo River alluvial fan groundwater, inverse distance weighting interpolation was performed on the scores of the three factors, resulting in a spatial distribution map of sulfate pollution sources in the Hutuo River basin (Figure 7). We employed cross-validation techniques to assess the accuracy and reliability of our interpolation model. The results showed that the mean prediction errors for PC1, PC2, and PC3 were −0.018, 0.026 and 0.012, respectively, with root mean square errors of 0.938, 1.106, and 0.857. The results of the cross-validation demonstrated that our interpolation model performed well.

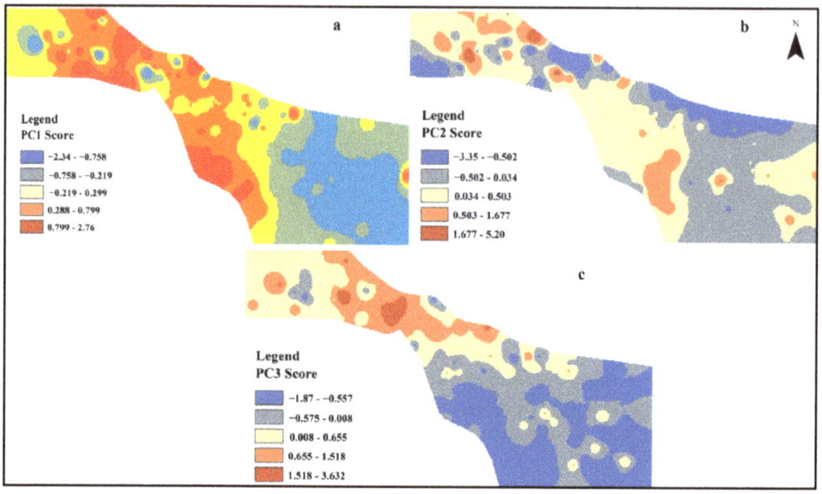

Figure 7. Spatial distribution map of sulfate sources in the groundwater of the Hutuo River alluvial fan. (**a**–**c**) represent the spatial interpolation maps of the scores for PC1, PC2, and PC3 in the principal component analysis, respectively.

As shown in Figure 7, high scores of PC1 are mainly located in the RVP and UAF (around power plants and reservoirs) (Figure 7a), indicating that SO_4^{2-} in groundwater in these regions primarily originates from industrial and domestic wastewater discharges. High scores of PC2 are mainly distributed in the RVP and sporadically in the study area (Figure 7b), suggesting that the dissolution of evaporites also contributes to sulfate in groundwater in these regions. High scores of PC3 are primarily located around the Huangbizhuang Reservoir in the river valley plain (Figure 7c), where coal-bearing strata are widely distributed. The oxidation of sulfide minerals is an important factor affecting SO_4^{2-} concentrations in groundwater in this region.

Based on the above analysis, the elevated SO_4^{2-} concentrations in the Hutuo River alluvial fan groundwater are mainly influenced by industrial activities around the Huangbizhuang Reservoir and coal mining in the Ye River basin. Intensive groundwater extraction has accelerated the dissolution of evaporites and the oxidation of sulfide minerals in the strata. Therefore, to control the continuous increase in SO_4^{2-} concentrations in groundwater in the Hutuo River alluvial fan area, strict measures should be implemented to regulate the discharge of untreated wastewater and prevent the overexploitation of groundwater.

4. The Limitations of This Study

This study provides a comprehensive analysis of the pollution characteristics and mechanisms of sulfate in groundwater in over-exploited areas, which can offer scientific evidence for the prevention and control of sulfate pollution in groundwater in similar regions both domestically and internationally. However, there are some limitations in this study. For example, the identification of sulfate sources in groundwater mainly relied on hydrochemical methods, which can only qualitatively recognize the sources of sulfate in the water environment, without quantifying the contribution rates of each sulfate pollution source. Therefore, future research should focus on improving the accuracy of sulfate source apportionment in over-exploited areas. Additionally, the study area is a thick multi-layered aquifer system in the North China Plain. Due to limitations in research objectives and funding, this study only collected shallow groundwater samples once and conducted a preliminary exploration of the pollution mechanisms of sulfate in groundwater in over-exploited areas. Future research should involve high-frequency sampling of groundwater at different depths and times in this region to deeply analyze the pollution mechanisms and evolution mechanisms of sulfate in over-exploited areas.

5. Conclusions

This study provides a comprehensive analysis of the spatial distribution and sources of sulfate (SO_4^{2-}) in the groundwater of the Hutuo River alluvial fan, an intensively exploited region where groundwater quality is influenced by complex natural and anthropogenic interactions. Our innovative integration of multivariate statistical techniques, including principal component analysis (PCA), with advanced geostatistical methods has revealed unprecedented insights into the heterogeneity of SO_4^{2-} distribution across distinct hydrogeological units. A key novel finding is the demonstrated relationship between groundwater depth and SO_4^{2-} concentration, establishing groundwater depth as a critical factor in sulfate distribution. This discovery enhances our understanding of how groundwater dynamics influence solute transport and distribution in overexploited aquifers.

Our spatially explicit analysis identifies industrial activities as the primary driver of elevated sulfate levels in the river valley plain and upper alluvial fan, contrasting with the central alluvial fan where natural processes dominate. Beyond anthropogenic sources, we have uncovered the significant contribution of natural processes—specifically evaporite dissolution and sulfide mineral oxidation—to groundwater sulfate pollution. The methodological framework developed in this study offers a cutting-edge approach for identifying and mapping pollution sources in complex hydrogeological settings. By combining PCA with geostatistical techniques, we created detailed spatial distributions of sulfate sources, providing a clearer picture of pollution dynamics than previously possible. These results highlight the need for a paradigm shift in groundwater management, emphasizing that effective pollution control requires both source-specific mitigation and systemic hydrogeological considerations.

The outcomes of this research not only advance the scientific understanding of sulfate pollution dynamics in groundwater systems but also deliver actionable knowledge for policymakers and water resource managers. By providing a robust scientific basis for sustainable groundwater management, this study serves as a model for addressing groundwater pollution challenges in other regions facing similar threats from overexploitation and contamination. This work represents a significant step forward in the development of science-based solutions for groundwater protection and sustainable water resource management.

Supplementary Materials: The following supporting information can be downloaded at: https://www.mdpi.com/article/10.3390/w17101530/s1. Figure S1: Hydrogeological cross-section of study area; Table S1: Hydrochemical parameters, analytical method, equipment and detection limits.

Author Contributions: L.W.: investigation, methodology, software, data curation, writing—original draft preparation. Q.W.: investigation, software, methodology. W.L.: investigation, software. Y.L.: investigation, data curation. Q.Z.: supervision, methodology, writing—reviewing and editing. All authors have read and agreed to the published version of the manuscript.

Funding: This work was supported by the Natural Science Foundation of Hebei Province of China (grant no. D2022504015) and the Nanyang Institute of Technology Doctoral Research Startup Fund Project (NGBJ-510069).

Data Availability Statement: The data are available from the authors upon reasonable request.

Acknowledgments: The authors gratefully acknowledge the editor and anonymous reviewers for their valuable comments on this manuscript.

Conflicts of Interest: The authors declare no conflicts of interest.

References

1. Zhang, Q.; Wang, H.; Lu, C. Tracing sulfate origin and transformation in an area with multiple sources of pollution in northern China by using environmental isotopes and Bayesian isotope mixing model. *Environ. Pollut.* **2020**, *265*, 115105. [CrossRef] [PubMed]
2. Liu, Y.; Ma, R. Human health risk assessment of heavy metals in groundwater in the luan river catchment within the north China Plain. *Geofluids* **2020**, *2020*, 8391793. [CrossRef]
3. Jung, H.; Kim, J.; Lee, J. Seasonal and spatial contributions of sulfate and trace elements in river water in mining districts: Insights from hydrogeochemical and isotopic analysis based on statistical models. *J. Hazard. Mater.* **2025**, *488*, 137246. [CrossRef]
4. Yustika, R.D.; Handayani, C.O.; Dewi, T.; Yanti, D.; Dariah, A. Risk assessment for non-carcinogenic effect posed by sulfates in water on the health of residents around The Sumpur River, West Sumatra-Indonesia. *Toxicol. Rep.* **2025**, *14*, 101921.
5. Ren, B.; Shi, X.; Guo, J. Interaction of sulfate-reducing bacteria and methanogenic archaea in urban sewers, leads to increased risk of proliferation of antibiotic resistance genes. *Environ. Pollut.* **2025**, *368*, 125777. [CrossRef]
6. Wang, H.; Zhang, Q. Research advances in identifying sulfate contamination sources of water environment by using stable isotopes. *Int. J. Environ. Res. Public Health* **2019**, *16*, 1914. [CrossRef]
7. Tostevin, R.; Turchyn, A.V.; Farquhar, J.; Johnston, D.T.; Eldridge, D.L.; Bishop, J.K.; McIlvin, M. Multiple sulfur isotope constraints on the modern sulfur cycle. *Earth Planet. Sci. Lett.* **2014**, *396*, 14–21. [CrossRef]
8. Li, X.; Liu, Y.; Zhou, A.; Zhang, B. Sulfur and oxygen isotope compositions of dissolved sulfate in the Yangtze River during high water period and its sulfate source tracing. *Earth Sci.* **2014**, *39*, 1647–1654.
9. Qiong, X.I.A.O.; Wen, L.I.U.; Fang, G.U.O.; Moucheng, P.A.N.; Shi, Y.U. Using $\delta^{34}S\text{-}SO_4{}^{2-}$and $\delta^{15}N\text{-}NO_3{}^{-}$, $\delta^{18}O\text{-}NO_3{}^{-}$ to trace the sources of sulfur and nitrate in Lihu lake undergound water, Guangxi. *China Environ. Sci.* **2015**, *36*, 2877–2886.
10. Otero, N.; Soler, A. Sulphur isotopes as tracers of the influence of potash mining in groundwater salinization in the Llobregat Basin (NE Spain). *Water Res.* **2002**, *36*, 3989–4000. [CrossRef]
11. Lu, Y.; Tang, C.; Chen, J.; Song, X.F.; Li, F.D.; Sakura, Y. Spatial characteristics of water quality, stable isotopes and tritium associated with groundwater flow in the Hutuo River alluvial fan plain of the North China Plain. *Hydrogeol. J.* **2008**, *16*, 1003–1015. [CrossRef]
12. Zhang, Q.; Miao, L.; Wang, H.; Hou, J.; Li, Y. How rapid urbanization drives deteriorating groundwater quality in a provincial capital of China. *Pol. J. Environ. Stud.* **2020**, *29*, 441–450. [CrossRef]
13. Ye, H.; Liu, C.L.; Jiang, J.M.; Pei, L.X.; Guo, J. Risk assessments of groundwater pollution in the shijiazhuang reach of the hutuo river, China. *Geol. Bull. China* **2008**, *27*, 1065–1070.
14. Zhang, Q.; Wang, H. Assessment of sources and transformation of nitrate in the alluvial-pluvial fan region of north China using a multi-isotope approach. *J. Environ. Sci.* **2020**, *89*, 9–22. [CrossRef]
15. Kannel, P.R.; Lee, S.; Kanel, S.R.; Khan, S.P. Chemometric application in classification and assessment of monitoring locations of an urban river system. *Anal. Chim. Acta* **2007**, *582*, 390–399. [CrossRef] [PubMed]
16. *GB/T14848-2017*; Standard for Groundwater Quality. General Administration of Quality Supervision, Inspection and Quarantine of the People' s Republic of China: Beijing, China, 2017.
17. Li, Y.; Zhang, Z.; Fei, Y.; Chen, H.; Qian, Y.; Dun, Y. Investigation of quality and pollution characteristics of groundwater in the Hutuo River Alluvial Plain, North China Plain. *Environ. Earth Sci.* **2016**, *75*, 581. [CrossRef]
18. Martín del Campo, M.A.; Esteller, M.V.; Expósito, J.L.; Hirata, R. Impacts of urbanization on groundwater hydrodynamics and hydrochemistry of the Toluca Valley aquifer (Mexico). *Environ. Monit. Assess.* **2014**, *186*, 2979–2999. [CrossRef]
19. Zhang, Q.; Wang, H.; Wang, Y.; Yang, M.; Zhu, L. Groundwater quality assessment and pollution source apportionment in an intensely exploited region of northern China. *Environ. Sci. Pollut. Res.* **2017**, *24*, 16639–16650. [CrossRef]
20. Xiong, Y.; Abdulraheem, M.I.; Li, L.; Moshood, A.Y.; Zhang, W.; Zhang, Y.Y.; Hu, J.D. Spatial analysis techniques for assessing groundwater vulnerability: A strategic approach to proactive conservation and mitigation strategies. *Environ Earth Sci.* **2024**, *83*, 621. [CrossRef]

21. Jakóbczyk-Karpierz, S.; Ślósarczyk, K. Isotopic signature of anthropogenic sources of groundwater contamination with sulfate and its application to groundwater in a heavily urbanized and industrialized area (Upper Silesia, Poland). *J. Hydrol.* **2022**, *612*, 128255. [CrossRef]
22. Jeong, C.H. Effect of land use and urbanization on hydrochemistry and contamination of groundwater from Taejon area, Korea. *J. Hydrol.* **2001**, *253*, 194–210. [CrossRef]
23. Gao, Y.; Qian, H.; Ren, W.; Wang, H.; Liu, F.; Yang, F. Hydrogeochemical characterization and quality assessment of groundwater based on integrated-weight water quality index in a concentrated urban area. *J. Clean. Prod.* **2020**, *260*, 121006. [CrossRef]
24. Qin, R.; Wu, Y.; Xu, Z.; Xie, D.; Zhang, C. Assessing the impact of natural and anthropogenic activities on groundwater quality in coastal alluvial aquifers of the lower Liaohe River Plain, NE China. *Appl. Geochem.* **2013**, *31*, 142–158. [CrossRef]
25. Zhou, B.; Wang, H.; Zhang, Q. Assessment of the evolution of groundwater chemistry and its controlling factors in the Huangshui River Basin of northwestern China, using hydrochemistry and multivariate statistical techniques. *Int. J. Environ. Res. Public Health* **2021**, *18*, 7551. [CrossRef]
26. Samantara, M.K.; Padhi, R.K.; Satpathy, K.K.; Sowmya, M.; Kumaran, P. Groundwater nitrate contamination and use of Cl/Br ratio for source appointment. *Environ. Monit. Assess.* **2015**, *187*, 50. [CrossRef]
27. Murgulet, D.; Tick, G.R. Understanding the sources and fate of nitrate in a highly developed aquifer system. *J. Contam. Hydrol.* **2013**, *155*, 69–81. [CrossRef]
28. Cao, W.; Zhang, Z.; Guo, H.; Fu, Y.; Gao, Z.; Nan, T.; Li, Z. Spatial distribution and controlling mechanisms of high fluoride groundwater in the coastal plain of Bohai Rim, North China. *J. Hydrol.* **2023**, *617*, 128952. [CrossRef]
29. Liu, L.; Wang, G.; Li, Y.; Wei, Z.; Lin, W.; Qin, X.; Long, X. Hydrogeochemistry of geothermal water in Huangshadong and adjacent areas of Guangdong province: Implications for water-rock interaction. *Geothermics* **2024**, *122*, 103084. [CrossRef]
30. Niu, C.; Zhang, Q.; Wang, H. Spatiotemporal variation in groundwater quality and source apportionment along the ye river of north China using the PMF model. *Int. J. Environ. Res. Public Health* **2022**, *19*, 1779. [CrossRef]
31. Wang, L.; Yang, N.; Zhao, Y.; Zhang, Q. Research on the features and driving factors of shallow groundwater quality in arid areas, northwest China. *Water* **2025**, *17*, 934. [CrossRef]
32. Wang, P.; Zhang, W.; Zhu, Y.; Liu, Y.; Li, Y.; Cao, S.; Li, B. Evolution of groundwater hydrochemical characteristics and formation mechanism during groundwater recharge: A case study in the Hutuo River alluvial–pluvial fan, North China Plain. *Sci. Total Environ.* **2024**, *915*, 170159. [CrossRef] [PubMed]
33. Kong, J.; Zhou, Z.; Cao, W. Tracing the source and behaviour of sulphate in karst reservoirs, using stable isotopes and Bayesian isotopic-mixing models. *Sci. Total Environ.* **2025**, *958*, 177994. [CrossRef] [PubMed]
34. Huang, P.; Zhang, Y.; Li, Y.; Gao, H.; Cui, M.; Chai, S. A multiple isotope (S, H, O and C) approach to estimate sulfate increasing mechanism of groundwater in coal mine area. *Sci. Total Environ.* **2023**, *900*, 165852. [CrossRef] [PubMed]

water

MDPI

Article

Spillovers and State Politics: Explaining Variation in U.S. Water Quality Permit Stringency

Neal D. Woods

Department of Political Science, University of South Carolina, Columbia, SC 29208, USA; neal.woods@sc.edu; Tel.: +1-803-777-7389

Abstract: Why do environmental regulators allow some facilities to discharge more water pollution than similar facilities in other locations? Drawing on general theories of regulatory decisionmaking, this study assesses four possible reasons: (1) variation in governments' ability to export pollution to other political jurisdictions, (2) variation in the demographic composition of the neighborhood surrounding the facility, (3) variation in local communities' ability to mobilize for effective collective action, and (4) variation in subnational political context. Analyses of effluent discharge limits imposed by the U.S. states on two common classes of water pollution indicate that state regulators allow greater discharges when the receiving river crosses state lines, and that they are responsive to a variety of other demographic and political factors. The specific factors, however, often vary across the two pollutants assessed. These results suggest that subnational political considerations may shape the conditions imposed in water pollution permits, and, ultimately, water quality outcomes.

Keywords: water quality; point-source water pollution; Clean Water Act; free riding; environmental justice; U.S. state environmental policy

1. Introduction

Few regulatory decisions impact water quality outcomes, including water contamination and its impact on public health, aquatic life, and ecosystems, as directly as how much pollution to allow a facility to discharge into a waterway. In U.S. clean water regulation, these discharge limits are set by regulators who possess a significant amount of discretion [1], and the limits themselves have been shown to vary substantially for the same pollutant across similar facilities [2,3]. Understanding the sources of this variation thus has important implications for efforts to improve surface water quality in the U.S.

Yet although other aspects of U.S. clean water policy, such as funding for wastewater treatment and drinking water infrastructure [4–8], and regulatory compliance monitoring and enforcement [9–13], have generated substantial empirical literatures, the discharge limits imposed in water quality permits have to date received scant scholarly attention. Moreover, existing studies that do assess variation in these limits largely restrict themselves to examining state-level political factors [2]. However, the broader literature on regulation suggests that, in addition to these state-level considerations, a number of local level influences may affect the restrictiveness of water quality permits, including (1) local demographic factors, (2) local level capacity to engage in political action, and (3) whether the receiving waterbody will transport the pollution out of state.

This study assesses these potential explanations in the context of state implementation of the U.S. Clean Water Act (CWA). It examines the discharge limits imposed in facility permits for two widespread and important classes of pollution. Using Geographical Information Systems (GIS) to precisely locate facilities and isolate characteristics of the surrounding communities and receiving waterbodies, it finds that (1) facilities that discharge their effluent into rivers where the pollution is likely to flow out of state systematically have more permissive permits, (2) local demographic characteristics also impact permit stringency, though their influence is less consistent across pollutants, and (3) permit stringency is related to a variety of characteristics of the state political environment, including government ideology and environmental interest group strength. In sum, these findings point to the need for a greater understanding of the role that subnational political factors play in determining the amount of pollution that facilities are allowed to discharge into U.S. waterways.

2. Water Quality Regulation in the United States

The primary law governing water quality protection in the United States is the CWA of 1972. Originating as an amendment to the 1948 Federal Water Pollution Control Act, it empowers the U.S. Environmental Protection Agency (EPA) to limit the discharge of pollutants into navigable waterways from point sources, including both industrial facilities and publicly owned treatment facilities. These facilities must obtain a legally binding National Pollutant Discharge Elimination System (NPDES) permit that specifies the frequency, quantity, and location of allowable discharges into waterbodies. The permit also establishes a set of reporting requirements, which are designed to help monitor compliance with permit obligations.

The discharge limits contained in NPDES permits are based on federal standards for the best available technology (BAT) for pollution abatement applicable to the specific industrial process involved, though stricter limits may be mandated if the abatement technology would not assure sufficient water quality [9,14]. The process of translating the general technological standard to emissions limits for a particular point source gives regulators substantial discretion [1]. Consequently, these limits are often the product of negotiations between polluting firms and government officials [9], and have been shown to vary substantially for the same pollutant across similar facilities [2]. Generally speaking, these discharge limits are the primary factor determining the overall stringency of a permit. Other factors that could impact permit stringency in some circumstances include monitoring requirements and special conditions [2].

As with many other U.S. environmental protection laws, the CWA is implemented through a system of partial preemption that allows state governments to assume authority to implement key portions of the act, including the NPDES permit program [15,16]. State environmental agencies may assume this authority by demonstrating to the EPA that they have a program that is at least as stringent as that required by applicable federal law, and that they have sufficient capacity to run the program. If states do not seek primary implementation authority, or primacy, for a program, the EPA runs it through one of its ten regional offices. By now, almost all states have received primacy for the NPDES program, which means that NPDES permits are generally being written by regulators in various state agencies [1]. One consequence of this dispersion of regulatory authority appears to be significant heterogeneity in permitted discharge limits; a study by the U.S. General Accounting Office, for instance, found that the allowable pollution for similarly sized facilities emitting the same pollutant could vary by more than an order of magnitude across states [3].

3. Explanations for Variation in Regulatory Policy Implementation

From the social science literature on regulatory policy implementation, one can draw four broad classes of potential explanations for geographic variation in environmental regulatory stringency, which may then be employed to explain variation in permitted effluent discharge limits.

The first set of explanations revolves around incentives to free-ride on the pollution control efforts of downstream states. State regulators, this argument suggests, may be less concerned with pollution that predominantly impacts other states. In such situations, they may regulate pollutants less stringently, thereby creating pollution spillover effects [17–21]. Thus, the expectation is that facilities located where the environmental consequences of their pollution are more likely to be felt by out-of-state residents should have less stringent permits.

The second set of explanations has to do with local demographic conditions. This argument suggests that the size and composition of the local affected population should affect permit stringency. In particular, a large empirical literature on environmental justice investigates the possibility that areas with lower income populations or a higher percentage of racial minority residents may receive lower levels of environmental regulatory protection [10,22–24]. The results of such studies have been mixed, but in general, recent studies have suggested that there are meaningful disparities in environmental regulation across race and income levels. The expectation is, therefore, that facilities located near lower-income and minority populations will receive more lenient permits.

A third, related set of explanations pertains to local political capacity. The literature on Not in My Backyard (NIMBY) politics suggests that facilities in areas that have greater capacity to overcome barriers to collective action and organize to pursue common interests may receive higher levels of environmental protection. Political capacity may be derived from greater education, homeownership, and residential stability [25–27]. This argument leads to the expectation that facilities located in areas with stronger political capacity should have more stringent permits.

The final set of explanations pertains to state politics. A long stream of literature suggests that state regulators are responsive to political pressures [2,9,28–33]. Thus, the expectation is that environmental agencies in states with more pro-environment mass publics and elected officials and a more active organized environmental movement may face stronger pressure to write permits with more stringent effluent discharge limits, all else equal, than those in states in which these pressures are weaker.

4. Materials and Methods

4.1. Dependent Variables: Permitted Effluent Discharge Limits

The research question motivating this study is as follows: Why do environmental regulators allow some facilities to discharge more water pollution than similar facilities in other locations? To address this question, this study assesses the above theoretical expectations as applied to NPDES permit conditions for point sources of water pollution regulated under the federal CWA. Specifically, it focuses on sources the EPA designates as being "major" facilities. These consist of those private industrial dischargers and publicly owned facilities that are the largest water-polluting facilities in the country, and are utilized here because they are the primary producers of effluent discharges. For the time period covered by the study, there were about 6400 active major facilities in the United States.

The permit conditions examined in these analyses set limits governing the allowable discharge of two major sources of water pollution: biochemical oxygen demand (BOD) and total suspended solids (TSS). BOD is an indicator of the amount of organic pollution, as represented by the amount of dissolved oxygen needed by microorganisms to break down organic matter in water. TSS is a measure of the concentration of particles in water, both organic and inorganic, exceeding 2 microns in size. For the purposes of this study, these have the advantage of being traditional pollution parameters that are fairly commonly discharged by NPDES facilities, and which have relatively easily interpretable standards [2,34].

For wastewater treatment facilities, national technology-based standards have been set by the EPA for both BOD and TSS at a monthly average of 30 milligrams per liter (mg/L) per source. Nonetheless, as discussed above, state regulators have substantial discretion in applying these standards to individual plants due to differences in the water quality of the receiving waterbody and other factors, and observed standards for wastewater treatment facilities vary significantly [2]. For other types of facilities, the limits for both BOD and TSS may also vary across industries. In the analyses reported below, the dependent variable is the discharge limit imposed on BOD or TSS in the facility's NPDES permit. These data come from the EPA's Envirofacts database.

These facility-specific permit data are then merged with a separate dataset created by Konisky and Woods [11], which contains a rich set of facility-level information regarding facility size, type, and location, as well as the surrounding area's demographic conditions, the facility's free riding potential, and state-level political factors. This dataset was created using detailed location information provided by the EPA's Geospatial Data Access Project to precisely locate each major NPDES permit holder in geographic space, which the authors then overlayed with a map of U.S. Census tracts using Geographic Information Systems (GIS). These GIS data were used in three ways: (1) to develop measures related to the opportunity to free ride on the pollution control efforts of other state governments, (2) to extract a set of demographic information regarding potentially affected local populations, and (3) to create indicators of local political capacity. These data were combined with both facility-level characteristics and a variety of state-level political variables to form the Konisky and Woods [11] dataset.

4.2. Independent Variables: Free-Riding Potential

The potential of a state to free ride by exporting its water pollution to other states depends on the features of the environmental resource receiving the pollution from the NPDES facilities. The analyses include the two types of resources included in the Konisky and Woods [11] dataset: rivers and watersheds. For rivers, information on the receiving stream of each NPDES facility's pollution discharges was obtained from the EPA's Permit Compliance System database, and GIS software was used to determine whether the receiving stream was fully intrastate or whether it crossed two or more states. Cases in which a river composed the border between two or more states, such as the Missouri River and the Ohio River, were coded as interstate (rather than intrastate) resources.

For watersheds, the receiving watershed was classified using the unique hydrologic unit code (HUC) assigned by the U.S. Geological Survey (USGS). The analysis uses 8-digit HUC codes, which are the smallest level of hydrologic units known as cataloging units, which are the geographic area representing part or all of a surface drainage basin, a combination of drainage basins, or a distinct hydrologic feature. As with rivers, a GIS intersection utility was utilized to identify whether these watersheds crossed two or more states.

4.3. Independent Variables: Local Demographic Composition

One important set of facility-level attributes relates to the demographic composition of the neighborhood surrounding the facility, which has been shown to impact regulatory inspection and enforcement behavior [10,13,17,23,35–37]. The analysis includes several attributes of the local population, including population density and the percentage of the population living around facilities that are poor, African-American, and Hispanic. In each case, Konisky and Woods [11] measured the composition of the population living around CWA facilities using an areal apportionment method [38,39]. The scope of the potentially affected local populations was delimited by a concentric circular buffer around each facility with a radius of 1 mile, which was intersected with spatial data from the census maps. These intersections were used as weights for each demographic attribute, where the weight was the proportion of each census unit contained within the circular buffers.

4.4. Independent Variables: Local Political Capacity

A second set of facility-level factors pertains to the political capacity of the local population. Communities with high levels of political capacity are more likely to be able to overcome organizational constraints and collective action problems in order to effectively advocate for stronger environmental regulation [25–27,40]. The analyses, therefore, include three common indicators of political capacity created using the same areal apportionment method discussed above: the percentage of the population with a college education, the percentage of owner-occupied housing, and residential stability, as defined below.

4.5. Independent Variables: State Political Factors

In addition to local factors, a variety of state political conditions may be expected to impact the stringency of NPDES permits. In this respect, it is reasonable to assume that permit stringency is driven by the same political and economic factors that have been shown to influence regulatory monitoring and enforcement behavior [9,13,28,32,33,41]. Accordingly, the analyses include measures of citizen and government ideology developed by Berry et al. [42], and the average League of Conservation Voters' scores of the state's House delegation as an indicator of congressional ideology. Higher values on these measures indicate greater liberalism. The analyses also include state spending on water quality as an indicator of state commitment to environmental protection in this area. In order to reflect the state's interest group environment, the analyses include a measure of the number of environmental interest groups registered to lobby in the state [43].

Regulatory stringency may also be a function of state economic conditions, with poorer economic conditions putting greater political pressure on environmental agencies to loosen regulatory standards. Unemployment rate, in particular, has been found to influence government regulatory behavior [44], and the analysis includes the state unemployment rate as a covariate.

4.6. Independent Variables: Facility Level Controls

The analyses also include a variety of other facility-level characteristics as control variables. To control for the amount of pollution emitted by the facility, the analyses incorporate a variable capturing the facility's flow capacity, which some studies have shown to be correlated with regulatory decisions [35,36]. The analyses also include a variable indicating whether the facility is publicly or privately owned. Prior research suggests that publicly owned facilities are more lightly regulated than privately owned ones [45]. Finally, a set of dummy variables for 2-digit Standard Industrial Classification (SIC) codes is included to control for the type of industry the facility is in. These variables capture the fact that there is substantial variation in the amount and type of pollution discharged by facilities in different industrial sectors.

4.7. Dataset Construction

To ensure temporal comparability, permit conditions for BOD and TSS were measured as of 2005, which is the last year covered by the Konisky and Woods [11] dataset. That dataset contains complete data for 5129 major NPDES facilities nationwide. Of these, data could be matched with permit data from the EPA Envirofacts dataset that contains BOD discharge limits for 1382 facilities and TSS discharge limits for 2905 facilities.

Although the permit conditions were measured as of 2005, they may have been established earlier. NPDES permits are effective for a period not exceeding five years, and the CWA stipulates that they must be renewed prior to expiration in order to continue to allow pollutant discharges. Despite this requirement, some states have allowed facilities to continue to operate for some period of time on expired permits [2]. When permit holders apply for renewal, all permit conditions are reviewed, and data collected under the previous permit are used to determine new conditions. Existing permit conditions for a given facility have thus generally been established within the prior five years (though it may at times be longer if it is operating on an expired permit). In recognition of this, each of the independent variables in the dataset represents averages over the five-year period between 2001 and 2005, inclusive. For residential stability, the variable represents the percentage of the local population that has lived in the same residence since 1995. Since these variables are slowly changing, choosing an individual year during this period, or even a year shortly before the period, would generally lead to comparable results.

The dataset is limited to states that have been delegated authority to implement the federal CWA, since the logic underlying many of the expected relationships pertains to state governments. By 2005, all but four states had been delegated authority by the U.S. EPA to write NPDES permits. In the remaining states (Idaho, Massachusetts, New Hampshire, and New Mexico), the relevant EPA regional office retained this authority.

The descriptive statistics for the variables in each model are presented in Table 1. Worth particular attention are the statistics for BOD and TSS discharge limits contained in NPDES permits. The mean of each is close to the general baseline EPA standard for wastewater treatment plants of 30 mg/L (26.2 for BOD and 29.6 for TSS). There is substantial variation, however, with standard deviations of 14.32 and 17.81, respectively. The most stringent permitted discharge limits in the dataset are around 1 mg/L for each, while the most permissive are several hundred mg/L. These differences are related in part to differences in technologies and production processes across differing industrial sectors, which are controlled for, to the extent possible, via SIC code dummies in the analyses to follow. These analyses assess the extent to which the remaining variation can be explained by the four sets of explanatory variables discussed above.

Table 1. Descriptive statistics.

Variable	Mean	SD	Min.	Max.
BOD	26.20	14.32	1.2	344.33
TSS	29.59	17.81	1	473
Interstate river	0.27	0.45	0	1
Interstate watershed	0.39	0.49	0	1
Pop. density	1.11	2.38	0	73.55
% Black	10.60	16.31	0	97.45
% Hispanic	7.73	13.99	0	98.11
% Poverty	12.63	8.32	0	60.87
% Owner-occupied	64.43	14.27	0	96.34
% College-educated	19.31	12.15	0	80.92
% Same residence	56.57	12.22	0	100
Govt ideology	47.89	18.50	8.15	85.12
Citizen ideology	49.94	10.56	26.59	87.24
Cong. ideology	42.49	17.95	5.23	95.43
Water quality spending	42.15	56.89	2.11	260.74

Table 1. *Cont.*

Variable	Mean	SD	Min.	Max.
Env. interest groups	17.68	12.38	2	46
Unemployment rate	5.42	0.69	3.44	7.12
Public facility	0.76	0.43	0	1
Flow capacity	38.70	181.57	0	3325

5. Results

The empirical models are estimated via ordinary least squares (OLS) regression, with state-clustered standard errors. Clustering the standard errors at the state level allows for potential correlation in errors across facilities within each state, which, if not accounted for, may lead to a violation of the standard OLS assumption of independently and identically distributed (i.i.d) errors [46]. This modeling approach is appropriate for these analyses because the discharge limits for the facilities in this dataset are set by state environmental agencies, suggesting that dependence across facility observations is likely to be caused by state-level factors. Clustering the standard errors has been argued to be a more straightforward and practical approach for dealing with correlated errors in many state policy contexts than an alternative approach, hierarchical linear modeling, which makes heavier demands on theory and data [46].

The results are presented in Table 2. The results in the first column indicate that facilities that discharge effluent into rivers that ultimately cross state lines receive systematically higher BOD discharge limits. All else constant, these facilities have permits that allow about 2.9 mg/L greater BOD discharges than those whose effluent remains within the state's boundaries. This finding provides support for the free-riding explanation for permit stringency.

Table 2. The determinants of NPDES permit stringency across two pollution parameters.

Variable	BOD	BOD	TSS	TSS
Interstate river	2.873 ** (1.183)		2.602 ** (1.277)	
Interstate watershed		0.302 (1.286)		−0.855 (0.591)
Pop. density	0.433 (0.346)	0.573 (0.381)	−0.507 * (0.272)	−0.401 (0.274)
% Black	−0.043 (0.040)	−0.048 (0.041)	−0.011 (0.042)	−0.013 (0.042)
% Hispanic	−0.060 (0.035)	−0.057 (0.037)	−0.039 (0.070)	0.035 (0.072)
% Poverty	0.138 ** (0.065)	0.131 ** (0.062)	0.140 (0.089)	0.140 (0.089)
% Owner−occupied	−0.030 (0.026)	−0.034 (0.026)	−0.059 (0.039)	−0.061 (0.040)
% College-educated	−0.029 (0.030)	−0.037 (0.032)	−0.019 (0.039)	0.008 (0.037)
% Same residence	0.067 * (0.034)	0.073 * (0.034)	0.091 ** (0.042)	0.096 ** (0.042)

Table 2. *Cont.*

Variable	BOD	BOD	TSS	TSS
Govt ideology	−0.149 ** (0.042)	−0.143 ** (0.045)	−0.068 (0.052)	−0.061 (0.049)
Citizen ideology	0.239 (0.148)	0.248 (0.154)	−0.253 (0.267)	−0.251 (0.265)
Cong. ideology	−0.050 (0.091)	−0.038 (0.096)	0.143 (0.164)	0.140 (0.161)
Water quality spending	0.007 (0.012)	0.006 (0.012)	0.041 ** (0.014)	0.038 ** (0.014)
Env. interest groups	−0.036 (0.072)	−0.016 (0.077)	−0.336 ** (0.073)	−0.334 ** (0.070)
Unemployment rate	−0.575 (1.209)	−0.671 (1.194)	−0.495 (1.074)	−0.687 (1.064)
Public facility	−3.392 ** (1.533)	−4.152 ** (1.456)	1.306 (1.193)	0.861 (1.159)
Flow capacity	−0.000 (0.001)	−0.000 (0.001)	−0.000 (0.001)	−0.001 (0.001)
Constant	12.177 (8.67)	12.936 (8.802)	35.93 ** (8.14)	37.94 ** (7.88)
R^2	0.12	0.11	0.11	0.11
N	1382	1382	2905	2905

Notes: OLS coefficient estimates, with state-clustered standard errors in parentheses. For purposes of clarity, coefficient estimates for SIC code fixed effects are not presented. * $p < 0.1$, ** $p < 0.05$, two-tailed test.

In addition, facilities that are located in poorer communities have less stringent permits, raising questions of environmental justice. Each additional ten percent of the local population that is in poverty is associated with an increase of about 1.4 mg/L in the allowable BOD discharge. States with more liberal state governments, in contrast, have significantly more stringent permits, suggesting that state politics may play an important role in the stringency levels of NPDES permits. Each ten-point increase in government liberalism (on a 100-point scale) is associated with a 1.5 mg/L increase in the discharge limit. Residential stability is associated with higher allowable discharges, all else constant, with a ten percent increase in the percentage of the local population living in the same residence since 1995, leading to about a 0.7 mg/L higher discharge limit. This effect is directionally opposite to expectations, though the magnitude of the effect is not as large as that of other variables in the model. Finally, public facilities have significantly lower limits, with a strong substantive effect of nearly 3.4 mg/L. As with residential stability, the direction of this effect is the opposite of expectations.

The second model uses interstate watersheds as an alternative measure of free-riding potential. This measure does not have a significant effect. The other significant variables, however, mirror those in the first model, with similar substantive effects.

The third model shifts its attention to the TSS discharge limits imposed in water quality permits. Again, the results provide evidence for environmental free riding in permit stringency, with allowable discharges into rivers that cross state lines being about 2.6 mg/L higher, ceteris paribus.

As with the BOD findings, greater residential stability is associated with higher TSS discharge limits, all else constant, which is the opposite of expectations. The other significant variables in this model depart from those of the BOD models, however. The results indicate that allowable discharges into areas with a denser population are lower, with each 100-person increase in population per square mile leading to a reduction about a 0.51 mg/L reduction in effluent discharge limits, suggesting that regulators may take the size of the potentially affected population into account in setting these limits. States with greater water quality expenditures also tend to have higher allowable discharges, with each USD 10,000 of additional expenditures per facility being associated with 0.04 mg/L higher allowable discharges. Causality is a bit difficult to disentangle here: although water quality spending was hypothesized to be an indicator of state commitment to water quality protection (and thus be negatively associated with allowable pollution discharges in state permits) it may also be the case that greater effluent discharges lead to greater state expenditures on water quality, due to a more substantial surface water pollution problem. Future research may examine this relationship temporally to assess whether water pollution spending is causally prior to changes in pollution levels, or vice versa. Finally, greater numbers of environmental interest groups are associated with more stringent permits, as expected, with each additional interest group being associated with a 0.33 mg/L decrease in allowable discharges.

The final model assesses TSS permits using the interstate watershed measure of free riding potential. As with the second BOD model, this alternative measure is not significant, and the other significant independent variables are unchanged from the previous model.

6. Conclusions

The stringency of NPDES permits has significant implications for water quality outcomes [47]. Among the four classes of explanations for variation in permit stringency examined here, the empirical results provide the broadest support for the effect of free riding potential. Both BOD and TSS permits are systematically more lenient if the recipient river ultimately flows across state lines. This suggests that state regulators may use their permitting authority as a mechanism to export their pollution, capturing the economic benefits of industrial production in their own state while forcing downriver states to absorb at least some of the environmental costs. Such spillover effects, however, are not found for interstate watersheds.

Several aspects of the state political environment also appear to impact discharge limits, though the specific variables that were significant in the analyses varied across the two classes of pollution studied. More liberal state governments (for BOD) and more organized environmental interest groups (for TSS) were found to lead to more restrictive permits, findings that were consistent with expectations. In addition, greater water quality spending was associated with more lenient permits for TSS, a finding that is less easily interpretable and deserves further investigation.

Some variables representing local demographic characteristics were likewise found to be significantly associated with permit stringency, although many were not statistically significant, and the significant variables were not consistent across the two pollution parameters. One finding that stands out for BOD, in particular, is that of higher allowable pollution discharges in higher poverty areas, which may reinforce the concerns of environmental justice advocates, and provide support for policy reforms that explicitly target environmental justice considerations, such as government monitoring and permit audits.

The only class of explanations studied that received no support in these findings was that of local political capacity. Factors related to the ability of communities to mobilize politically were either not significant predictors of permit stringency, or, in the case of residential stability, consistently significant in a direction opposite to what theory would predict. This is consistent with the mixed results these variables have produced in other analyses of government regulatory activity [10,11], though it should be noted that some of the reasons for the lack of support in these analyses may simply have to do with variable classification. Specifically, percent poverty, though treated here as an environmental justice variable (as is typical in the literature), could also be considered an indicator of (lack of) political capacity, and it is a significant predictor of permit stringency for BOD.

On the whole, these findings provide suggestive evidence that a variety of political considerations may be at play when regulators make decisions about allowable discharges in water quality permits. These conclusions are tempered, though, but the fact that much of the variation in these permits remains unexplained in these empirical models. Moreover, although some consistent findings emerged, many of the significant findings varied across the two pollution parameters examined. This suggests that the role of political factors may vary across pollutants, and other factors that are specific to particular pollutants may play a large role in the permitting process.

Ultimately, therefore, this study merely represents a first step in assessing the factors that lead to variation in discharge limits. Future research may expand upon this work by assessing permit conditions imposed for other types of pollutants, evaluating the temporal stability of these findings across time, and seeking to develop datasets that contain a richer set of facility level control variables that allow for greater understanding of the role that differences in production processes play in determining permit conditions. Such research could also extend this inquiry cross-nationally, investigating the role that the factors assessed here play in determining allowable pollution levels in different legal and institutional settings. Further work on this issue holds the promise of leading to significantly greater understanding of what causes differences in permit stringency, and, ultimately, water quality.

Funding: This research received no external funding.

Data Availability Statement: The raw data supporting the conclusions of this article will be made available by the author on request.

Conflicts of Interest: The author declares no conflicts of interest.

Abbreviations

The following abbreviations are used in this manuscript:

BAT	Best Available Technology
BOD	Biochemical Oxygen Demand
CWA	Clean Water Act
EPA	Environmental Protection Agency
HUC	Hydrologic Unit Code
NPDES	National Pollution Discharge Elimination System
TSS	Total Suspended Solids
USGS	United States Geological Survey

References

1. Woods, N.D. Regulatory competition, administrative discretion, and environmental policy implementation. *Rev. Policy Res.* **2022**, *39*, 486–511. [CrossRef]
2. Hoornbeek, J.A. *Water Pollution Policies in the American States: Runaway Bureaucracies or Congressional Control?* SUNY Press: Albany, NY, USA, 2011.
3. United States General Accounting Office. *Water Pollution: Differences Among the States in Issuing Permits Limiting the Discharge of Pollutants*; Government Printing Office: Washington, DC, USA, 1996.

4. Aistrup, J.A.; Morris, J.C. Does institutional location matter? The effect of agency type on resource distribution for nonpoint water pollution. *Water Policy* **2024**, *26*, 1039–1055. [CrossRef]
5. Daley, D.M.; Mullin, M.; Rubado, M.E. State agency discretion in a delegated federal program: Evidence from drinking water investment. *Publius* **2014**, *44*, 564–586. [CrossRef]
6. Morris, J.C. *Clean Water Policy and State Choice: Promise and Performance in the Water Quality Act*; Cambridge University Press: Cambridge, UK, 2022.
7. Morris, J.C.; Aistrup, J.; Nemerever, Z.; Zhang, X. Explaining State Allocation Shares Under the Water Quality Act of 1987: Politics, Incrementalism, or Need? *Publius* **2024**, *55*, 353–378. [CrossRef]
8. Mullin, M.; Daley, D.M. Multilevel instruments for infrastructure investment: Evaluating state revolving funds for water. *Policy Stud. J.* **2018**, *46*, 629–650. [CrossRef]
9. Hunter, S.; Waterman, R.W. *Enforcing the Law: The Case of the Clean Water Acts*; M. E. Sharpe: Armok, NJ, USA, 1996.
10. Konisky, D.M.; Schario, T.S. Examining environmental justice in facility-level regulatory enforcement. *Soc. Sci. J.* **2010**, *91*, 835–855. [CrossRef]
11. Konisky, D.M.; Woods, N.D. Environmental free riding in state water pollution enforcement. *State Polit. Policy Q.* **2012**, *12*, 227–252. [CrossRef]
12. Magat, W.A.; Viscusi, W.K. Effectiveness of the EPA's regulatory enforcement: The case of industrial effluent standards. *J. Law Econ.* **1990**, *33*, 331–360. [CrossRef]
13. Scholz, J.T.; Wang, C.L. Cooptation or transformation? Local policy networks and federal regulatory enforcement. *Am. J. Pol. Sci.* **2006**, *50*, 81–97. [CrossRef]
14. Freeman, A.M. Water Pollution Policy. In *Public Policies for Environmental Protection*, 2nd ed.; Portney, P., Stavins, R.N., Eds.; RFF Press: Washington, DC, USA, 2000; pp. 97–149.
15. Fowler, L.; Birdsall, C. Does the primacy system work? State versus federal implementation of the Clean Water Act. *Publius* **2021**, *51*, 131–160. [CrossRef]
16. Woods, N.D. Primacy implementation of environmental policy in the U.S. states. *Publius* **2006**, *36*, 259–276. [CrossRef]
17. Gray, W.B.; Shadbegian, R.J. "Optimal" pollution abatement—Whose benefits matter, and how much? *J. Environ. Econ. Manag.* **2004**, *47*, 510–534. [CrossRef]
18. Helland, E.; Whitford, A.B. Pollution incidence and political jurisdiction: Evidence from TRI. *J. Environ. Econ. Manag.* **2003**, *46*, 403–424. [CrossRef]
19. Konisky, D.M.; Woods, N.D. Exporting air pollution? Regulatory enforcement and environmental free riding in the United States. *Polit. Res. Q.* **2010**, *63*, 771–782. [CrossRef]
20. Monogan, J.E., III; Konisky, D.M.; Woods, N.D. Gone with the wind: Federalism and the strategic location of air polluters. *Am. J. Pol. Sci.* **2017**, *61*, 257–270. [CrossRef]
21. Sigman, H. Transboundary spillovers and decentralization of environmental policies. *J. Environ. Econ. Manag.* **2005**, *50*, 82–101. [CrossRef]
22. Bae, J.; Lynch, M.J. Ethnicity, Poverty, Race, and the Unequal Distribution of US Safe Drinking Water Act Violations, 2016–2018. *Sociol. Q.* **2023**, *64*, 274–295. [CrossRef]
23. Konisky, D.M. Inequities in enforcement? Environmental justice and government performance. *J. Policy Anal. Manag.* **2009**, *28*, 102–121. [CrossRef]
24. Liang, J.; Park, S.; Zhao, T. Representative bureaucracy, distributional equity, and environmental justice. *Public Adm. Rev.* **2020**, *80*, 402–414. [CrossRef]
25. Hamilton, J.T. Politics and social costs: Estimating the impact of collective action on hazardous waste facility. *Rand J. Econ.* **1993**, *24*, 101–125. [CrossRef]
26. Hamilton, J.T. Testing for environmental racism: Prejudice, profits, political power? *J. Policy Anal. Manag.* **1995**, *14*, 107–132. [CrossRef]
27. Hamilton, J.T.; Viscusi, W.K. *Calculating Risks? The Spatial and Political Dimensions of Hazardous Waste Policy*; MIT Press: Cambridge, MA, USA, 1999.
28. Cline, K.D.; Davis, C. Assessing the influence of EPA regional offices on state hazardous waste enforcement decisions. *Soc. Sci. J.* **2007**, *44*, 349–358. [CrossRef]
29. Fowler, L.; Kettler, J.J. Are Republicans bad for the environment? *State Polit. Policy Q.* **2021**, *21*, 195–219. [CrossRef]
30. Fowler, L. *Environmental Federalism: Old Legacies and New Challenges*; Routledge: New York, NY, USA, 2020.
31. Hoornbeek, J.A. The promises and pitfalls of devolution: Water pollution policies in the American states. *Publius* **2005**, *35*, 7–114. [CrossRef]

32. Hopper, J.S. The regulation of combination: The implications of combining natural resource conservation and environmental protection. *State Polit. Policy Q.* **2017**, *17*, 105–124. [CrossRef]
33. Woods, N.D. Structuring bureaucratic performance? Assessing the policy impact of environmental agency design. *Sustainability* **2024**, *16*, 7505. [CrossRef]
34. Magat, W.A. *Rules in the Making: A Statistical Analysis of Regulatory Agency Behavior*; RFF Press: Washington, DC, USA, 1986.
35. Earnhart, D. Regulatory factors shaping environmental performance at publicly-owned treatment plants. *J. Environ. Econ. Manag.* **2004**, *48*, 655–681. [CrossRef]
36. Earnhart, D. The effects of community characteristics on polluter compliance levels. *Land Econ.* **2004**, *80*, 408–432. [CrossRef]
37. Mennis, J.L. The distribution and enforcement of air polluting facilities in New Jersey. *Prof. Geogr.* **2005**, *57*, 411–422. [CrossRef]
38. Downey, L. Environmental racial inequality in Detroit. *Soc. Forces* **2006**, *85*, 771–796. [CrossRef]
39. Mohai, P.; Saha, R. Reassessing racial and socioeconomic disparities in environmental justice research. *Demography* **2006**, *43*, 383–399. [CrossRef] [PubMed]
40. Hird, J.A.; Reese, M. The distribution of environmental quality: An empirical analysis. *Soc. Sci. Q.* **1998**, *79*, 693–716.
41. Helland, E. Environmental protection in the federalist system: The political economy of NPDES inspections. *Econ. Inq.* **1998**, *36*, 305–319. [CrossRef]
42. Berry, W.; Ringquist, E.J.; Fording, R.; Hansen, R. Measuring citizen and government ideology in the American states: 1960-93. *Am. J. Pol. Sci.* **1998**, *42*, 337–348. [CrossRef]
43. Gray, V.; Lowery, D. *The Population Ecology of Interest Representation: Lobbying Communities in the American States*; University of Michigan Press: Ann Arbor, MI, USA, 1996.
44. Helland, E. The revealed preferences of state EPAs: Stringency, enforcement, and substitution. *J. Environ. Econ. Manag.* **1998**, *35*, 242–261. [CrossRef]
45. Konisky, D.M.; Teodoro, M.P. When governments regulate governments. *Am. J. Pol. Sci.* **2016**, *60*, 559–574. [CrossRef]
46. Primo, D.M.; Jacobsmeier, M.L.; Milyo, J. Estimating the impact of state policies and institutions with mixed-level data. *State Polit. Policy Q.* **2007**, *7*, 446–459. [CrossRef]
47. Earnhart, D. The influence of facility characteristics and permit conditions on the effectiveness of environmental regulatory deterrence. *J. Regul. Econ.* **2009**, *36*, 247–273. [CrossRef]

water

MDPI

Article

Spatiotemporal Distribution Characteristics and Removal Efficiency of Microplastics in a Wastewater Treatment Plant

Xudong Chen, Yang Li *, Keyi Lu, Xishu Liang, Kaibo Jin, Tianyu Ao, Lei Zhang, Jingjing Lv, Yanyan Dou * and Xuejun Duan

School of Smarts Energy and Environment, Zhongyuan University of Technology, Zhengzhou 450007, China
* Correspondence: liyang@zut.edu.cn (Y.L.); 6259@zut.edu.cn (Y.D.)

Abstract: This study examined the removal efficiency of microplastics (MPs) in a wastewater treatment plant (WWTP) in Zhengzhou, China. A three-point sampling approach (influent, process effluent, and final effluent) was employed, with samples collected across three seasons (summer, winter, and autumn) to investigate seasonal variations in MPs. The abundance of MPs in influent ranged from 184.3 $\pm$ 4.0 to 145.3 $\pm$ 24.0 n/L, while in the process effluent it decreased to 79.3 $\pm$ 18.7 to 62.3 $\pm$ 15.0 n/L. Furthermore, in final effluent it was further reduced to 26.0 $\pm$ 7.0 to 38.7 $\pm$ 5.1 n/L. Fragments and granule-shaped MPs predominated (>80%), with polypropylene (PP, 42.6%) and polyethylene terephthalate (PET, 31.8%) emerging as the dominant polymer types. The removal efficiency of MPs in the WWTP was 86%, 81%, and 73% in summer, autumn, and winter, respectively. Additionally, the plant exhibited differing removal efficiencies for MPs of varying sizes. Notably, residual sludge retained substantial MPs loads, with seasonal abundances measuring 22.3 $\pm$ 3.2, 14.2 $\pm$ 2.4, and 29.1 $\pm$ 6.7 n/g in summer, autumn, and winter samples, respectively. The findings underscore the importance of implementing effective management strategies and interventions in wastewater systems to mitigate MP pollution.

Keywords: microplastics; wastewater treatment plant; removal efficiency; spatiotemporal distribution characteristics

1. Introduction

Plastics have emerged as one of the most essential materials for the survival and development of human society. Microplastics (MPs), which are defined as plastic fragments and particles smaller than 5 mm in diameter, have increasingly become a significant global pollutant and a focal point of research in recent years [1]. They have been universally recognized as persistent emerging contaminants due to their high specific surface areas. Their small size facilitates the adsorption and transport of toxic substances, thereby posing significant threats to both environmental and human health [2]. They pose a greater risk to the environment and human health than biodegradable plastics. This significant gap between plastic production and recycling efficiency results in plastic waste infiltrating ecosystems, where oxidation and mechanical abrasion cause them to fragment into MPs [3]. Owing to their indigestible nature, MPs can accumulate in biological organisms over prolonged periods, resulting in diseases and mortality among animals [4]. These persistent particles pose significant threats to respiratory, reproductive, and metabolic functions in organisms, thereby introducing unpredictable risks to human health [5,6]. Furthermore, in wastewater, MPs exhibit synergistic interactions with a variety of contaminants, such as contaminants of emerging concern, pathogenic bacteria, antibiotic-resistant bacteria, and antibiotic resistance genes [7].

The emergence and persistence of MPs are primarily attributed to terrestrial anthropogenic activities, with wastewater treatment plants (WWTPs) playing a crucial role in MP management [8]. These facilities serve as dual interfaces by simultaneously receiving and potentially discharging MPs, thereby forming critical components in urban water systems, particularly in water-scarce regions where treated effluents are reused. WWTPs' effluents and biosolids serve as indirect pathways for MPs to enter the environment and the majority of MPs intercepted during treatment accumulate in sludge matrices, which are often repurposed as agricultural fertilizers [9,10]. Although conventional wastewater treatment plants do not have dedicated MP filtration mechanisms, they can have a positive effect on the removal of MPs from wastewater [11]. Globally, approximately 300 million metric tons of plastics are produced annually, yet only 21% are recycled, leaving the remaining 79% to enter the environment [12]. This practice sustains the environmental impact of MPs through agricultural applications, a phenomenon observed globally across treatment facilities [13]. These realities position MPs in WWTPs as a critical environmental issue garnering increasing scientific and regulatory attention in recent years [14].

Our understanding of the presence and removal efficiency of MPs in different units of WWTPs remains limited [15,16]. Previous studies have confirmed that seasonal variations in wastewater characteristics significantly influence MP abundance [17,18]. However, research aiming to establish clear correlations between MP distribution and seasonal weather patterns necessitates further data exploration and analysis [19]. Given the divergent outcomes observed regarding seasonal and precipitation effects, enhanced investigations remain essential for clarifying inter-seasonal variations.

Existing research on MPs in WWTPs has predominantly focused on MP abundance, removal efficiency, and transport dynamics within urban, suburban, and riverine systems [20–22], with comparatively less emphasis on cyclical seasonal variations [23,24]. This knowledge gap underscores the necessity for deeper insights into seasonal variations in MP characteristics, removal performance, and occurrence patterns in municipal WWTPs. This study investigates the removal efficiency of MPs in two treatment processes (secondary and tertiary) during summer, autumn, and winter in a WWTP located in Zhengzhou, China. The spatiotemporal distribution characteristics and removal efficiency of MPs in this WWTP were investigated. Experimental data reveal significant seasonal differences in polymer-specific MP abundance in secondary and tertiary treatment processes. Different sizes of microplastics have different removal efficiencies. Microplastics larger than 1 mm can be completely removed in the secondary treatment process. By analyzing the abundance and characterization of MPs in this WWTP, the study provides robust data to support seasonal assessments of MP removal efficiency and offers empirical evidence for drawing conclusions regarding seasonal correlations. Numerous studies have identified WWTPs as critical hotspots for the release of MPs into the environment [25,26]. These findings emphasize the pressing need to enhance monitoring efforts and implement targeted intervention measures to mitigate the role of WWTPs in MPs dissemination. Furthermore, they highlight the significance of optimizing treatment processes and adopting effective mitigation strategies to address this pressing environmental challenge.

2. Materials and Methods

2.1. Sampling

Wastewater and sludge samples were collected from a WWTP in Zhengzhou, China. The WWTP treats the sewage of 440,000 inhabitants of the area, with a capacity of

100,000 m^3/d, and a service area of 74 km^2. The plant is equipped with a three-stage treatment unit. The primary treatment consists of a grille and rotational flow grit chamber, the secondary treatment includes an anaerobic–anoxic–oxic (AAO) oxidation ditch and radial-flow-sedimentation tank, and the tertiary treatment comprises a fiber rotary filter and ultraviolet disinfection tank. Water samples were collected from three key locations: the inlet (S1), the process effluent (S2), and the final effluent (S3) during three distinct seasons: summer (July 2023), autumn (November 2023), and winter (January 2024). Sludge samples were taken at the sludge dewatering plant room. These sampling points are illustrated in Figure 1. At each sampling site, three separate 1 L wastewater samples were collected in stainless steel buckets and subsequently transferred to brown glass bottles for parallel experiments. For the sludge sample, approximately 500 g was collected using a stainless steel shovel and sealed in an aluminum foil bag for storage. Upon the completion of sample collection, the samples were transported back to the laboratory for further processing.

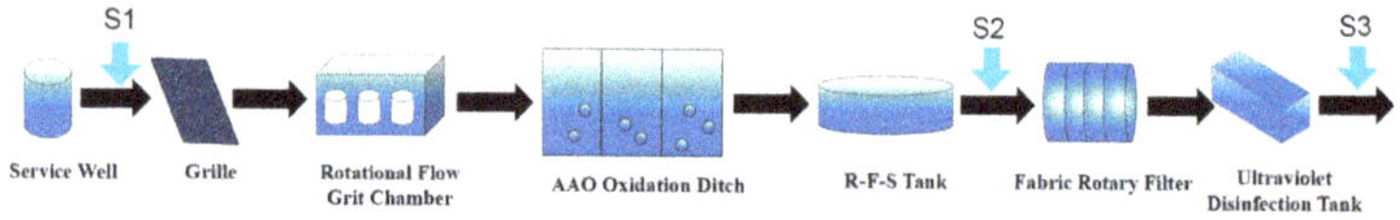

Figure 1. Location of water sampling sites in the WWTP (AAO: anaerobic–anoxic–oxic; R-F-S: radial-flow-sedimentation).

2.2. Sample Processing

Experimental pretreatment protocols followed some literature sources [27–29]. Firstly, three sets of parallel samples were prepared for both wastewater and sludge. Each set of wastewater comprised 1 L, while each set of sludge weighed 50 g. The wastewater samples were filtered through a stainless steel sieve with a mesh aperture of 0.0385 mm. At the same time, they were rinsed multiple times with ultra-pure water into the labelled beaker. For sludge samples, we accurately weighed 50 g and transferred it to a clean labelled beaker. Additionally, we weighed 100 g of sludge, dried it in a 105 °C oven for 24 h, and then reweighed it to determine the moisture content of the sludge. Density-based separation used a NaCl (Sinopharm Group Chemical reagent Co., LTD, Shanghai, China) solution (1.3 g/cm^3) with a solid-to-liquid ratio of 1:10—an economical salt solution selected based on density differences for polymer separation. After 8 h of settling, the sample was floated, and this was repeated three times. Organic matter in wastewater and sludge was removed by digestion with 30% H_2O_2 (volume ratio 2:1, Sinopharm Group Chemical reagent Co., LTD, Shanghai, China). The mixtures were homogenized and the reaction was performed at 60 °C in a water bath for 48–72 h. Post-digestion samples were vacuum-filtered through a filtration apparatus. Ultrapure water rinses ensured the complete transfer of residual particles from separatory funnels onto filter membranes. Final MP-bearing membranes (0.45 μm of pore size, Delvstlab Co., LTD, Haining, Zhejiang, China) were preserved in sterilized Petri dishes under controlled laboratory conditions. Air-dried filters were preserved for subsequent microscopic and spectroscopic analyses.

2.3. Detection Method

The filtered mixed cellulose membranes were examined under a digital optical microscope (Motic, BA310Digital, Motic Industrial Group Ltd., Xiamen, China). Samples were photographed through the integrated camera system (Motic Images Plus 3.0 ML, Motic Industrial Group Ltd., Xiamen, China) to visually identify the morphology, color, and size of the suspected MPs. Suspected MP particles were manually counted and marked on the filtered membranes. A confocal Raman microscopy spectrometer (gora-Lite, Ideaoptics

Co. Ltd., Shanghai, China) with an excitation wavelength of 785 nm, coupled with the Gora Dawn software (gora. Dawn v1.0) was used to identify the MPs (spectral range: 60–3500/cm, resolution: <5/cm). The obtained spectra were compared against standard Raman spectral libraries for polymer identification. The identified images of MPs are shown in Figure S1. Due to the high particle density, particularly in influent water samples, 30% of each filtered membrane's area was systematically analyzed by random field selection to ensure representative quantification. The Raman detection results for various types of polymers are summarized in Figure S2.

2.4. Data Analysis

The removal efficiency (RE) of MPs in the WWTP was calculated using the following equation:

$$\text{RE (\%)} = 100 - (\text{MPs abundance in } S_{n+1} \times 100/\text{MPs abundance in } S_n) \quad (1)$$

MP samples were collected at three sampling sites, S1, S2, and S3, at the WWTP to assess the effect of WWTP structures on MP abundance. A one-way analysis of variance (ANOVA) was conducted using SPSS 26.0 software to assess differences in MPs abundance across different seasons. Spearman's rank correlation test was employed to evaluate the relationship between MP removal efficiency and particle characteristics, including size, shape, and color.

2.5. Quality Control

Before each sampling event, the sampling equipment was sterilized using ethanol to minimize the risk of contamination. Personnel involved in the sampling process were required to wear gloves and lab coats to prevent potential skin-contact contamination. Sample collection bottles were subjected to thorough cleaning and sterilization procedures, followed by rinsing with ultrapure water to ensure purity. All sludge or treated water samples were protected from environmental exposure during handling and storage by being covered with aluminum foil or Petri dishes.

3. Results and Discussion

3.1. Abundance and Removal Efficiency of MPs

The abundance of MPs across wastewater treatment stages during monitored months are displayed in Table 1. The abundance of MPs in S1 fluctuated between 145.3 ± 24.0 n/L and 184.3 ± 4.0 n/L (mean ± standard deviation, S.D.). In contrast, the data in S2 exhibited MP levels ranging from 62.3 ± 15.0 n/L to 79.3 ± 18.7 n/L, while the abundance in S3 varied between 26.0 ± 7.0 n/L and 38.7 ± 5.1 n/L. Seasonal analysis revealed higher MP abundance in S1 during summer (184.3 ± 4.0 n/L), compared to autumn levels (184.0 ± 8.9 n/L), and minimal winter values (145.3 ± 24.0 n/L). Final effluent showed no marked seasonal variation, with summer–autumn–winter measurements at 26.0 ± 7.0 n/L, 35.0 ± 11.5 n/L, and 38.9 ± 5.1 n/L, respectively, ANOVA tests showed that there were significant seasonal differences in MP abundance in S1 ($P < 0.05$). Observational data indicated that the abundance of MPs in S1 during summer and autumn was 20–30% higher than in winter. This seasonal disparity may be attributed to increased plastic packaging demand during summer due to weather-related consumption patterns and summer clothes being washed more frequently [30], coupled with higher rainfall intensity compared to autumn and winter. In addition, elevated precipitation enhances surface runoff, mobilizing greater quantities of MPs into wastewater systems [31].

Regarding MP abundance in S2, concentrations measured 49.7 ± 10.1 n/L (summer), 79.3 ± 18.7 n/L (autumn), and 62.3 ± 15.0 n/L (winter), translating to the removal efficiencies of 73%, 57%, and 57%, respectively (Figure 2). The superior summer removal performance potentially relates to enhanced biological process efficiency under warmer thermal conditions. Tertiary treatment demonstrated MP removal efficiencies of 48%, 56%, and 38% across different seasons, while secondary treatment significantly outperformed tertiary processes in terms of overall efficiency. The diminished winter tertiary removal efficiency likely results from the low-temperature inhibition of specialized purification mechanisms. Cumulative seasonal removal efficiency reached maximums of 86% (summer), 81% (autumn), and 73% (winter), aligning with global municipal WWTPs' MP removal benchmarks (Table 2). Sludge analysis (dry weight) revealed retained MP abundances of 22.3 ± 3.2 n/g (summer), 14.2 ± 2.4 n/g (autumn), and 29.1 ± 6.7 n/g (winter), confirming biosolids as major reservoirs of MPs in wastewater systems [12]. These findings emphasize sludge management's critical role in environmental MP control.

Table 1. Abundance of MPs in different seasons.

Sampling Time	Mean Abundance of MPs			
	S1 (n/L)	S2 (n/L)	S3 (n/L)	Sludge Sample (n/g)
Summer	184.3 ± 4.0	49.7 ± 10.1	26.0 ± 7.0	22.3 ± 3.2
Autumn	184.0 ± 8.9	79.3 ± 18.7	35.0 ± 11.5	14.2 ± 2.4
Winter	145.3 ± 24.0	62.3 ± 15.0	38.9 ± 5.1	29.1 ± 6.7

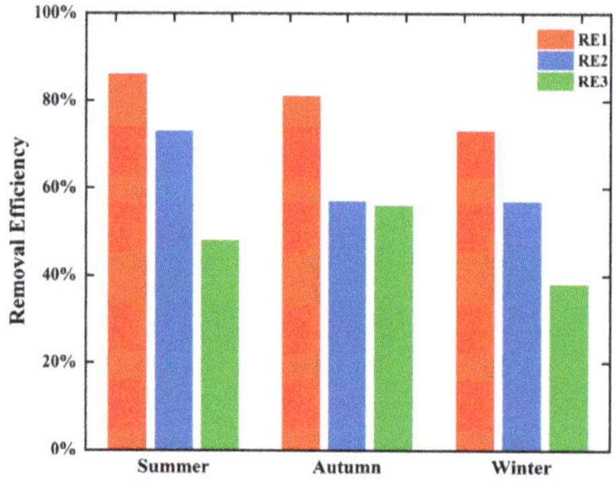

Figure 2. Removal efficiency of MPs in different seasons (RE 1 represents the total removal efficiency of MPs; RE 2 represents the removal efficiency of MPs at S2; and RE 3 represents the removal efficiency of MPs at S3).

Table 2. Removal efficiency of MPs in sewage plants in different regions.

Region	Scale of Treatment (m^3/d)	Digestion Reagent	Total RE (%)	Influent Abundance (n/L)	Effluent Abundance (n/L)	Reference
Lithuania	225,000	H_2O_2	97.0	1964 ± 50–2982 ± 54	744 ± 13–1244 ± 21	[32]
Mikelli, Finland	10,000	Fenton reagent	98.3	57.6 ±12.4	1.05 ± 0.4	[33]
Astana, Kazakhstan	253,900	H_2O_2	88.6–93.0	47.1 ± 37.6–69.4 ± 41.0	4.1 ± 3.1–5.4 ± 3.5	[34]
Kathmandu, Nepal	32,400	Fenton reagent	72.5	31.2 ± 17.3	8.5 ± 5.6	[35]
Wuhan, China	70,000	H_2O_2	66.1	23.3 ± 2.0	7.9 ± 1.1	[36]
Hohhot, China	180,000	Fenton reagent	80.8 ± 12.1	73.0 ± 5.0	14.0 ± 2.0	[37]
Nanjing, China	80,000	Fenton reagent	96.0	44.1 ± 3.2	2.0 ± 0.3	[38]
Zhengzhou, China	100,000	H_2O_2	73–86	171.3 ± 22.3	33.3 ± 6.6	This study

Table 2 summarizes the digestion reagents and removal efficiencies of MPs in various WWTPs with differing treatment scales. There are some differences in the characterization of MPs in the studies of different wastewater plants. It can be seen that the removal efficiencies of MPs in foreign WWTPs are maintained at a high level. Comparative analysis reveals generally superior MP removal efficiency in foreign WWTPs, with this study's cumulative MP elimination exceeding some domestic counterparts while remaining below international benchmarks. As shown in Table 2, there are significant differences in the relationship between pretreatment reagents and total removal efficiency. This may be attributed to the strong oxidative action of hydroxyl radicals in Fenton reagents, which efficiently degrade complex pollutants. In contrast, the effect of H_2O_2 as a pretreatment reagent varies significantly, potentially due to its weaker oxidative capacity, its influence on water pH, and variations in organic matter concentration.

It had been calculated that this WWTP, from the total effluent, will discharge many MP particles into nature in summer (19.0 to 33.0 × 10^8 n/d), autumn (23.5 to 46.5 × 10^8 n/d), and winter (33.8 to 4.0 × 10^8 n/d), which will be discharged into the river and will cause different kinds of impacts on the water environment. It was shown that PA, PE, and PP polymer types were the most common polymer types found in fish samples and because of this, ingested organisms and associated food chains were threatened due to the high distribution of PP [39,40]; studies have shown that these MPs were not only ecologically hazardous, but also acted as carriers of dissolved pollutants, thus exacerbating environmental pollution [41].

3.2. *Size of MPs*

MPs exhibit distinct characteristics that provide valuable insights into their sources and potential environmental impacts. This study classified MPs based on their size, shape, color, and polymer type. As presented in Table 3, to systematically investigate the seasonal variations in removal efficiencies of MPs of various sizes by this WWTP, this study compared and analyzed the retention patterns of different treatment units for specific MPs sizes across the summer, autumn, and winter seasons. After verification, there were no significant seasonal differences in the MP abundance of various particle sizes in S1 of the WWTP in the three seasons ($P > 0.05$), significant correlation between MP removal for sizes 20–50 μm and >1000 μm ($P < 0.05$), and no correlation between 50 and 1000 μm removals across seasons ($P > 0.05$).

Table 3. Removal efficiency of size-specific MPs in different seasons.

Season	Size	RE2 (%)	RE3 (%)
Summer	20–50 μm	79%	59%
	50–100 μm	48%	44%
	100–500 μm	50%	31%
	500–1000 μm	−200%	−100%
	>1000 μm	100%	0%
Autumn	20–50 μm	67%	59%
	50–100 μm	61%	44%
	100–500 μm	71%	31%
	500–1000 μm	67%	−100%
	>1000 μm	100%	0%
Winter	20–50 μm	45%	72%
	50–100 μm	64%	58%
	100–500 μm	100%	−138%
	500–1000 μm	100%	−20%
	>1000 μm	100%	0%

Figure 3a–c and Table 3 collectively demonstrate the removal efficiency of MPs across various size ranges at different treatment stages and seasons. Summer influent exhibited the highest abundance of MPs in the 20–50 μm size fraction (153.7 ± 44.2 n/L, accounting for 83% of the total), followed by the 50–100 μm size fraction (21.7 ± 29.7 n/L, accounting for 12% of the total). Larger size fractions (100–500 μm, 500–1000 μm, >1000 μm) collectively represented only 5.4% of the total MPs. The treatment sequence exhibited significantly higher removal efficiency for MPs in the 50–500 μm range compared to finer MPs in the 20–50 μm range, primarily due to gravitational settling mechanisms in the primary treatment process [38]. The results in S2 indicated a significant reduction in the distribution of MPs across all particle sizes, with particularly pronounced removal for MPs in the range of 20–50 μm. The abundance decreased to 33.0 ± 18.5 n/L, corresponding to a removal rate of 79%. Due to the high adsorption capacity of sludge, MPs with small particle sizes are more likely to be adsorbed and retained in it [25].

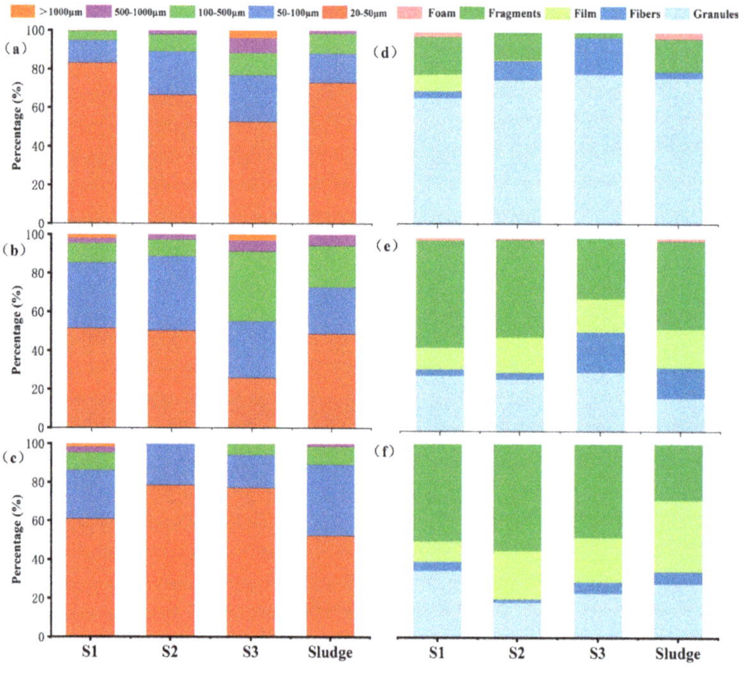

Figure 3. Percentages of MPs' size and shape in different seasons (**a**) size in summer; (**b**) size in autumn; (**c**) size in winter; (**d**) shape in summer; (**e**) shape in autumn; and (**f**) shape in winter.

The removal of MPs in the particle size range of 50–100 μm and 100–500 μm was 48% and 50%, respectively. Previous studies had demonstrated that in the secondary treatment process, MPs were either adsorbed by bioflocs or encapsulated by biofilm, eventually leading to their deposition into the sludge [38,42]. In addition, the removal effect of the secondary treatment significantly reduced the amount of MPs in the effluent at this stage [43]. The experimental results showed that at the S3 sampling site, the highest removal rate of MPs was 59% for particle sizes 20–50 μm, and the lowest rate was 30% for 100–500 μm. As a result, a large number of MPs with particle sizes of 20–50 μm were treated in the sludge, and the largest proportion of MPs with this size was detected in the sludge samples in this quarter with 73%, followed by 50–100 μm with 15%. Table 3 showed that S2 had a higher removal efficiency

(79%) for MPs of small particle size 20–50 μm. However, as particle size increased, the removal efficiency decreased, and particles with a particle size of 500–1000 μm even exhibited negative values. In the tertiary treatment stage, the removal efficiency of MPs with the same particle size was relatively stable, except for those in the >1000 μm range, which exhibited greater variability. MPs in the >1000 μm range had no retention capacity (0%), and the removal effect was poor. In summer, the removal efficiency of large-particle MPs in S2 was more pronounced compared to S3. Meanwhile, it could be observed that during the tertiary treatment stage, there were limitations in intercepting large-particle MPs.

Autumn data revealed distinct MP distribution patterns, with the influent predominantly consisting of 20–50 μm (94.7 ± 9.3 n/L, 51.4%) and 50–100 μm (62.7 ± 7.1 n/L, 34.1%). The results showed that the distribution of MPs in S2 was significantly reduced in all particle sizes. Specifically, the removal rates in the 20–50 μm, 50–100 μm, and 100–500 μm particle size ranges were similar, in the range of 60–70%. In S3, the removal of MPs in the 20–50 μm particle size range was significantly increased by 71%, while the removal of MPs in the 50–100 μm particle size range was 58%. Findings indicated that the abundance of MPs with particle sizes greater than 100 μm increased compared to autumn. This may be attributed to the input of polymers from the WWTP or experimental limitations observed during the study. In this season, the highest proportion of MPs with a particle size of 20–50 μm in the sludge reached 49%. The removal efficiency of MPs smaller than 500 μm in S2 remained stable at 67–71%, with the majority of MPs being retained within the biofilm carriers during this stage [44], which was lower compared to the summer season. It was found that the high temperatures characteristic of the summer season positively influence the biological reactions occurring during this stage. Additionally, high-temperature conditions enhance the ability of biofilms to retain MPs more effectively. However, the RE 3 was significantly lower, between 31 and 44%, and there may be particle fragmentation or a re-suspension phenomenon [45]. The secondary treatment process was still more effective than the tertiary treatment process in treating large-sized MPs in this season.

The winter data showed that the abundance of MPs was mainly composed of particles in the size ranges of 20–50 μm (61%) and 50–100 μm (25%). The experimental results at the S2 sampling site showed that the removal of MPs in the size ranges of 20–50 μm and 50–100 μm was 45% and 64%, respectively. It is interesting to note that the MPs in the 100–500 μm and 500–1000 μm size ranges showed complete removal (100% removal) during the winter treatment, indicating an enhanced ability to remove larger particles under cold conditions. In the effluent, the removal of MPs in the 20–50 μm, 50–100 μm, and 100–500 μm size ranges was 39% and 50%, respectively, and no MPs with particle sizes larger than 500 μm were detected. In sludge samples, 20–50 μm still accounted for the largest proportion of 52%. The physical retention effect of the secondary treatment process demonstrated a significantly higher removal efficiency for MPs > 100 μm, as evidenced by the experimental results, particularly when compared to the summer and autumn seasons. This can be attributed to the low-temperature environment in winter, which inhibits microbial activity in activated sludge, thereby enhancing the retention effect at this stage. However, the physical retention effect during the tertiary treatment process showed abnormal negative values for the removal of MPs within the particle size range of 100–500 μm.

Overall, the removal efficiency of small-sized MPs ranging from 20 to 500 μm was primarily attributed to the activated sludge in the secondary treatment process. In this process, a significant amount of extracellular polymeric substances secreted by microorganisms within the activated sludge adsorbed the MPs, forming larger flocs [46]. The increased relative densities and settling velocities of these flocs facilitated the efficient removal of MPs through sedimentation in the secondary treatment process [45,47]. Seasonal performance variations showed winter flocculation efficiency exhibited stronger temperature dependence compared to summer, while the fiber disc filtration maintained consistent performance. Notably, the secondary treatment process exhibited a superior removal of large MPs (>500 μm) during winter through enhanced flocculation and sedimentation. This contrasts with the negative removal efficiency observed in summer and autumn, which may be attributed to increased apparent concentrations caused by sludge recirculation. In the tertiary treatment process, the effective removal of large MPs could not be achieved due to the absence of a targeted process. Specifically, fiber disc filtration tanks failed to achieve a satisfactory removal of large-sized MPs, likely because high suspended solids loads led to clogging, thereby impairing the retention of large particle-sized MPs.

3.3. *Shape of MPs*

All of the detected MPs were categorized into four main forms: particles, fibers, fragments, and films. Fibers, which were elongated and thread-like particles, likely originate from the washing of synthetic fabrics and the degradation of textiles such as cloth, linen, and carpets during their aging process [48]. Fragments, typically characterized by broken or fragmented pieces of larger plastic items, were likely the result of the degradation or disintegration of plastic products. Film, which manifested as soft flakes of MPs, typically originates from items such as plastic bags or packaging materials. The statistical results showed that there were no significant seasonal differences in the MP abundance of different shapes in S1 ($P > 0.05$); there was no significant correlation between the shape of the MPs and the removal efficiency in different seasons ($P > 0.05$). As illustrated in Figure 3d–f, which displays the percentages of MP shapes at each stage across the three seasons, Figure 3d indicates that this WWTP had the highest percentage of particles in the influent water during the summer months, accounting for 66% with an abundance of 121.3 ± 28.5 n/L. Many types of facial washes, facial scrubs, or exfoliating products contain MP particles, and these were considered to be one of the sources of particles [49]. The hot weather in summer leads to an increased usage of household products, such as facial cleansers, which are prone to releasing MPs. Fragmented MPs accounted for the second highest percentage at 19%, with an average abundance of 35.3 ± 9.0 n/L. The results from S2 demonstrated that particles, thin-film, and fragmented MPs were largely removed at this stage, with removal efficiencies of 69%, 98%, and 80%, respectively. Their abundances decreased to 37.3 ± 11.0 n/L, 1.0 ± 0.3 n/L, and 7.0 ± 2.0 n/L, respectively. Thin-film shaped MPs, characterized by their relatively low density and large surface area, tend to float on the water surface, facilitating their removal along with flotsam [50]. In the effluent, all other shapes of MPs were removed in large quantities, and the removal effect of fibrous MPs was not obvious in this stage relative to other shapes of MPs. The largest percentage of particle-shaped MPs was 76%, followed by debris, at 17%, in sludge samples.

Figure 3e indicates that in autumn, fragmented MPs accounted for the largest share of MPs, reaching 55% with an abundance of 40.0 ± 17.3 n/L in S1. Irregular fragmented shapes usually come from fragments of larger plastic objects, such as tires, bottles, and plastic bags [51], and particles accounted for the next largest share of MPs, reaching an abundance of 29% with an abundance of 21.3 ± 3.2 n/L. In this season, the secondary treatment process demonstrated greater effectiveness in removing MPs across all shapes, except for film-like MPs. The removal efficiency exceeded 60% in all cases, with the exception of films. Among the various MP shapes, fragmented MPs, which were the most abundant, exhibited the most significant removal rate in S3. This is likely due to their faster settling velocity compared to fibers, spheres, and films, leading to their eventual adsorption in the sludge [52]. Consequently, fragments constituted the highest proportion (46%) of the mud samples collected during the fall quarter, followed by films at 20%. Figure 3f indicated that during the winter season, the fragmented MPs in this WWTP accounted for the largest proportion, approximately 50%, followed by particle and film, which each constituted 20% to 30%. The most significant removal occurred for particle and fibrous MPs during the S2 phase of the season, achieving removal rates of 78% and 88%, respectively. Overall, the removal efficiency for particles was the highest during the winter phase at 82%, followed by fragmented MPs at 74%. Films and fragments represented the higher proportions during the winter phase, accounting for 37% and 29%, respectively.

3.4. Color of MPs

The analysis of the MPs collected from the wastewater samples revealed a diverse range of colors and the distribution of different color groups (grey, black, blue, red, yellow, and transparent) is presented in Figure 4. Among the identified colors, gray was the most prevalent, accounting for 36–51% of the total observed MPs. This finding indicated that gray MPs dominated in wastewater, and their presence may be associated with various consumer products, such as packaging materials or plastic products. Transparent MPs ranged from 37% to 38%, while the color distribution varied across other studies. Several studies had demonstrated that transparent MPs constitute a significant proportion of effluent [36,53]. This range indicated a relatively consistent distribution of transparent MPs in wastewater. Additionally, both grey and transparent MPs exhibited a relatively uniform total removal rate of approximately 80% compared to other colors in this study, with grey MPs achieving the highest removal rate of 97% in summer. While transparent MPs exhibited the highest treatment efficiency in the autumn season, with a removal rate of 87%, summer biosolids consisted of 49% grey and 36% transparent particles, shifting to 40% grey and 34% transparent particles in autumn. Winter samples displayed distinct coloration patterns, with yellow (25%) and grey (35%) MPs forming the predominant fractions. Yellow MPs showed variable prevalence ranging from 11% to 24%, with peak removal efficiencies observed in summer (86%), followed by autumn (59%) and winter (75%). The sources of these colors are diverse and may include packaging materials, plastic film fragments, or other yellow and transparent plastic items. The proportions of blue and red MPs were relatively low, ranging from 0% to 0.8%. While these colors were consistently detected, their prevalence is significantly lower compared to grey and clear MPs. The presence of blue and red MPs suggested the existence of plastic debris and its specific pigments in wastewater. These colors might also originate from items such as bottles, containers, or synthetic fibers released during washing machine operations.

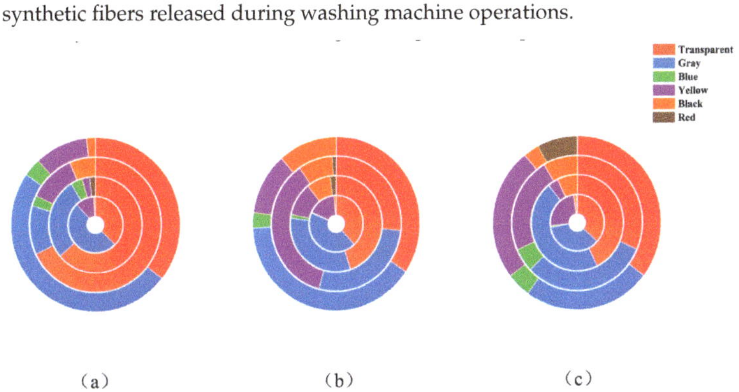

Figure 4. Percentages of MP colors in different seasons (from the inner ring to the outer ring are S1, S2, S3, and sludge samples, respectively; (**a**) summer, (**b**) autumn, and (**c**): winter).

The color distribution of MPs offers critical insights into the types and sources of plastic fragments entering wastewater systems. For instance, the incorporation of heavy metals, such as chromium (Cr), copper (Cu), cobalt (Co), selenium (Se), lead (Pb), and cadmium (Cd), during manufacturing processes results in the production of vividly colored polymers. These metals are often added as pigments or stabilizers, which not only influence the polymer's coloration but also enhance its durability and resistance to environmental degradation [54]. Pale yellow to reddish-brown coloration potentially signals environmental weathering through photo-oxidative degradation [55,56]. Seasonal chromatic shifts in MPs exhibited distinct patterns. Grey MPs were predominant in summer influent (51% at site S1), with decreased proportions observed in autumn (43%) and winter (36%). Transparent MPs showed higher proportions in summer and autumn influent streams. Winter data indicate potential correlations between color profiles and increased outdoor plastic usage or climatic changes, warranting further investigation. By analysis, there were no significant seasonal differences ($P > 0.05$) in MP abundance by season for different colors in S1 among the three seasons, and there was no significant correlation between the color and removal efficiency of MPs between seasons ($P > 0.05$). This finding underscores the necessity for systematic investigations to ascertain whether seasonal variations affect the removal efficiency of differently colored MPs. Specifically, controlled experiments were required to isolate chromatic properties from other physicochemical factors.

3.5. *Polymer Types of MPs*

In the results of this experiment, polypropylene (PP) and polyethylene terephthalate (PET) were identified as the two polymer types with the highest abundance percentages during the study period (Figure 5). These two polymers are widely used in food packaging and food containers (such as beverage bottle caps), as they were extensively processed and produced in the plastic supply market through traditional processes (such as plastic processing, film processing, or plastic molding) [57]. The detailed experimental results are presented in the Supplementary Materials (Tables S1–S3). The standardization of priorities and methods for future polymer investigations can be guided by these conclusions drawn from WWTP data. Moreover, the detection of material information on MPs indicated that the majority of MPs in wastewater are predominantly derived from daily human activities [58].

Polymer-specific analysis revealed distinct seasonal variations in polymer composition across different stages of WWTP. In summer influent (S1), PP was the predominant polymer, accounting for 101.7 ± 7.1 n/L (55%), followed by PET at 45.7 ± 2.3 n/L (29%) and polyphenylene oxide (PPO) at 18.7 ± 13.8 n/L (10%). High-density PET removal occurred through gravity sedimentation in the secondary treatment process, demonstrating density-dependent elimination patterns [59,60]. In S2, the abundance of PP was reduced to 19.0 ± 4.4 n/L, corresponding to a removal rate of 81%. Low-density MPs, such as PE and PP, tend to naturally float on the surface of the effluent [61]. Meanwhile, medium-density polymers, like polyamide (PA) and PS, were transported to the effluent surface by bubbles generated during the air flotation process [62]. Other polymers, including PVC, PET, and PS, were removed at rates of 85%, 82%, and 96%, respectively. In S3, the polymer PET exhibited the highest removal percentage, while polymers PP and PPO demonstrated the most significant removal efficiencies of 88% and 66%, respectively.

In autumn, the influent of the WWTP contained relatively high levels of the polymers PP and PET, accounting for 32% and 36%, respectively. Based on the process effluent test results, the removal efficiencies for the polymers PP, PET, PE, and PA were 61%, 71%, 94%, and 80%, respectively, indicating that the treatment effects for these polymers were the most significant. The content of PS and polytetrafluoroethylene (PTFE) showed no significant changes or abnormalities, which might be attributed to observational errors or measurement limitations. In S3, the main polymers PP, PPO, and POM were more significantly removed, with removal efficiencies of 90%, 83%, and 75%, respectively. In contrast, the removal efficiency of the polymer PET was only 22% in this stage. Notably, while PE was completely removed (100%) during the test, its abundance at S1 was relatively low, making it difficult to objectively conclude that it was entirely removed.

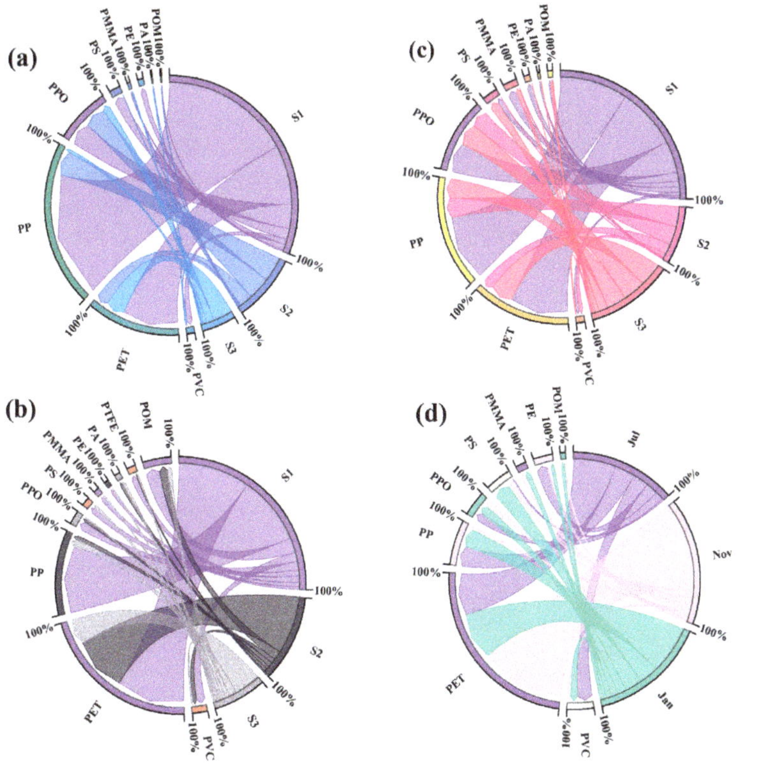

Figure 5. Percentages of MP polymer types in different seasons, (**a**) summer, (**b**) autumn, and (**c**) winter, and (**d**) in sludge.

In addition, PA and PTFE showed consistently increasing abundance in the observations, which might contribute to local sampling errors. During winter at site S1, the abundances of PET and PP were measured as 16.3 ± 4.0 n/L and 15.7 ± 8.1 n/L, respectively, accounting for 37% and 36% of the total polymer content. The treatment effects on polymers PVC, PP, PPO, and PTFE were more pronounced at site S2, with removal efficiencies reaching 50%, 94%, 67%, and 75%, respectively. Based on these results, the removal rates for PS, polymethyl methacrylate (PMMA), and PE were relatively low, at 20% and 40%, while the concentrations of PPO, PS, PMMA, PE, and PTFE remained at 1.0 ± 0.3 n/L. The main component of the wastewater was PET, with an abundance of 6.0 ± 1.0 n/L and 5.0 ± 1.7 n/L, respectively. The abundance of PVC, PPO, PS, PMMA, PE, PA, and PTFE was approximately 1.0 ± 0.3 n/L. It is evident that the removal of polymers PVC, PET, PPO, and POM at this stage was relatively significant, with removal efficiencies of 67%, 54%, 50%, and 50%, respectively. In contrast, there were no significant changes in the abundance of PP, PS, PMMA, and PE. PET was identified as the most prevalent polymer type across all three seasons of the mud samples. In summer, PET constituted 42% of the MPs, followed by PP at 20%. During autumn, PET accounted for 74% of the sludge samples, while other polymer types, such as PP and PPO, collectively represented only 5% on average. In winter, PET made up 47% of the polymers in the sludge samples, with PP following at 19%.

4. Limitations

The composition of various types of wastewater, local lifestyles, population structure, sampling techniques (e.g., sampling duration and mesh size), extraction methods (e.g., digestion and density separation), and identification techniques, as well as sampling locations and times, are among the key factors influencing the abundance of MPs in WWTPs within a study area. Subjective misjudgments by observers during testing and potential errors in pre-processing may lead to inaccuracies in the results. Furthermore, the randomness of sampling failed to account for seasonal variations across the three sampled seasons, contributing to uncertainty in quantifying MPs discharged from the WWTP. In this study, it was observed that the abundance of larger MPs was relatively low, with flocculation and gravity settling identified as effective primary treatment methods for intercepting larger particles. Larger MPs tend to settle at the bottom of the pool or adsorb onto suspended solids, potentially introducing error into the sampling results. Current sampling tools and techniques may not fully capture all sizes and forms of MP particles, particularly those smaller than 20 microns. The distribution of MPs in WWTP is typically uneven, and human errors during sampling may affect result accuracy. During data processing, the diversity and complexity of MPs, combined with the limitations of different testing techniques, may introduce subjectivity into result interpretation, complicating inter-laboratory comparisons and validations. Additionally, continuous observation and more meticulous sampling are required to better understand the effects of seasonal variations and to accurately characterize MP abundance in WWTPs.

5. Conclusions

This study investigated the abundance, characterization, and removal efficiency of MPs in the influent, process effluent, and final effluent of the WWTP and analyzed the results through experimental testing. The main findings of this study revealed that the total removal rate of the WWTP was highest at 86% during the summer, followed by 81% in autumn, and lowest at 73% in winter. Specifically, secondary treatment removed 73% of MPs in summer, similar to the efficiency in autumn, while achieving only 57% removal in winter. Tertiary treatment demonstrated removal efficiencies of 48%, 56%, and 38% for summer, autumn, and winter, respectively. The overall removal rate is highest in summer, higher than in autumn and winter, indicating that overall treatment efficiency is better during the hot season. Secondary treatment shows seasonal fluctuations, with a secondary removal rate of 73% in summer, which is significantly different from autumn and winter. Tertiary treatment shows non-linear seasonal characteristics, with tertiary removal rates showing "summer < autumn < winter". The high retention efficiency of secondary treatment in summer reduces the tertiary load. These results demonstrate the WWTP's efficacy in removing the majority of MPs; however, substantial quantities are still released into the environment on a daily basis, underscoring the urgent need for enhanced environmental protection measures. The shape of MPs in the effluent from this WWTP was predominantly fragments and particles across the three seasons. Specifically, granular MPs constituted the largest proportion in summer, whereas fragmented MPs were more prevalent in autumn and winter. PET and PP were identified as the predominant components of MPs in the effluent. Specifically, PP exhibited the highest concentration during summer observations, whereas both PP and PET accounted for more than one-third of the total concentration in autumn and winter. Transparent and grey were the most frequently observed colors across all three seasons. The most prevalent sizes during the three seasons ranged from 20 to 50 μm, followed by 50 to 100 μm. In addition, the study revealed distinct seasonal trends in MP abundance, with a higher abundance observed during specific seasons, particularly summer and autumn. This correlation with seasonal weather patterns indicates that hydrologically-driven human activities and weather-related changes may play a crucial role in influencing MP levels. This study offers several recommendations and suggestions for future research. First, it is crucial to broaden the investigation of MPs in wastewater by incorporating a more extensive range of sampling locations across the region. Such an approach would yield a more comprehensive understanding of the extent and origins of MP contamination. Furthermore, long-term monitoring studies are essential for elucidating spatiotemporal trends and seasonal variations in MPs concentrations with greater precision. Consequently, it is imperative to examine the potential impacts of MP pollution on aquatic ecosystems. The findings of this study underscore the importance of implementing effective management strategies and interventions in wastewater systems to mitigate MP pollution.

Supplementary Materials: The following supporting information can be downloaded at https://www.mdpi.com/article/10.3390/w17172614/s1: Figure S1: Electron micrograph of MPs (fiber: a; fragment: b, f, g, and j; film: c; and granules: d, e, and h); Figure S2: Raman spectra of various polymers (PS: polystyrene; PET: polyethylene terephthalate; PPO: polyphenylene oxide; PE: polyethylene; PMMA: polymethyl methacrylate; PVC: polyvinyl chloride; PP: polypropylene; POM: polyoxymethylene; PTFE: polytetrafluoroethylene; and PA: polyamide); Table S1: Percentages of polymer types in summer; Table S2: Percentages of polymer types in autumn; and Table S3: Percentages of polymer types in winter.

Author Contributions: Data curation, X.C. and Y.L.; Formal analysis, X.C., K.L. and X.L.; Funding acquisition, Y.L. and Y.D.; Investigation, X.C., K.L., X.L., K.J. and T.A.; Methodology, Y.L. and J.L.; Project administration, X.D.; Resources, Y.L.; Software, X.C., K.L. and L.Z.; Supervision, Y.D. and X.D.; Validation, K.J. and T.A.; Visualization, L.Z.; Writing—original draft, X.C. and Y.L.; Writing—review and editing, J.L. and Y.D. All authors have read and agreed to the published version of the manuscript.

Funding: This research was supported by the Natural Science Foundation of Henan (No. 242300421666), Postgraduate Education Reform and Quality Improvement Project of Henan Province (No. YJS2023JD17, YJS2024KC16).

Data Availability Statement: The original contributions presented in this study are included in the article and Supplementary Materials. Further inquiries can be directed to the corresponding authors.

Conflicts of Interest: The authors declare no conflict of interest.

References

1. Liu, X.; Wei, W.; Chen, Z.; Wu, L.; Duan, H.; Zheng, M.; Wang, D.; Ni, B.-J. The threats of micro- and nanoplastics to aquatic ecosystems and water health. *Nat. Water* **2025**, *3*, 764–781. [CrossRef]
2. Leslie, H.A.; van Velzen, M.J.; Brandsma, S.H.; Vethaak, A.D.; Garcia-Vallejo, J.J.; Lamoree, M.H. Discovery and quantification of plastic particle pollution in human blood. *Environ. Int.* **2022**, *163*, 107199. [CrossRef]
3. Wu, T.; Ding, J.; Wang, S.; Pang, J.-W.; Sun, H.-J.; Zhong, L.; Ren, N.-Q.; Yang, S.-S. Insight into effect of polyethylene microplastic on nitrogen removal in moving bed biofilm reactor: Focusing on microbial community and species interactions. *Sci. Total. Environ.* **2024**, *932*, 173033. [CrossRef]
4. Ragusa, A.; Svelato, A.; Santacroce, C.; Catalano, P.; Notarstefano, V.; Carnevali, O.; Papa, F.; Rongioletti, M.C.A.; Baiocco, F.; Draghi, S.; et al. Plasticenta: First evidence of microplastics in human placenta. *Environ. Int.* **2021**, *146*, 106274. [CrossRef]

5. Bhatt, P.; Pathak, V.M.; Bagheri, A.R.; Bilal, M. Microplastic contaminants in the aqueous environment, fate, toxicity consequences, and remediation strategies. *Environ. Res.* **2021**, *200*, 111762. [CrossRef] [PubMed]
6. Eze, C.G.; Nwankwo, C.E.; Dey, S.; Sundaramurthy, S.; Okeke, E.S. Food chain microplastics contamination and impact on human health: A review. *Environ. Chem. Lett.* **2024**, *22*, 1889–1927. [CrossRef]
7. Zhao, K.; Li, C.; Li, F. Research progress on the origin, fate, impacts and harm of microplastics and antibiotic resistance genes in wastewater treatment plants. *Sci. Rep.* **2024**, *14*, 1–19. [CrossRef] [PubMed]
8. Talbot, R.; Chang, H. Microplastics in freshwater: A global review of factors affecting spatial and temporal variations. *Environ. Pollut.* **2022**, *292*, 118393. [CrossRef] [PubMed]
9. Freeman, S.; Booth, A.M.; Sabbah, I.; Tiller, R.; Dierking, J.; Klun, K.; Rotter, A.; Ben-David, E.; Javidpour, J.; Angel, D.L. Between source and sea: The role of wastewater treatment in reducing marine microplastics. *J. Environ. Manag.* **2020**, *266*, 110642. [CrossRef]
10. Schell, T.; Hurley, R.; Buenaventura, N.T.; Mauri, P.V.; Nizzetto, L.; Rico, A.; Vighi, M. Fate of microplastics in agricultural soils amended with sewage sludge: Is surface water runoff a relevant environmental pathway? *Environ. Pollut.* **2022**, *293*, 118520. [CrossRef]
11. Kruglova, A.; Muñoz-Palazón, B.; Gonzalez-Martinez, A.; Mikola, A.; Vahala, R.; Talvitie, J. The dangerous transporters: A study of microplastic-associated bacteria passing through municipal wastewater treatment. *Environ. Pollut.* **2022**, *314*, 120316. [CrossRef]
12. Enfrin, M.; Dumée, L.F.; Lee, J. Nano/microplastics in water and wastewater treatment processes—Origin, impact and potential solutions. *Water Res.* **2019**, *161*, 621–638. [CrossRef]
13. Lv, X.; Dong, Q.; Zuo, Z.; Liu, Y.; Huang, X.; Wu, W.-M. Microplastics in a municipal wastewater treatment plant: Fate, dynamic distribution, removal efficiencies, and control strategies. *J. Clean. Prod.* **2019**, *225*, 579–586. [CrossRef]
14. Jani, V.; Wu, S.; Venkiteshwaran, K. Advancements and Regulatory Situation in Microplastics Removal from Wastewater and Drinking Water: A Comprehensive Review. *Microplastics* **2024**, *3*, 98–123. [CrossRef]
15. Jiang, J.; Wang, X.; Ren, H.; Cao, G.; Xie, G.; Xing, D.; Liu, B. Investigation and fate of microplastics in wastewater and sludge filter cake from a wastewater treatment plant in China. *Sci. Total. Environ.* **2020**, *746*, 141378. [CrossRef] [PubMed]
16. Tagg, A.S.; Sapp, M.; Harrison, J.P.; Sinclair, C.J.; Bradley, E.; Ju-Nam, Y.; Ojeda, J.J. Microplastic Monitoring at Different Stages in a Wastewater Treatment Plant Using Reflectance Micro-FTIR Imaging. *Front. Environ. Sci.* **2020**, *8*, 145. [CrossRef]
17. Bayo, J.; Olmos, S.; López-Castellanos, J. Microplastics in an urban wastewater treatment plant: The influence of physicochemical parameters and environmental factors. *Chemosphere* **2020**, *238*, 124593. [CrossRef] [PubMed]
18. Cheung, P.K.; Hung, P.L.; Fok, L. River Microplastic Contamination and Dynamics upon a Rainfall Event in Hong Kong, China. *Environ. Process.* **2018**, *6*, 253–264. [CrossRef]
19. Constant, M.; Ludwig, W.; Kerhervé, P.; Sola, J.; Charrière, B.; Sanchez-Vidal, A.; Canals, M.; Heussner, S. Microplastic fluxes in a large and a small Mediterranean river catchments: The Têt and the Rhône, Northwestern Mediterranean Sea. *Sci. Total. Environ.* **2020**, *716*, 136984. [CrossRef]
20. Hongprasith, N.; Kittimethawong, C.; Lertluksanaporn, R.; Eamchotchawalit, T.; Kittipongvises, S.; Lohwacharin, J. IR microspectroscopic identification of microplastics in municipal wastewater treatment plants. *Environ. Sci. Pollut. Res.* **2020**, *27*, 18557–18564. [CrossRef]
21. Huang, Q.; Liu, M.; Cao, X.; Liu, Z. Occurrence of microplastics pollution in the Yangtze River: Distinct characteristics of spatial distribution and basin-wide ecological risk assessment. *Water Res.* **2022**, *229*, 119431. [CrossRef]
22. Iordachescu, L.; Vest Nielsen, R.; Papacharalampos, K.; Barritaud, L.; Denieul, M.-P.; Plessis, E.; Baratto, G.; Julien, V.; Vollertsen, J. Point-source tracking of microplastics in sewerage systems. Finding the culprit. *Water Res.* **2024**, *257*, 121696. [CrossRef]
23. Can, T.; Üstün, G.E.; Kaya, Y. Characteristics and seasonal variation of microplastics in the wastewater treatment plant: The case of Bursa deep sea discharge. *Mar. Pollut. Bull.* **2023**, *194*, 115281. [CrossRef]
24. Bengüsu, I.K.; Gökhan, E.Ü.; Tuğba, C. Characteristics, and seasonal change of microplastics in organized industrial zone wastewater treatment plant. J. *Environ. Chem. Eng.* **2025**, *13*, 115516. [CrossRef]
25. Hamidian, A.H.; Ozumchelouei, E.J.; Feizi, F.; Wu, C.; Zhang, Y.; Yang, M. A review on the characteristics of microplastics in wastewater treatment plants: A source for toxic chemicals. *J. Clean. Prod.* **2021**, *295*, 126480. [CrossRef]
26. Tadsuwan, K.; Babel, S. Microplastic abundance and removal via an ultrafiltration system coupled to a conventional municipal wastewater treatment plant in Thailand. *J. Environ. Chem. Eng.* **2022**, *10*, 107142. [CrossRef]
27. Turan, N.B.; Erkan, H.S.; Engin, G.O. Microplastics in wastewater treatment plants: Occurrence, fate and identification. *Process. Saf. Environ. Prot.* **2021**, *146*, 77–84. [CrossRef]
28. Zhang, B.; Wu, Q.; Gao, S.; Ruan, Y.; Qi, G.; Guo, K.; Zeng, J. Distribution and removal mechanism of microplastics in urban wastewater plants systems via different processes. *Environ. Pollut.* **2023**, *320*, 121076. [CrossRef]

29. Mathew, J.; Pulicharla, R.; Rezai, P.; Brar, S.K. Microplastics in wastewaters: Pretreatment to detection trail. *J. Water Process. Eng.* **2024**, *64*, 105702. [CrossRef]
30. Liu, W.; Zhang, J.; Liu, H.; Guo, X.; Zhang, X.; Yao, X.; Cao, Z.; Zhang, T. A review of the removal of microplastics in global wastewater treatment plants: Characteristics and mechanisms. *Environ. Int.* **2021**, *146*, 106277. [CrossRef]
31. Kittipongvises, S.; Phetrak, A.; Hongprasith, N.; Lohwacharin, J. Unravelling capability of municipal wastewater treatment plant in Thailand for microplastics: Effects of seasonality on detection, fate and transport. *J. Environ. Manag.* **2022**, *302*, 113990. [CrossRef]
32. Uogintė, I.; Pleskytė, S.; Pauraitė, J.; Lujanienė, G. Seasonal variation and complex analysis of microplastic distribution in different wastewater treatment plant treatment stages in Lithuania. *Environ. Monit. Assessm.* **2022**, *194*, 829. [CrossRef]
33. Lares, M.; Ncibi, M.C.; Sillanpää, M.; Sillanpää, M. Occurrence, identification and removal of microplastic particles and fibers in conventional activated sludge process and advanced MBR technology. *Water Res.* **2018**, *133*, 236–246. [CrossRef]
34. Zhaxylykova, D.; Alibekov, A.; Lee, W. Seasonal variation and removal of microplastics in a central Asian urban wastewater treatment plant. *Mar. Pollut. Bull.* **2024**, *205*, 116597. [CrossRef]
35. Bastakoti, S.; Adhikari, A.; Thaiba, B.M.; Neupane, B.B.; Gautam, B.R.; Dangi, M.B.; Giri, B. Characterization and removal of microplastics in the Guheshwori Wastewater Treatment Plant, Nepal. *Sci. Total. Environ.* **2024**, *935*, 173324. [CrossRef] [PubMed]
36. Tang, N.; Liu, X.; Xing, W. Microplastics in wastewater treatment plants of Wuhan, Central China: Abundance, removal, and potential source in household wastewater. *Sci. Total. Environ.* **2020**, *745*, 141026. [CrossRef] [PubMed]
37. Mi, J.H.; Lu, J.P.; Liu, T.X.; Liu, Y.; Shi, X.H.; Zhang, X.J.; Shi, Z.Y.; Liu, Y.H. Whole Process Analysis and Fate Behavior of mi-croplastics in Urban Wastewater Treatment Plants, Including their Occurrence Forms, Components, and Removal Efficiency. *Environ. Sci.* **2024**, *45*, 4052–4062. [CrossRef]
38. Zhang, Y.; Wang, H.; Xu, J.; Su, X.; Lu, M.; Wang, Z.; Zhang, Y. Occurrence and Characteristics of Microplastics in a Wastewater Treatment Plant. *Bull. Environ. Contam. Toxicol.* **2021**, *107*, 677–683. [CrossRef] [PubMed]
39. Djekoun, M.; Gaaied, S.; Romdhani, I.; Rida, A.M.; Missaoui, Y.; Boubekeur, M.S.; Trea, F.; Lakbar, C.; Ouali, K.; Banni, M. Abundance and distribution of environmental microplastic in edible fish and mussels from the south Mediterranean coasts. *Mar. Pollut. Bull.* **2024**, *206*, 116705. [CrossRef]
40. Oo, P.Z.; Boontanon, S.K.; Boontanon, N.; Tanaka, S.; Fujii, S. Horizontal variation of microplastics with tidal fluctuation in the Chao Phraya River Estuary, Thailand. *Mar. Pollut. Bull.* **2021**, *173*, 112933. [CrossRef]
41. Saud, S.; Yang, A.; Jiang, Z.; Ning, D.; Fahad, S. New insights in to the environmental behavior and ecological toxicity of microplastics. *J. Hazard. Mater. Adv.* **2023**, *10*, 100298. [CrossRef]
42. Xiao, L.; Wen, K.; Ming, X.; Zhen, L.; Jun, W. Transfer and fate of MPs during the conventional activated sludge process in one wastewater treatment plant of China. *Chem. Eng. J.* **2019**, *362*, 176–182. [CrossRef]
43. Lofty, J.; Muhawenimana, V.; Wilson, C.; Ouro, P. Microplastics removal from a primary settler tank in a wastewater treatment plant and estimations of contamination onto European agricultural land via sewage sludge recycling. *Environ. Pollut.* **2022**, m. [CrossRef]
44. Jose, S.; Lonappan, L.; Cabana, H. Prevalence of microplastics and fate in wastewater treatment plants: A review. *Environ. Chem. Lett.* **2024**, *22*, 657–690. [CrossRef]
45. Li, L.; Liu, D.; Song, K.; Zhou, Y. Performance evaluation of MBR in treating microplastics polyvinylchloride contaminated polluted surface water. *Mar. Pollut. Bull.* **2020**, *150*, 110724. [CrossRef] [PubMed]
46. Xu, Z.; Bai, X.; Ye, Z. Removal and generation of microplastics in wastewater treatment plants: A review. *J. Clean. Prod.* **2021**, *291*, 125982. [CrossRef]
47. Ali, I.; Ding, T.; Peng, C.; Naz, I.; Sun, H.; Li, J.; Liu, J. Micro- and nanoplastics in wastewater treatment plants: Occurrence, removal, fate, impacts and remediation technologies – A critical review. *Chem. Eng. J.* **2021**, *423*, 130205. [CrossRef]
48. Napper, I.E.; Thompson, R.C. Release of synthetic microplastic plastic fibres from domestic washing machines: Effects of fabric type and washing conditions. *Mar. Pollut. Bull.* **2016**, *112*, 39–45. [CrossRef] [PubMed]
49. Cheung, P.K.; Fok, L. Evidence of microbeads from personal care product contaminating the sea. *Mar. Pollut. Bull.* **2016**, *109*, 582–585. [CrossRef]
50. Vardar, S.; Onay, T.T.; Demirel, B.; Kideys, A.E. Evaluation of microplastics removal efficiency at a wastewater treatment plant discharging to the Sea of Marmara. *Environ. Pollut.* **2021**, *289*, 117862. [CrossRef]
51. Wang, W.; Ndungu, A.W.; Li, Z.; Wang, J. Microplastics pollution in inland freshwaters of China: A case study in urban surface waters of Wuhan, China. *Sci. Total. Environ.* **2017**, *575*, 1369–1374. [CrossRef]
52. Bilgin, M.; Yurtsever, M.; Karadagli, F. Microplastic removal by aerated grit chambers versus settling tanks of a municipal wastewater treatment plant. *J. Water Process. Eng.* **2020**, *38*, 101604. [CrossRef]

53. Hidayaturrahman, H.; Lee, T.-G. A study on characteristics of microplastic in wastewater of South Korea: Identification, quantification, and fate of microplastics during treatment process. *Mar. Pollut. Bull.* **2019**, *146*, 696–702. [CrossRef]
54. Razzak, S.A.; Faruque, M.O.; Alsheikh, Z.; Alsheikhmohamad, L.; Alkuroud, D.; Alfayez, A.; Hossain, S.M.Z.; Hossain, M.M. A comprehensive review on conventional and biological-driven heavy metals removal from industrial wastewater. *Environ. Adv.* **2022**, *7*, 100168. [CrossRef]
55. Abaroa-Pérez, B.; Ortiz-Montosa, S.; Hernández-Brito, J.J.; Vega-Moreno, D. Yellowing, Weathering and Degradation of Marine Pellets and Their Influence on the Adsorption of Chemical Pollutants. *Polymers* **2022**, *14*, 1305. [CrossRef] [PubMed]
56. Liu, P.; Zhan, X.; Wu, X.; Li, J.; Wang, H.; Gao, S. Effect of weathering on environmental behavior of microplastics: Properties, sorption and potential risks. *Chemosphere* **2020**, *242*, 125193. [CrossRef] [PubMed]
57. Gritsch, L.; Breslmayer, G.; Rainer, R.; Stipanovic, H.; Tischberger-Aldrian, A.; Lederer, J. Critical properties of plastic packaging waste for recycling: A case study on non-beverage plastic bottles in an urban MSW system in Austria. *Waste Manag.* **2024**, *185*, 10–24. [CrossRef] [PubMed]
58. Sun, J.; Dai, X.; Wang, Q.; van Loosdrecht, M.C.; Ni, B.-J. Microplastics in wastewater treatment plants: Detection, occurrence and removal. *Water Res.* **2019**, *152*, 21–37. [CrossRef]
59. Xu, X.; Zhang, L.; Jian, Y.; Xue, Y.; Gao, Y.; Peng, M.; Jiang, S.; Zhang, Q. Influence of wastewater treatment process on pollution characteristics and fate of microplastics. *Mar. Pollut. Bull.* **2021**, *169*, 112448. [CrossRef]
60. Long, Z.; Pan, Z.; Wang, W.; Ren, J.; Yu, X.; Lin, L.; Lin, H.; Chen, H.; Jin, X. Microplastic abundance, characteristics, and removal in wastewater treatment plants in a coastal city of China. *Water Res.* **2019**, *155*, 255–265. [CrossRef]
61. Petroody, S.S.A.; Hashemi, S.H.; van Gestel, C.A. Transport and accumulation of microplastics through wastewater treatment sludge processes. *Chemosphere* **2021**, *278*, 130471. [CrossRef] [PubMed]
62. Ngo, P.L.; Pramanik, B.K.; Shah, K.; Roychand, R. Pathway, classification and removal efficiency of microplastics in wastewater treatment plants. *Environ. Pollut.* **2019**, *255*, 113326. [CrossRef] [PubMed]

water

MDPI

Article

Sources and Transport of Dissolved Organic Matter (DOM) in Surface and Groundwater Within a Dominated Greenhouse Agriculture Catchment: Insights from Multi-Tracer

Haoyang Wang [1], Shuang Song [2], Wei Xu [2,*] and Fu-Jun Yue [1,*]

[1] Institute of Surface-Earth System Science, School of Earth System Science, Tianjin University, Tianjin 300072, China
[2] Center of Eco-Environmental Monitoring and Scientific Research, Administration of Ecology and Environment of Haihe River Basin and Beihai Sea Area, Ministry of Ecology and Environment, Tianjin 300170, China
* Correspondence: tjhw525@163.com (W.X.); fujun_yue@tju.edu.cn (F.-J.Y.)

Abstract: Intensive greenhouse agriculture significantly alters dissolved organic matter (DOM) dynamics in aquatic ecosystems, but related research remains scarce. To address this knowledge gap, this study employed an integrated approach combining Excitation–Emission Matrix Parallel Factor Analysis (EEM-PARAFAC), Two-Dimensional Correlation Spectroscopy (2D-COS), and Self-Organizing Map (SOM) analyses with hydrochemical and stable water isotopes ($\delta^{18}O$ and δD) to investigate the dynamic characteristics of DOM in surface water and groundwater in an intensive greenhouse agriculture catchment (XER) in northern China. Water chemistry and isotope results consistently demonstrated mixing between surface water and groundwater, which was attributed to irrigation pumping. Four fluorescent components were identified via EEM-PARAFAC (C1 and C4 are humic components, while C2 and C3 are tryptophan components), with microbial decomposition of organic fertilizers and domestic wastewater discharge being important sources. Compared with tryptophan components, terrestrial humic substances in groundwater preferentially change in the parallel river direction, while microbial humic substances are more sensitive in the vertical direction, as confirmed by 2D-COS. SOM analysis validated the EEM-PARAFAC results through component plane visualization, demonstrating both DOM inter-component relationships and their correlations with inorganic ions. These results provide critical scientific support for developing sustainable water resource management strategies and optimizing organic fertilizer use in greenhouse agricultural systems, with important practical implications for protecting groundwater quality in intensively cultivated catchments.

Keywords: greenhouse agriculture catchment; dissolved organic matter (DOM); fluorescence; self-organizing network (SOM); surface–groundwater interactions

1. Introduction

China's agricultural sector is transitioning from traditional extensive farming to modern intensive systems, with facility-based agriculture becoming the predominant model for high-value crop production [1]. While this shift has significantly boosted agricultural productivity, particularly through year-round greenhouse cultivation of vegetables, fruits, and ornamental plants [2,3], it has also introduced environmental challenges. Intensive greenhouse practices characterized by excessive fertilizer application (e.g., poultry/pig manure and nitrogen fertilizers) and groundwater overexploitation for irrigation have substantially altered aquatic ecosystems near cultivation areas [4,5]. Compared to groundwater, surface water systems are particularly vulnerable to anthropogenic disturbances from agricultural inputs [6]. However, shallow groundwater and surface water form an interconnected hydrological system with continuous quality and quantity exchange [7,8]. Excessive groundwater extraction creates hydraulic gradients that enhance pollutants from surface water leaching into shallow aquifers [9], potentially degrading groundwater quality in intensive greenhouse farming regions [10].

Dissolved organic matter (DOM), the most environmentally reactive fraction of natural organic matter (NOM), comprises a complex mixture of allochthonous and autochthonous organic compounds in aquatic systems [11,12]. At the catchment scale, DOM serves as a dynamic indicator of surface–groundwater interactions [13] and plays a crucial role in mediating the transport and bioavailability of nutrients (N, P), heavy metals, and organic pollutants [4,14,15]. In natural aquifers, the detailed impact of DOM depends not only on its concentration [6] but also on its source and composition [16,17], and it is significantly influenced by hydrological seasonality [18]. During high-flow periods, rainfall-induced soil erosion mobilizes terrestrial organic matter through surface runoff into rivers, making humic substances the predominant DOM source [19,20]. In contrast, baseflow conditions lead to groundwater-derived DOM, which exhibits lower aromaticity but higher microbial and protein-like signatures [21,22]. Notably, agricultural practices have altered DOM dynamics at the catchment scale, including concentrations, biodegradability, and total export flux [21,23,24]. In greenhouse agricultural systems, irrigation practices including groundwater extraction and return flow may modify DOM characteristics in both surface water and shallow groundwater through enhanced soil erosion and organic fertilizer transport [4,25]. While existing research has thoroughly documented DOM concentration variations in agricultural catchments, the dynamic response of DOM properties to agricultural-induced alterations in surface–groundwater interactions remains insufficiently characterized [25]. This study specifically aims to bridge this critical research gap.

Absorbance and fluorescence spectroscopy remain powerful and cost-effective tools for characterizing DOM sources in aquatic systems, despite the availability of various characterization methods [26]. Excitation–emission matrix (EEM) fluorescence spectroscopy coupled with parallel factor analysis (PARAFAC) effectively resolves complex fluorophore mixtures into distinct components, enabling robust DOM source discrimination [6,27]. While EEM-PARAFAC is particularly sensitive to land use and hydrological influences [26], it typically treats DOM components as static variables rather than dynamic, interacting systems. Self-organizing maps (SOM), as an unsupervised neural network, effectively characterize DOM component distributions and sources [28–30]. SOM combined with EEM-PARAFAC can clearly show changes in the overall fluorescence characteristics of all samples rather than examining isolated fluorophores [31,32]. Two-dimensional correlation spectroscopy (2D-COS) can enhance spectral resolution and detect subtle DOM transformations under environmental perturbations (flow direction and geographic latitude) [7,33]. However, few studies have applied EEM-PARAFAC combined with 2D-COS technology to characterize DOM and to reveal the transformation processes of different DOM components in natural water environments [34–36]. Therefore, integrating multiple technical methods (EEM-PARAFAC-SOM-2D-COS) can facilitate more comprehensive, in-depth, and complementary exploration and analysis of the sources and dynamic transformation of DOM components from surface water to shallow groundwater in small agricultural catchments.

Stable water isotopes ($\delta^{18}O$ and δD) serve as powerful natural tracers for identifying water sources and circulation mechanisms due to their distinct isotopic signatures resulting from various hydrological and geochemical processes [37,38]. Notably, current hydrological process research mainly focuses on inorganic parameters, whereas organic indicators, despite their potential application value, remain under-explored [39]. Combining geochemical information identified by water chemistry, $\delta^{18}O$, and δD with organic indicators may provide new insights into the origin and transformation of DOM [38,40].

Intensive greenhouse agriculture has a significant impact on surrounding water quality [4,5]. This study employed a systematic sampling approach during winter (minimizing the effects of rainfall and terrestrial inputs) in greenhouse-dominated agricultural catchments to investigate DOM dynamics under key agricultural practices, including organic fertilizer application and irrigation pumping. Our research objectives were to conduct the following: (1) hydrochemical and stable isotope analyses to elucidate surface–groundwater interactions; (2) determining of DOM sources and composition through EEM-PARAFAC analysis; (3) 2D-COS analysis of groundwater DOM fluorescence component variations using river orientation as a perturbation factor; and (4) establishing of DOM-hydrogeochemical relationships via SOM clustering analysis and Pearson's correlation analysis. By integrating multiple analytical approaches, this study provides a scientific foundation for the protection of sustainable water resources in greenhouse-intensive catchments.

2. Materials and Methods

2.1. Study Area

The Xier River catchment (XER), located in central Shandong Province, is a 105.56 km^2 tributary catchment within the Mi River Basin (118.36–119.8° N, 36.11.5–37.8° E, 3847.5 km^2) (Figure 1a). This intensive agricultural region features predominantly farmland with year-round greenhouse cultivation of vegetables, flowers, and fruits. Characterized by a warm-temperate semi-humid monsoon climate, the area exhibits distinct seasonal variations, with an average annual temperature of 12 °C. The dry season typically extends from December to March, followed by the flood season from June to August. Land use patterns and topographic features of both basins are presented in Figure 1a–c.

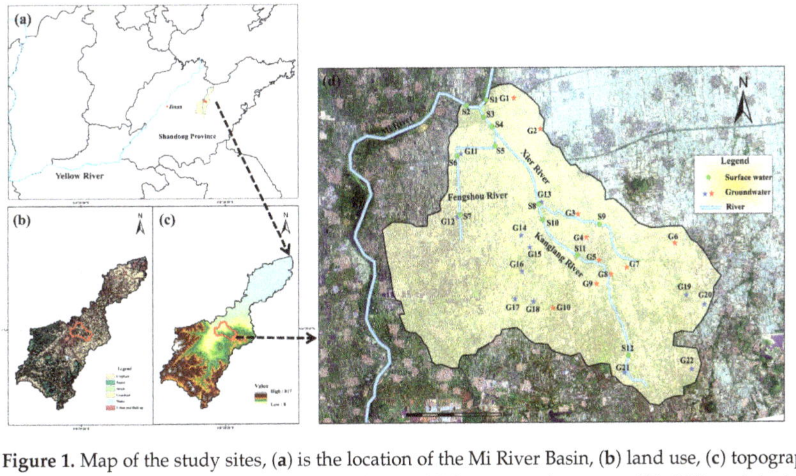

Figure 1. Map of the study sites, (**a**) is the location of the Mi River Basin, (**b**) land use, (**c**) topography, and the red line area is XER (**d**). Note: The letter "S" denotes surface water (S1–S12), and the letter "G" denotes groundwater (G1–G22). Among them, G1–G5 are parallel to the river direction, and G6–G10 are perpendicular to the river direction.

2.2. Field Sampling and Pretreatment

To investigate the characteristics of DOM under the influence of surface–groundwater interactions while minimizing interference from rapid hydrological variations such as rainfall events, we conducted a comprehensive water sampling campaign during the winter of 2023 (From 8th to 12th December). A total of 12 surface water (S-) and 22 groundwater (G-) samples were collected from the Mi River, Xier River, their tributaries (Kanglang River and Fengshou River), and adjacent villages (Figure 1d). Groundwater sampling covered densely populated areas and different agricultural zones (vegetables, flowers, and fruit fields; Table 1). Notably, the groundwater samples G1–G5 were collected parallel to the river direction (PR), while samples G6–G10 were collected perpendicular to the river direction (VR). In addition, based on the well depths obtained from field surveys (Table S1), groundwater was classified into deep groundwater (G4 and G22) and shallow groundwater (other sites). Surface water collection employed an extendable-pole bucket and stainless steel sampler with a bottom check valve, while groundwater was obtained through well pumping. All samples were collected in triplicate, uniformly mixed, and divided into unfiltered (plastic bottles) and filtered (0.45 μm membrane into 100 mL HDPE bottles) portions. Strict preservation protocols were followed, including overflow filling, Parafilm sealing, and light-protected cold storage during transport.

2.3. Analysis of Hydrochemical Indicators

Water quality parameters (temperature [T], pH, dissolved oxygen [DO], and electrical conductivity [EC]) were measured in situ using a calibrated EXO2 multiparameter probe (YSI, New York, NY, USA), with readings recorded after 5-min stabilization (Table 1). Laboratory analyses were completed within 7 days. Major ions (K^+, Na^+, Ca^{2+}, Mg^{2+}, Cl^-, NO_3^-, and SO_4^{2-}) were quantified by ion chromatography (Metrohm ECO IC, Herisau, Switzerland, Model: 940), while HCO_3^- concentrations were determined via 0.02 M HCl titration. Stable isotope ratios ($\delta^{18}O$ and δD) were analyzed using a Picarro L2140-i water isotope analyzer (Santa Clara, CA, USA), reported in δ-notation (‰) relative to V-SMOW (±0.1‰ for $\delta^{18}O$; ±0.8‰ for δD). DOC measurements employed an OI Analytical 1080 TOC analyzer (College Station, TX, USA) following acidification (0.5 M HCl) and high-purity oxygen purging to remove DIC, ensuring accurate non-purgeable organic carbon (NPOC) determination [18]. All analyses were performed in triplicate, with means calculated after outlier exclusion (±2% relative standard deviation).

2.4. Analysis of DOM

The DOM's UV-visible absorption spectra (200–800 nm, 1 nm resolution) were measured using a Shimadzu UV-2700 spectrophotometer with 1 cm quartz cuvettes. Milli-Q water (18.2 MΩ·cm resistivity) served as the reference blank, and all sample spectra were corrected by subtracting the corresponding blank spectrum. Three-dimensional excitation–emission matrices (3D-EEMs) were acquired on a Hitachi F-4700 fluorescence spectrophotometer (Tokyo, Japan) with the following parameters: excitation (Ex) 220–480 nm, emission (Em) 280–550 nm (5 nm intervals for both), 2.5 nm slit widths, and 12,000 nm min^{-1} scanning speed [16,41].

2.5. Data Analysis Methods of DOM

2.5.1. EEM-PARAFAC and Optical Indices Analysis

The raw EEM spectra were preprocessed to remove optical interferences prior to modeling. The methods described by Zepp et al. [42] were used to remove most of the Rayleigh and Raman scattering effects from the dataset (subtracting the Milli-Q blank spectrum). The inner-filter effect was corrected using measured absorbance at 254 nm (<0.3). All spectra were normalized to Raman units (R.U.) based on the integrated Raman scatter peak (Ex = 350 nm) of Milli-Q water [43]. PARAFAC modeling was conducted using the DOMFluor toolbox in MATLAB R2025a [44] and validated by split-half analysis [45]. The components identification was performed using OpenFluor (http://www.openfluor.org (accessed on 26 June 2025)) by spectral matching against reference libraries (fitting coefficient = 0.95) [46]. For each component, the maximum fluorescence intensity (F_{max}, R.U.) was determined from peak Ex/Em wavelengths, and relative abundances were calculated as the ratio of individual Fmax to total fluorescence intensity.

Four fluorescence indices (FI, β:α, BIX, HIX) were derived from EEM spectra to characterize DOM composition [12,47]. Complementary UV-visible absorption parameters included $SUVA_{254}$ (aromaticity indicator) and E2:E3 ratio (molecular weight proxy) [12,48]. Detailed formulations and interpretations of these indices are provided in Table S3.

2.5.2. 2D-COS Analysis

The 2D-COS has become an essential analytical tool for examining dynamic interactions and sequential responses among DOM components under environmental perturbations (e.g., temperature, pH gradients, or metal concentrations) [49,50]. Heterogeneous 2D-COS (hetero-2DCOS) extends this capability by enabling comparative analysis of spectral responses across different techniques under identical perturbation conditions, thereby providing enhanced characterization of system dynamics [33,51]. While synchronous correlation maps identify co-varying components, asynchronous maps reveal response sequences [52,53]. In this study, we implemented hetero-2DCOS analysis using 2Dshige v13 software (Kansai-Gakuin University, Nishinomiya, Japan) to process PARAFAC Ex load data, with groundwater sampling orientations (PR and VR to Xier River flow) serving as defined perturbations [7,33]. The analysis characterized DOM component dynamics relative to river orientation and established response sequences through Noda's rule [54]. Detailed methodological specifications are available in Zhu et al. [55].

2.5.3. SOM Analysis

SOM is an unsupervised neural network that projects high-dimensional data (e.g., 3D-EEM) onto a low-dimensional grid through a nonlinear transformation [31,56]. This method preserves topological relationships while enabling pattern classification, feature extraction, and dimensionality reduction, with minimal quantification error. The SOM algorithm operates through iterative weight initialization and adjustment based on input vectors and neighborhood interactions, with similarity quantification achieved via best-matching unit (BMU) distance metrics [57]. Spatially proximate neurons represent similar samples, with relationships refined through continuous learning. In this study, we implemented SOM analysis using MATLAB's somtoolbox, constructing a 6 × 6 neuron grid (Figure S1a) sized according to the heuristic equation $5\sqrt{n}$ (where n = 34 sampling points), followed by K-means clustering optimization [58]. Our SOM analysis incorporated five distinct parameter categories: conventional water quality indices, hydrochemical parameters, DOM components, spectral indices, and stable water isotope data. The U-matrix (unified distance matrix) visually represents sample similarity relationships through spatial distances between adjacent map units, employing an Euclidean distance-based color gradient scheme: warm hues (e.g., yellow to red) delineate cluster boundaries corresponding to larger inter-neuron distances; cool tones (e.g., blue) indicate regions of high sample similarity with minimal topological separation [31].

2.6. Statistical Analysis

Geospatial analysis was conducted using ArcGIS 10.2 for sampling location mapping and Kriging interpolation. All statistical analyses were performed with Microsoft Office 2021, while Pearson's correlation analysis was carried out using Origin 2024, examining relationships among DOC, DOM components, optical indices (FI, $\beta:\alpha$, HIX, BIX, $SUVA_{254}$, E2:E3), and stable water isotopes ($\delta^{18}O$ and δD).

3. Results and Discussion

3.1. Surface–Groundwater Interaction Based on Water Chemistry and Isotopes (δD and $\delta^{18}O$)

Hydrochemical characterization revealed a striking convergence between surface water and groundwater compositions in the study area, as evidenced by their comparable total dissolved solids (TDS) concentrations (surface water: 1166 $\pm$ 356 mg/L, $n = 12$; groundwater: 1196 $\pm$ 589 mg/L, $n = 22$) (Table S2). Generally speaking, the TDS in groundwater substantially exceeds surface water due to prolonged water–rock interactions [59,60], suggesting significant hydraulic connectivity potentially mediated by intensive irrigation practices in the study area [25]. The aqueous matrices exhibited distinct ionic fingerprints: surface waters were characterized by dominant K^+-HCO_3^--SO_4^{2-} assemblages, while groundwaters showed marked enrichment in Na^+-NO_3^--Cl^--Ca^{2+}-Mg^{2+} components (Table S2). Notably, high concentrations of Cl^- and NO_3^- in groundwater are particularly diagnostic and can serve as reliable water chemical tracers for non-point source pollution from greenhouse agriculture [9]. In this study, the observed DOC depletion in groundwater, especially at deep monitoring points G4 and G22 (Table S1), reflects progressive attenuation of organic compounds along flow paths through adsorption processes [23,61]. Piper trilinear analysis classified the waters primarily as Ca-Mg-HCO_3 and Cl-SO_4 types (Figure 2a), indicating dynamic recharge conditions in shallow aquifers that are likely enhanced by irrigation return flows [62]. Gibbs diagram analysis confirmed evaporation as the dominant process controlling water chemistry (Figure 2b,c), while cation ratio plots (Ca^{2+}/Na^+ vs. Mg^{2+} vs. HCO_3^-/Na^+) revealed dual contributions from silicate and carbonate weathering (Figure 2d,e) [63]. All in all, these collective hydrochemical signatures (Figure 2) demonstrate how anthropogenic perturbations, particularly irrigation activities, can alter natural hydrogeochemical regimes by modifying hydrodynamic conditions and solute transport pathways [59,60], ultimately leading to the observed convergence between surface and groundwater chemistries [38].

Stable water isotopes ($\delta^{18}O$ and δD) serve as effective natural tracers in hydrological studies, revealing distinct spatiotemporal variations across water cycle components [37,38]. Our measurements showed $\delta^{18}O$ values ranging from -9.2‰ to -6.5‰ and δD values from -71.8‰ to -49.7‰, plotted near both the Global and Local Meteoric Water Lines (GMWL and LMWL) (Figure 3a), confirming their precipitation origin. The significant isotopic overlap observed between surface water and groundwater samples (purple elliptical region) indicates intensive mixing processes within the XER [61,64], primarily driven by two mechanisms: groundwater recharges to surface water during low-precipitation periods [23] and irrigation returns flow from agricultural pumping activities [65]. The pronounced evaporation effect on surface water is evidenced by the rightward shift of several sampling points (particularly S2) from the GMWL and LMWL (Figure 2a) [61]. Spatial analysis revealed downstream depletion of $\delta^{18}O$ in surface water (Figure 3b), contrasting with groundwater's generally uniform isotopic distribution (Figure 3c), suggesting elevation-dependent fractionation. Notably, the deep groundwater sites G4 (vegetable field, 70 m) and G22 (residential area, 150 m) exhibited anomalously depleted $\delta^{18}O$ values (Table S1), which align with previous reports of more depleted isotopic signatures in deeper aquifers

(Figure 3a) [64]. These findings collectively highlight the complex interplay between natural hydrological processes and anthropogenic influences in the XER.

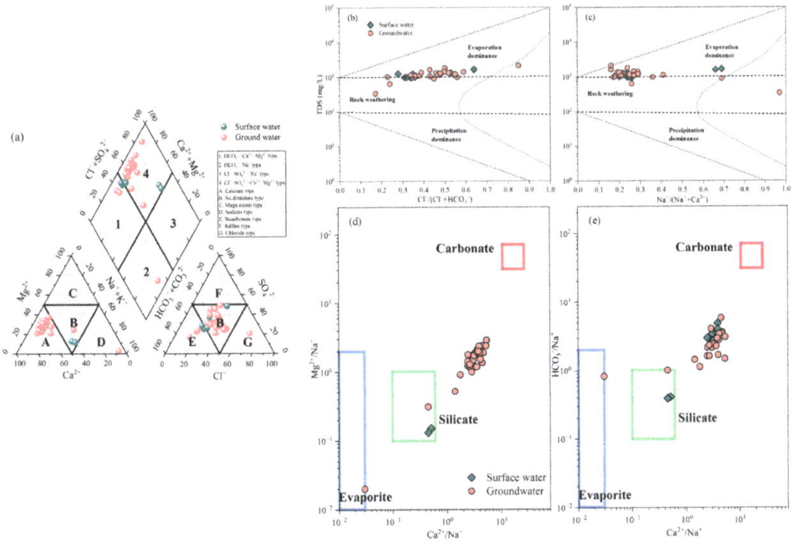

Figure 2. (**a**) Piper diagrams, (**b**,**c**) Gibbs diagrams, and (**d**,**e**) molar ratio bivariate plots of Mg^{2+} and HCO_3^-/Na^+ vs. Ca^{2+}/Na^+ of surface water and groundwater.

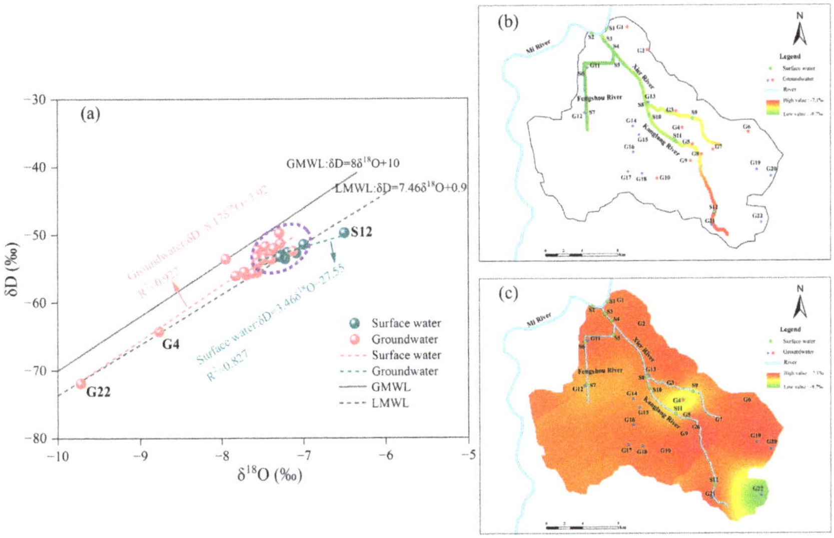

Figure 3. (**a**) The relationship between δD and $\delta^{18}O$ in the study area. The spatial distribution of (**b**) $\delta^{18}O$ of surface water (**c**) and groundwater in XER by Kriging interpolation. Note: The LMWL in this study for Eastern Monsoon China follows the established relationship: $\delta D = 7.46\ \delta^{18}O + 0.9$ [66].

3.2. Variations in DOM Components and Optical Indices Characteristics

This study identified four DOM components using EEM-PARAFAC analysis (Figure 4, Table 1), classified into two categories: (1) humic-like substances (C1 and C4) and (2) protein-like substances (C2 and C3). C1 displays maximum excitation/emission wavelengths at 245(325)/400 nm, representing a microbial humic substance typically found in groundwater [22,67]. This low molecular weight component originates from biological activity [68] and shows sensitivity to agricultural influences [9]. In contrast, C4 in surface water is a terrestrial humic component characterized by high molecular weight and aromaticity, strongly associated with agricultural land use [7,68]. Yang et al. [64] indicated that the relatively high abundance of C4 may be associated with terrestrial organic matter brought by monsoon season rainfall, while the non-monsoon season sees greater C1 production. C2 and C3 both exhibit Ex/Em peaks typical of tryptophan-like proteins, yet demonstrate different origins. While C2 primarily indicates domestic wastewater input [18], C3 reflects microbial metabolic activity in groundwater systems [13]. Studies demonstrate strong correlations between tryptophan-like proteins and biodegradable organic loads, such as BOD and COD in wastewater. In contrast, humic-like substances, which are less readily biodegradable, influence the formation and long-term persistence of disinfection byproducts [7,13,26]. Therefore, distinguishing between these two components holds significant importance for water quality monitoring and assessment.

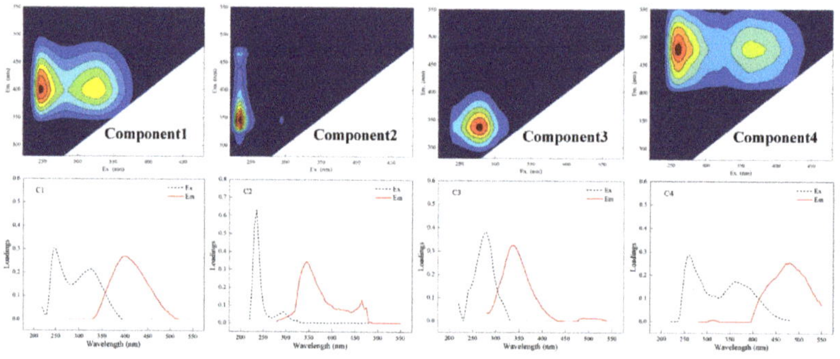

Figure 4. The fluorescent DOM composition and loading maps identified by EEM-PARAFAC analysis.

Table 1. The Ex and Em of DOM components (C1–C4) in XER water samples identified by EEM-PARAFAC analysis.

Component	Maximum Ex(nm)	Maximum Em(nm)	Description	Previous Studies	Number of Matches [b]
C1	245 (325) [a]	400	Microbial humic-like	C2 [68] C2 [9]	94
C2	235	345	Tryptophan-like, produced by human activities	C2 [7] C3 [18]	14
C3	275	335	Tryptophan-like, produced by microbial metabolism	C4 [68] C3 [13]	96
C4	260 (365)	480	Terrestrial humic-like	C4 [7] C1 [68]	80

Note: [a] Ex wavelengths in parentheses represent secondary peaks. [b] http://www.openfluor.org (conducted on 26 June 2025).

The PARAFAC model quantified the maximum fluorescence intensities (F_{max}, R.U.) of four DOM components in surface water and groundwater systems [4,43]. Component abundance varied significantly among sampling sites (Figure 5a). S1 and S12 showed exceptionally high levels of tryptophan-like components (C2 and C3) despite not exhibiting the highest DOC concentrations, a pattern potentially linked to localized material inputs (Table S1). Groundwater samples from flower cultivation areas (G10, G13, G14, Table S1) contained relatively high concentrations of C1, likely derived from microbial degradation of organic fertilizers (chicken/pig manure) used in these greenhouses [69]. Harjung et al. [70] demonstrated through regional chromatographic modeling that microbial processing of bioavailable organic matter plays a pivotal role in DOM transformation during its subsurface transport. Their findings indicate that groundwater-infiltrating DOM predominantly comprises hydrophilic low-molecular-weight compounds enriched with microbial metabolic byproducts. Relative abundance analysis (Figure 5b) showed a significant increase in C2 concentration in G16 (residential area), suggesting the input of untreated sewage rich in tryptophan-like compounds [71].

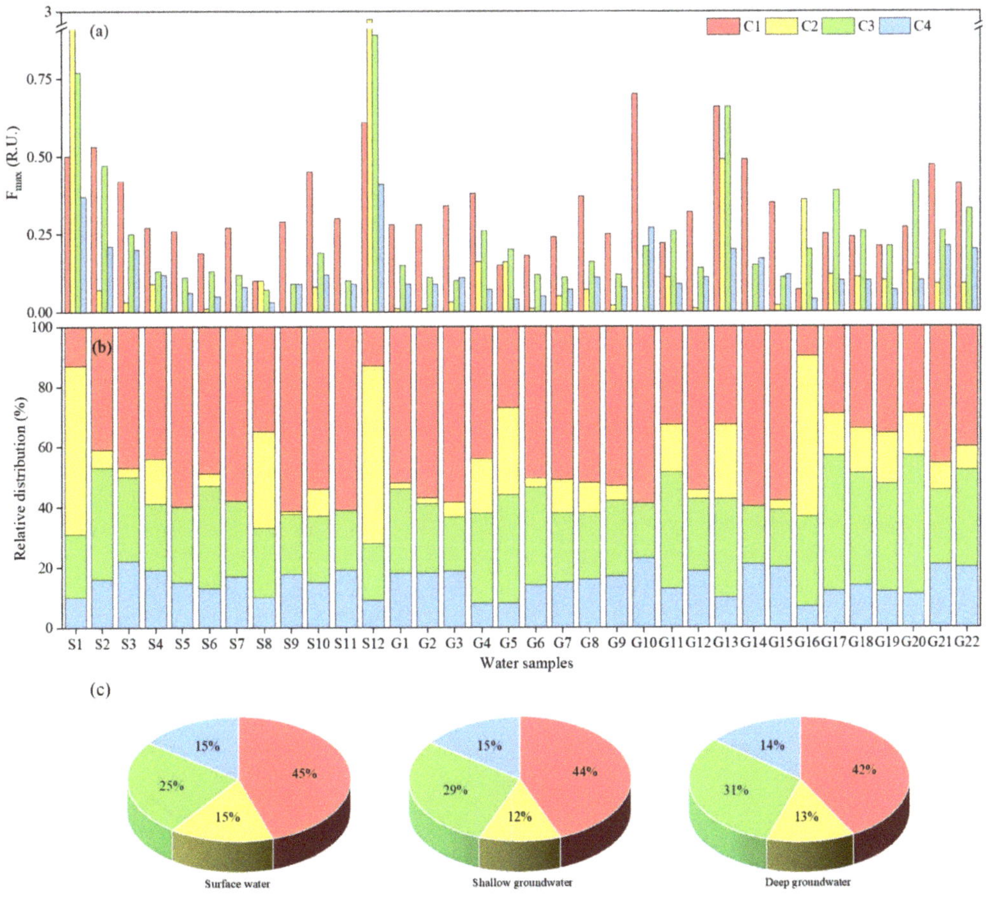

Figure 5. Comparisons of fluorescence components (C1–C4) of surface water and groundwater. (**a**) The PARAFAC fluorescence component loadings (F_{max}, R.U.) and (**b**) the relative abundance (%) and (**c**) total relative abundance (%) pie chart.

As shown in Figure 5c, the relative abundance of fluorescence components in both surface water and groundwater (shallow and deep) follows the order C1 > C3 > C4 ≥ C2. Specifically, C1 exhibits significantly higher abundance in surface water compared to groundwater, with particularly pronounced differences observed in deep aquifers. This distribution pattern is strongly associated with intensive organic fertilizer application in agricultural regions, as microbial degradation of these fertilizers generates substantial quantities of C1 [25,38]. In addition, pumped irrigation further introduces C1 from groundwater into rivers, causing mixing of DOM components between surface water and groundwater (e.g., at sites S5–S7 on the Fengshou River). Notably, C1 in surface water can also enter shallow aquifers through vertical infiltration. C3, a tryptophan-like component indicative of microbial DOM degradation [23], dominates in groundwater, likely due to enhanced organic matter breakdown—especially in deep aquifers during non-monsoon periods [64,68]. In contrast, C2 (anthropogenically influenced tryptophan-like material) is more prevalent in surface water. The lower C4 levels in deep groundwater suggest either removal in shallow aquifers or isolation from surface inputs [23,38]. In the intensively farmed arid regions of northwestern China, irrigation return flows promote the downward migration and shallow aquifer (<50 m) accumulation of terrestrial humics [25]. Notably, some humic substances, such as fulvic acid, have been shown to resist degradation, which makes it possible to use them as stable tracers for exogenous DOM inputs [13,68].

Comparative analysis of spectral indices between surface water and groundwater (Figure 6) revealed distinct DOM characteristics. FI>1.9 in both water types indicated predominant microbial origins (Table S3) [4], consistent with reduced external inputs during winter's low rainfall and flow conditions. Wang et al. [22] found that FI values in groundwater ranged from 1.28 to 1.65, indicating that groundwater systems in summer contain more DOM of terrestrial origin than DOM of microbial origin. The comparable β:α ranges (0.96/0.97–1.26/1.26) reflected similarly constrained biological decomposition under low temperatures, aligning with Du et al. [72] observation of less fresh winter DOM. BIX (>0.8) and HIX (<4) values demonstrated strong microbial influence and limited humification [9], further supported by regional parameter distribution patterns (Figure S2, Table S3). Notably, Yang et al. [64] found that low HIX values in surface water may be attributed to the mixing of groundwater with surface water, which contains fresher, microbially derived, and insufficiently humified organic matter originating from domestic sewage and other wastewater inputs. UV spectral analysis provided additional compositional insights [12]. Groundwater exhibited greater E2:E3 variability than surface water, indicating smaller DOM molecular weights resulting from preferential adsorption of humic substances by aquifer minerals and biodegradation [13,70]. Higher $SUVA_{254}$ in groundwater (~6 vs. surface water's ~4) confirmed greater aromaticity, likely due to microbial activity-driven humification of aquifer sediments [68]. Studies have found that winter water bodies exhibit higher aromaticity and double bonds compared to summer water bodies, with significantly enhanced resistance to degradation [72].

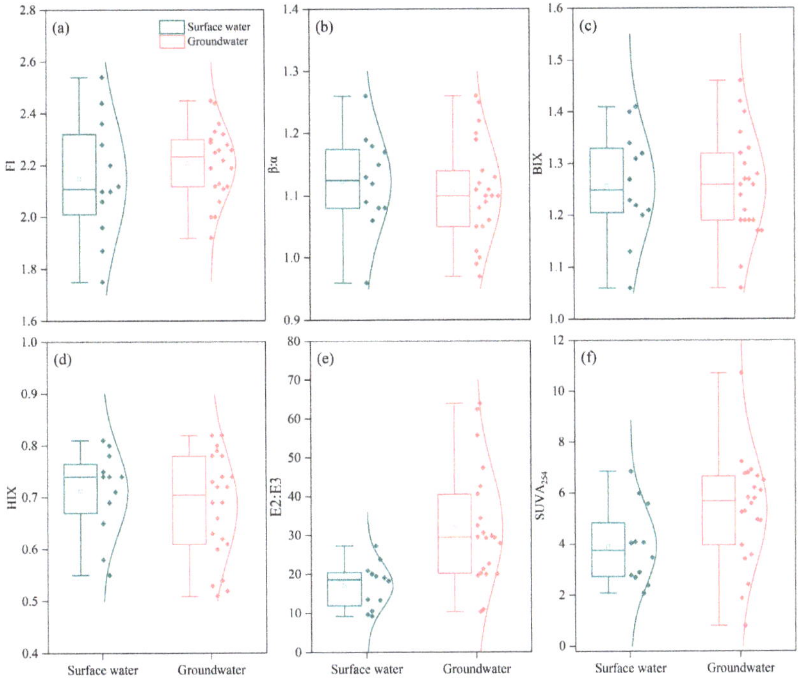

Figure 6. The optical indices of DOM in surface water and groundwater in the XER. (**a–d**) represents the fluorescence indexs: FI, β:α, BIX and HIX, and (**e**,**f**) represents UV-visible absorption parameters: E2:E3 and $SUVA_{254}$. The boxplot displays the data distribution through five key elements: maximum (top whisker), minimum (bottom whisker), upper quartile (75th percentile, box top), lower quartile (25th percentile, box bottom), and median (central line). The mean value is indicated by a central square marker. Adjacent scatter plots with trend lines illustrate sample point distributions and their variation patterns.

3.3. DOM Components Variation Sequences

While traditional methods like PARAFAC can characterize the static distribution of DOM components, they lack the capability to resolve changes in fluorescence signatures under external perturbations (e.g., river supply gradients) [7]. To address this limitation, we employed hetero-2DCOS to track the sensitivity and sequential variations of fluorescence components along both PR and VR river flow directions, aiming to reveal the key impacts of hydrological processes on DOM transformation under agricultural practices [7,33]. In the PR direction (Figure 7), synchronous and asynchronous 2D-COS revealed two key transformation sequences [54]: (1) C1→C3 (positive correlation) and C2→C3 and C4→C2 (negative correlation), and (2) C2→C1 and C4→C1 (inverse synchronous-asynchronous correlation patterns). These results establish the complete variability sequence as C4→C2→C1→C3 (Table 2), indicating preferential transformation of terrestrial humic substances over tryptophan-like and microbial humus, consistent with Ding et al. [7]. Xing et al. [61] showed that accelerated groundwater flow caused by irrigation pumping may lead to phase separation between sediment-adsorbed and dissolved components, thereby promoting the migration of DOM in aquifer sediments, particularly terrestrial humic substances [73]. Studies have shown that compared to complex terrestrial humic substances, relatively simple microbial humic substances are preferentially produced and stored in aquatic ecosystems [74]. Notably, the preferential migration of terrestrial humic substances along the PR may promote the transport of adsorbed pollutants and influence the formation of disinfection byproducts in downstream water bodies [75].

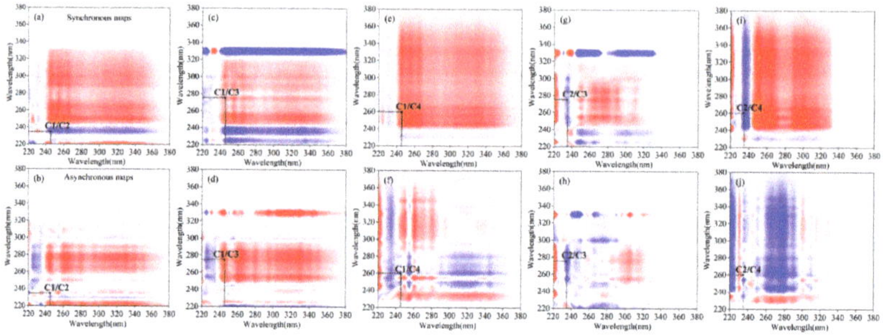

Figure 7. The 2D-COS of PARAFAC components in the PR direction: synchronous (top: (**a**,**c**,**e**,**g**,**i**)) and asynchronous (bottom: (**b**,**d**,**f**,**h**,**j**)) maps. Red and blue contours indicate positive and negative spectral responses, respectively.

Table 2. Symbols in the hetero-2DCOS synchronous and asynchronous (in parentheses) spectra in PR and VR.

Disturbance	Components	275 (C3)	260 (C4)	245 (C1)	235 (C2)
PR	275 (C3)	+	+(−)	+(+)	−(−)
	260 (C4)		+	+(−)	−(+)
	245 (C1)			+	−(+)
	235 (C2)				+
VR	275 (C3)	+	+(+)	+(+)	+(+)
	260 (C4)		+	+(+)	+(+)
	245 (C1)			+	+(+)
	235 (C2)				+

Note: + represents the positive cross peak; − represents the negative cross peak.

According to Noda's rule [54], the variability order of DOM components in the VR direction can also be determined as C1→C2→C3→C4 (Figure 8, Table 2). This indicates that protein- and microbial-related components in the VR direction are more prone to transformation due to their higher mobility and biodegradation potential [7,76,77]. Previous 2D-COS studies on the dynamics of DOM composition in groundwater around landfills demonstrated that rapid fluctuations in humic and protein-like components resulted from leachate-derived nutrients stimulating microbial activity, consequently elevating microbial byproduct concentrations (particularly proteinaceous substances) in groundwater [78]. Similarly, in our study area, intensive organic fertilizer application in agricultural zones has created optimal conditions for enhanced microbial metabolism in groundwater, potentially explaining the observed changes in fluorescent components. Therefore, optimizing fertilization timing and dosage in this catchment is essential to minimize the release of bioavailable organic matter and reduce groundwater pollution risks [4,26]. The above research results highlight the necessity of incorporating hydrological flow paths into pollution management strategies for small agricultural catchments, and have significant implications for ecological and water quality management [7].

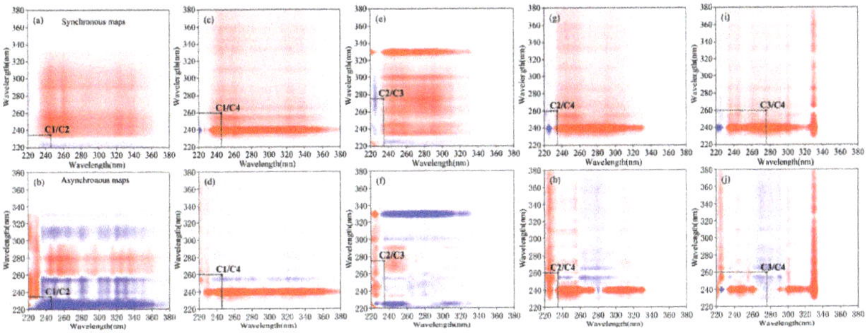

Figure 8. The 2D-COS of PARAFAC components in the VR direction: synchronous (top: (**a**,**c**,**e**,**g**,**i**)) and asynchronous (bottom: (**b**,**d**,**f**,**h**,**j**)) maps. Red and blue contours indicate positive and negative spectral responses, respectively.

3.4. Classification and Visualization

Kohonen's SOM, an artificial neural network algorithm, enables visualization of geometric correlations in fluorescence component data without prior assumptions [57]. As demonstrated in Figure 9a, the SOM network revealed significant dissimilarity between upper and lower neuron clusters [56], effectively differentiating sample characteristics. For enhanced sample distribution analysis, Figure 9b presents the hit histogram quantifying the frequency at which each neuron emerged as the winning neuron for water samples [32]. Neurons with higher hit counts correspond to water samples with similar characteristics (Figure S1b) [79]. Our results demonstrated that surface water exhibited peak representation (4 hits) and groundwater samples showed secondary prominence (3 hits). K-means clustering of neurons yielded six distinct clusters (Clusters 1–6, Figure 9c), with complete sample distribution mapping [80]. Groundwater samples predominantly occupied the upper-right SOM region (Clusters 1, 2, 4), while surface water samples concentrated in the lower-left area (Clusters 3, 5, 6). Interestingly, both the G4 and G22 deep groundwater samples were assigned to Cluster 5, while the S12 sample, which exhibits strong evaporation, was assigned to Cluster 4. This suggests that the groundwater in these clusters has a fairly direct connection to surface sources [70], although this was not captured in the present study. Building upon the analytical results presented in Section 3.2, which identified potential surface water–groundwater mixing at the Fengshou River (sites S5–S7), we hypothesize that comparable hydrological interactions may exist at other sampling locations within Cluster 3. However, additional investigation is required to verify. Previous studies have shown that combining DOM fluorescence with SOM can serve as a rapid monitoring tool for identifying sensitive recharge areas of surface–groundwater interactions affected by human activities [70]. Cluster 6 (S1 and S2) grouping reflects DOC concentration patterns (Table S1), where downstream sample S2 showed reduced DOC relative to upstream S1, indicating dilution effects at the Xier-Mi River confluence. In a previous study, Zhou et al. [30] classified SOM clusters based on differences in precipitation dilution of nitrogen concentrations before and after the rainy season.

Figure 9d displays the SOM component plane analysis of 25 parameters following network training, revealing consistent color patterns that indicate positive correlations between parameters [32]. Our analysis demonstrates three key findings: (1) C2 and C3 show parallel color gradients (intensity increasing from top-left to bottom-right), confirming their strong positive correlation. Figure 10's correlation coefficients quantitatively confirm this intuitive observation, and both belong to the tryptophan components (Table 1). (2) C1 and C4 exhibited distinct gradient patterns corresponding to their different origins (microbial degradation vs. terrestrial input). (3) C2/C3 displayed correlated patterns with Cl^- (extremes at upper-left/lower-right), implicating human activities (e.g., domestic wastewater) as their common source [71]. In addition, concordant β:α and HIX gradients (upper-right extremes) confirmed the microbial origin of DOM components.

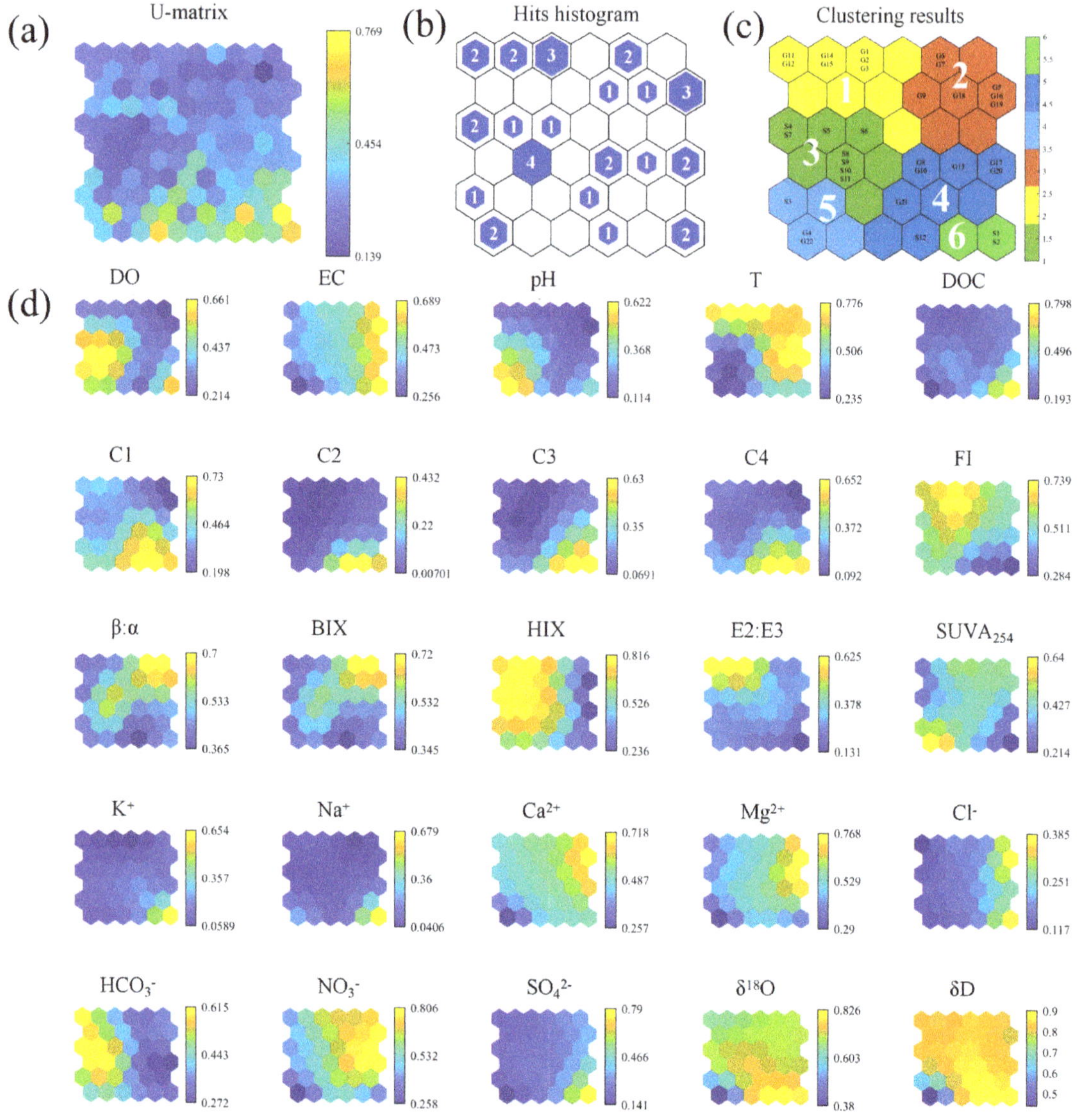

Figure 9. SOM visualization of surface water and groundwater samples. (**a**) U-matrix, (**b**) hit histogram, (**c**) clustering results, and (**d**) component planes for SOM.

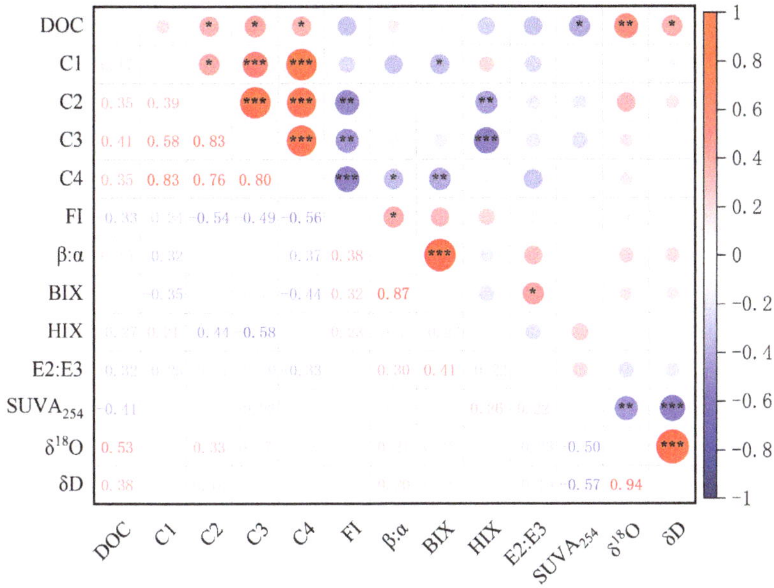

Figure 10. Pearson's correlation analysis of DOC, PARAFAC components, optical indices, and stable isotopes. Red indicates positive correlations, and blue indicates negative correlations. Asterisks indicated different significance levels: *** p-value < 0.001, ** p-value < 0.01, * p-value < 0.05.

3.5. Results of Pearson's Correlation Analysis

Figure 10 presents the Pearson's correlation analysis among DOM content, PARAFAC components, spectral indices, and stable water isotopes. DOC content was positively correlated with all four fluorescent components and δD ($p < 0.05$) and showed a stronger correlation with $\delta^{18}O$ ($p < 0.01$). Xing et al. [61] demonstrated a positive correlation between $\delta^{18}O$ values and DOC, suggesting that groundwater DOM composition is significantly influenced by mixing with surface water-derived DOM. Our findings further indicate that this mixing process is likely enhanced by irrigation pumping activities, which facilitate increased hydraulic connectivity between surface water and groundwater systems in the study area. While isotopic signatures partially reflect DOM content variations, their utility for tracing recharge source influences on DOM dynamics requires further investigation. Although $\delta^{18}O$ and δD generally showed weak correlations with DOM components and most spectral indices, they displayed significant negative correlations with $SUVA_{254}$ ($p < 0.01$ and $p < 0.001$, respectively), suggesting their potential utility as indicators of DOM aromaticity. Strong inter-component relationships were observed, with C1, C2, and C3 showing highly significant positive correlations with C4 ($p < 0.001$). The particularly strong positive correlation ($p < 0.001$) of C1 and C3 supports their shared microbial degradation/metabolism origin [73]. In addition, according to Yang et al. [81], some tryptophan compounds may also leach from agricultural fertilizers, which may be the same source as microbial humus. The β:α index was significantly positively correlated with BIX ($p < 0.001$), which is consistent with the conclusions drawn from SOM (Figure 9).

3.6. Environmental Implications

Groundwater is an important source of drinking water in water-scarce northern China [82]. In recent years, the development of greenhouse agriculture has inevitably altered the microenvironment of surrounding water bodies, particularly during winter, as winter cultivation of vegetables, fruits, and flowers offers higher economic profitability for producers [3]. A portion of the groundwater extracted for irrigation, which is not fully utilized, carries organic and nitrogen fertilizers into rivers, exacerbating water eutrophication. Another portion flows back into the shallow aquifer, thereby increasing the DOM content in groundwater [4]. Numerous studies have shown that DOM in water bodies contains various functional groups that can potentially bind to pollutants (such as toxic metals and organic pollutants), thereby significantly altering their mobility, bioavailability, and toxicity [10]. In this study, two tryptophan-containing proteins were identified that are precursors of disinfection byproducts (DBPs), which pose a potential threat to drinking water quality and human health [75]. The study results emphasize the importance of agricultural practices (irrigation pumping and organic fertilizer application) on the sources, migration pathways, and environmental behavior of DOM in water bodies surrounding intensive greenhouse agriculture, which is crucial for the effective prevention, monitoring, and management of water quality in agricultural catchments [8,83]. Notably, while this study provides insights into DOM characteristics under winter conditions, future work should incorporate multi-seasonal sampling to fully capture its annual cycle.

4. Conclusions

This study demonstrates that intensive groundwater extraction for irrigation in XER is a key driver of surface–groundwater mixing, significantly influencing DOM dynamics in the greenhouse agricultural catchment during winter. DOM composition was primarily characterized by humic substances from terrestrial inputs and microbial degradation of organic fertilizers, alongside tryptophan compounds derived from domestic sewage and microbial metabolism. Microbial activities strongly dominated the DOM characteristics with limited humification in winter conditions. Notably, humic substances exhibited greater sensitivity to parallel river flow directions, while tryptophan-based proteins showed higher mobility perpendicular to river flow. The integration of SOM with PARAFAC analysis revealed a positive correlation between tryptophan proteins and Cl^-, confirming their common origin from domestic wastewater discharge. These findings highlight the value of combining DOM parameters with traditional geochemical indicators and complementary analytical methods to elucidate DOM sources, migration pathways, and environmental impacts, thereby providing a robust theoretical foundation for water quality monitoring and management in agricultural catchments.

Supplementary Materials: The following supporting information can be downloaded at: https://www.mdpi.com/article/10.3390/w17182681/s1, Figure S1: (1) Neuron numbers in the SOM and (b) the sample distribution map of SOM analysis; Figure S2: Distribution patterns of DOM fluorescence parameters (FI vs HIX and BIX vs HIX) in different water samples. Table S1: Geographic location, land use, and water quality information of sampling points within the study area; Table S2: Water chemistry information of sampling points in the study area; Table S3: Optical measurement and the corresponding DOM characterization used in this study.

Author Contributions: H.W.: Investigation, data curation, methodology, formal analysis, visualization, writing—original draft. S.S.: Data curation, investigation, writing—review and editing. W.X.: Conceptualization, supervision, writing—review and editing. F.-J.Y.: Conceptualization, methodology, funding acquisition, formal analysis, investigation, supervision, writing—review and editing. All authors have read and agreed to the published version of the manuscript.

Funding: This research was funded by the Key Technologies Research and Development Program of China [2024YFD1701102].

Institutional Review Board Statement: Not applicable.

Informed Consent Statement: Not applicable.

Data Availability Statement: The datasets generated and/or analysed during the current study are available from the corresponding author on reasonable request.

Acknowledgments: The authors thank the editor and anonymous reviewers for their valuable comments and suggestions on this paper.

Conflicts of Interest: The authors declare that they have no known competing financial interests or personal relationships that could have appeared to influence the work reported in this paper.

References

1. Wang, X.Z.; Dou, Z.X.; Shi, X.J.; Zou, C.Q.; Liu, D.Y.; Wang, Z.Y.; Guan, X.L.; Sun, Y.X.; Wu, G.; Zhang, B.G.; et al. Innovative management programme reduces environmental impacts in Chinese vegetable production. *Nat. Food* **2020**, *2*, 47–53. [CrossRef]
2. Ni, B.; Zhang, W.; Xu, X.C.; Wang, L.G.; Bol, R.; Wang, K.Y.; Hu, Z.J.; Zhang, H.X.; Meng, F.Q. Exponential relationship between N_2O emission and fertilizer nitrogen input and mechanisms for improving fertilizer nitrogen efficiency under intensive plastic-shed vegetable production in China: A systematic analysis. *Agric. Ecosyst. Environ.* **2021**, *312*, 107353. [CrossRef]
3. Ti, C.P.; Luo, Y.X.; Yan, X.Y. Characteristics of nitrogen balance in open-air and greenhouse vegetable cropping systems of China. *Environ. Sci. Pollut. Res.* **2015**, *22*, 18508–18518. [CrossRef] [PubMed]
4. Ma, Y.; Liu, Z.-H.; Xi, B.-D.; He, X.-S.; Li, Q.-L.; Qi, Y.-J.; Jin, M.-Y.; Guo, Y. Characteristics of groundwater pollution in a vegetable cultivation area of typical facility agriculture in a developed city. *Ecol. Indic.* **2019**, *105*, 709–716. [CrossRef]
5. Wang, X.Z.; Zou, C.Q.; Gao, X.P.; Guan, X.L.; Zhang, Y.Q.; Shi, X.J.; Chen, X.P. Nitrate leaching from open-field and greenhouse vegetable systems in China: A meta-analysis. *Environ. Sci. Pollut. Res.* **2018**, *25*, 31007–31016. [CrossRef] [PubMed]
6. Tang, J.F.; Li, X.H.; Cao, C.L.; Lin, M.X.; Qiu, Q.L.L.; Xu, Y.Y.; Ren, Y. Compositional variety of dissolved organic matter and its correlation with water quality in peri-urban and urban river watersheds. *Ecol. Indic.* **2019**, *104*, 459–469. [CrossRef]
7. Ding, H.Y.; Zheng, M.X.; Yan, L.; Zhang, X.Y.; Liu, L.; Sun, Y.Y.; Su, J.; Xi, B.D.; Yu, H.B. Spectral and molecular insights into the variations of dissolved organic matter in shallow groundwater impacted by surface water recharge. *Water Res.* **2025**, *273*, 122978. [CrossRef]
8. Tong, H.; Gao, R.Z.; Yue, C.; Xie, L.M.; Duan, L.M.; Zhu, Y.; Wang, G.Q. Hydrochemical evolution and nitrate sources, migration, and transformation in surface water and groundwater of a typical tributary of the Yellow River. *J. Environ. Manag.* **2025**, *390*, 126218. [CrossRef]
9. Yu, H.; Feng, S.B.; Qiu, H.S.; Liu, J.Y. Interaction between the hydrochemical environment, dissolved organic matter, and microbial communities in groundwater: A case study of a vegetable cultivation area in Huaibei Plain, China. *Sci. Total Environ.* **2023**, *895*, 165166. [CrossRef]
10. Dong, L.; Zhang, J.; Guo, Z.Z.; Li, M.; Wu, H.M. Distributions and interactions of dissolved organic matter and heavy metals in shallow groundwater in Guanzhong basin of China. *Environ. Res.* **2022**, *207*, 112099. [CrossRef]
11. Hu, J.; Kang, L.Y.; Li, Z.L.; Feng, X.H.; Liang, C.F.; Wu, Z.; Zhou, W.; Liu, X.N.; Yang, Y.H.; Chen, L.Y. Photo-produced aromatic compounds stimulate microbial degradation of dissolved organic carbon in thermokarst lakes. *Nat. Commun.* **2023**, *14*, 3681. [CrossRef]
12. Zhang, H.Z.; Cheng, D.D.; Song, J.X.; Zhang, Y.X. Spatiotemporal fate of dissolved organic matter (DOM) in aquatic systems: Drivers, patterns and global implications. *J. Hydrol.* **2025**, *661*, 133637. [CrossRef]
13. Zheng, Y.X.; He, W.; Li, B.H.; Hur, J.; Guo, H.M.; Li, X.M. Refractory Humic-like Substances: Tracking Environmental Impacts of Anthropogenic Groundwater Recharge. *Environ. Sci. Technol.* **2020**, *54*, 15778–15788. [CrossRef]
14. Chen, J.Q.; Zhang, Y.W.; Guo, Y.Q.; Jiang, K.; Li, D.; Zheng, T.H. Differential Molecular Interactions of Imidacloprid with Dissolved Organic Matter in Citrus Soils with Diverse Planting Ages. *Agriculture* **2025**, *15*, 997. [CrossRef]
15. Yao, Y.R.; Zhang, J.Y.; Ma, K.; Li, J.; Hu, X.; Wang, Y.S.; Lin, Y.S.; Fang, F.M.; Li, S.Y. Combination Mechanism of Soil Dissolved Organic Matter and Cu^{2+} in Vegetable Fields, Forests and Dry Farmland in Lujiang County. *Agriculture* **2024**, *14*, 684. [CrossRef]
16. Ge, J.F.; Qi, Y.L.; Li, C.; Ma, J.F.; Yi, Y.B.; Hu, Q.Z.; Mostofa, K.M.G.; Volmer, D.A.; Li, S.-L. Fluorescence and molecular signatures of dissolved organic matter to monitor and assess its multiple sources from a polluted river in the farming-pastoral ecotone of northern China. *Sci. Total Environ.* **2022**, *837*, 154575. [CrossRef] [PubMed]
17. Yi, Y.B.; Zhong, J.; Bao, H.Y.; Mostofa, K.M.G.; Xu, S.; Xiao, H.-Y.; Li, S.-L. The impacts of reservoirs on the sources and transport of riverine organic carbon in the karst area: A multi-tracer study. *Water Res.* **2021**, *194*, 116933. [CrossRef]

18. Bai, Y.J.; Zhang, S.R.; Mu, E.L.; Zhao, Y.J.; Cheng, L.R.; Zhu, Y.; Yuan, Y.M.; Wang, Y.Y.; Ding, A.Z. Characterizing the spatiotemporal distribution of dissolved organic matter (DOM) in the Yongding River Basin: Insights from flow regulation. *J. Environ. Manag.* **2023**, *325*, 116476. [CrossRef]
19. Jeon, P.; Cho, S.; Hur, J.; Mun, H.; Chae, M.; Cho, Y.; Seok, K.; Hong, S. Close Association between Stream Water Quality and Fluorescence Properties of Dissolved Organic Matter in Agriculture-Dominated Watersheds. *Water* **2022**, *14*, 2459. [CrossRef]
20. Li, S.-L.; Zhang, H.; Yi, Y.B.; Zhang, Y.T.; Qi, Y.L.; Mostofa, K.M.G.; Guo, L.D.; He, D.; Fu, P.Q.; Liu, C.-Q. Potential impacts of climate and anthropogenic-induced changes on DOM dynamics among the major Chinese rivers. *Geogr. Sustain.* **2023**, *4*, 329–339. [CrossRef]
21. Bellmore, R.A.; Harrison, J.A.; Needoba, J.A.; Brooks, E.S.; Kent Keller, C. Hydrologic control of dissolved organic matter concentration and quality in a semiarid artificially drained agricultural catchment. *Water Resour. Res.* **2015**, *51*, 8146–8164. [CrossRef]
22. Wang, Y.H.; Tian, X.G.; Song, T.L.; Jiang, Z.; Zhang, G.L.; He, C.; Li, P. Linking DOM characteristics to microbial community: The potential role of DOM mineralization for arsenic release in shallow groundwater. *J. Hazard. Mater.* **2023**, *454*, 131566. [CrossRef] [PubMed]
23. Hu, Y.; Lu, Y.H.; Edmonds, J.W.; Liu, C.K.; Wang, S.; Das, O.; Liu, J.; Zheng, C.M. Hydrological and land use control of watershed exports of dissolved organic matter in a large arid river basin in northwestern China. *J. Geophys. Res. Biogeosci.* **2016**, *121*, 466–478. [CrossRef]
24. Wilson, H.F.; Xenopoulos, M.A. Effects of agricultural land use on the composition of fluvial dissolved organic matter. *Nat. Geosci.* **2008**, *2*, 37–41. [CrossRef]
25. Hu, Y.; Lu, Y.H.; Edmonds, J.; Liu, C.K.; Zhang, Q.; Zheng, C.M. Irrigation alters source-composition characteristics of groundwater dissolved organic matter in a large arid river basin, Northwestern China. *Sci. Total Environ.* **2021**, *767*, 144372. [CrossRef]
26. Zhang, Y.Z.; Liu, Y.D.; Zhou, A.G.; Zhang, L. Identification of groundwater pollution from livestock farming using fluorescence spectroscopy coupled with multivariate statistical methods. *Water Res.* **2021**, *206*, 117754. [CrossRef]
27. Zafar, R.; Lee, Y.K.; Oh, H.; Hur, J. Tracking microplastic-derived dissolved organic matter in the adsorption of its mixtures with natural organic matter via end-member mixing analysis. *Environ. Res.* **2025**, *283*, 122144. [CrossRef]
28. Cuss, C.W.; Guéguen, C. Relationships between molecular weight and fluorescence properties for size-fractionated dissolved organic matter from fresh and aged sources. *Water Res.* **2015**, *68*, 487–497. [CrossRef]
29. Li, M.; Wen, Q.X.; Zhang, Y.M.; Chen, Z.Q. New insights into the transformation of effluent organic matter during Fe(II)-assisted advanced oxidation processes: Parallel factor analysis coupled with self-organizing maps. *Water Res.* **2022**, *221*, 118789. [CrossRef]
30. Zhou, B.; Li, J.N.; Sun, Q.H.; Lei, K. Monsoon-driven DOM dynamics revealed by self-organizing map and end-member mixing analysis: Spatial partitioning of autochthonous and anthropogenic sources in plain river networks. *Environ. Res.* **2025**, *283*, 122160. [CrossRef]
31. Shi, K.; Zhao, Y.T.; Wu, C.B.; Geng, Y.T.; Zhou, S.L.; Chai, B.B. Revealing the distribution characteristics and key driving factors of dissolved organic matter in Baiyangdian Lake inflow rivers from different seasons and sources. *Sci. Total Environ.* **2024**, *951*, 175768. [CrossRef] [PubMed]
32. Zhang, Y.X.; Liang, X.Q.; Wang, Z.B.; Xu, L.X. A novel approach combining self-organizing map and parallel factor analysis for monitoring water quality of watersheds under non-point source pollution. *Sci. Rep.* **2015**, *5*, 16079. [CrossRef] [PubMed]
33. Jia, L.M.; Yang, Q.; Cui, H.Y. Insight into the dynamics of dissolved organic matter components under latitude change perturbation. *Ecotoxicol. Environ. Saf.* **2024**, *269*, 115734. [CrossRef] [PubMed]
34. Cui, H.Y.; Xie, L.N.; Zhang, G.G.; Zhao, Y.; Wei, Z.M. Revealing the Inner Changes of Component Composition Derived from DOM PARAFAC Based on Two-Dimensional Correlation Spectroscopy. *Molecules* **2022**, *27*, 7316. [CrossRef]
35. Lin, Z.Y.; Lee, K.H.; Lim, J.Y.; Kim, J.H.; Eun, B.J.; Lee, S.J.; Park, J.Y.; Oh, H.S.; Oh, J.-M. Revealing spatial-temporal impact of industrial effluent towards DOM in Riverine employing PARAFAC and MW-2D COS. *J. Environ. Chem. Eng.* **2024**, *12*, 113412. [CrossRef]
36. Zhang, X.L.; Nie, L.; Gao, H.J.; Yu, H.B.; Liu, D.P. Applying second derivative synchronous fluorescence spectroscopy combined with Gaussian band fitting to trace variations of DOM fractions along an urban river. *Ecol. Indic.* **2023**, *146*, 109872. [CrossRef]
37. Wang, J.J.; Liang, X.; Ma, B.; Liu, Y.F.; Jin, M.G.; Knappett, P.S.K.; Liu, Y.L. Using isotopes and hydrogeochemistry to characterize groundwater flow systems within intensively pumped aquifers in an arid inland basin, Northwest China. *J. Hydrol.* **2021**, *595*, 126048. [CrossRef]
38. Yi, B.; Liu, J.T.; He, W.; Lü, X.L.; Cao, X.; Chen, X.R.; Zeng, X.J.; Zhang, Y.X. Optical variations of dissolved organic matter due to surface water—Groundwater interaction in alpine and arid Datonghe watershed. *Sci. Total Environ.* **2023**, *864*, 161036. [CrossRef]
39. Martinez, J.L.; Raiber, M.; Cox, M.E. Assessment of groundwater–surface water interaction using long-term hydrochemical data and isotope hydrology: Headwaters of the Condamine River, Southeast Queensland, Australia. *Sci. Total Environ.* **2015**, *536*, 499–516. [CrossRef]

40. Zhang, S.R.; Bai, Y.J.; Wen, X.; Ding, A.Z.; Zhi, J.H. Seasonal and downstream alterations of dissolved organic matter and dissolved inorganic ions in a human-impacted mountainous tributary of the Yellow River, China. *Environ. Sci. Pollut. Res.* **2018**, *25*, 17967–17979. [CrossRef]
41. Hudson, N.; Baker, A.; Reynolds, D. Fluorescence analysis of dissolved organic matter in natural, waste and polluted waters—A review. *River Res. Appl.* **2007**, *23*, 631–649. [CrossRef]
42. Zepp, R.G.; Sheldon, W.M.; Moran, M.A. Dissolved organic fluorophores in southeastern US coastal waters: Correction method for eliminating Rayleigh and Raman scattering peaks in excitation–emission matrices. *Mar. Chem.* **2004**, *89*, 15–36. [CrossRef]
43. Huo, P.; Gao, P.C. Seasonal changes of dissolved CO_2 and its linkage with optical characteristics of DOM in groundwater in agricultural areas. *J. Hydrol.* **2024**, *643*, 131927. [CrossRef]
44. Stedmon, C.A.; Bro, R. Characterizing dissolved organic matter fluorescence with parallel factor analysis: A tutorial. *Limnol. Oceanogr. Methods* **2008**, *6*, 572–579. [CrossRef]
45. Zhang, Y.X.; Zhang, H.Z.; Song, J.X.; Cheng, D.D. More sensitive signatures of DOM are stimulated by socioeconomic-related anthropogenic influences in a typical loess watershed. *J. Hydrol.* **2025**, *657*, 133106. [CrossRef]
46. Murphy, K.R.; Stedmon, C.A.; Wenig, P.; Bro, R. OpenFluor-an online spectral library of auto-fluorescence by organic compounds in the environment. *Anal. Methods* **2014**, *6*, 658–661. [CrossRef]
47. Ni, M.F.; Jiang, S.H.; Li, S.Y. Spectroscopic indices trace spatiotemporal variability of dissolved organic matter in a river system with Karst characteristic. *J. Hydrol.* **2020**, *590*, 125570. [CrossRef]
48. Li, P.H.; Hur, J. Utilization of UV-Vis spectroscopy and related data analyses for dissolved organic matter (DOM) studies: A review. *Crit. Rev. Environ. Sci. Technol.* **2017**, *47*, 131–154. [CrossRef]
49. Chen, W.; Teng, C.-Y.; Qian, C.; Yu, H.-Q. Characterizing Properties and Environmental Behaviors of Dissolved Organic Matter Using Two-Dimensional Correlation Spectroscopic Analysis. *Environ. Sci. Technol.* **2019**, *53*, 4683–4694. [CrossRef]
50. Liu, M.X.; Han, X.K.; Liu, C.-Q.; Guo, L.D.; Ding, H.; Lang, Y.C. Differences in the spectroscopic characteristics of wetland dissolved organic matter binding with Fe^{3+}, Cu^{2+}, Cd^{2+}, Cr^{3+} and Zn^{2+}. *Sci. Total Environ.* **2021**, *800*, 149476. [CrossRef]
51. Li, K.; Yang, S.P.; Wang, H.Y.; Wu, Z.B.; Liang, Y.S.; Gong, X.M.; Peng, X.; Qin, P.F. Molecular spectra and docking simulations investigated the binding mechanisms of tetracycline onto E. coli extracellular polymeric substances. *Talanta* **2024**, *276*, 126231. [CrossRef]
52. Li, W.J.; Liu, G.B.; Lei, M.; Zhou, Y.Y.; Cui, H.J.; Du, H.H. Spectral fingerprints of DOM-tungsten interactions: Linking molecular binding to conformational changes. *J. Hazard. Mater.* **2025**, *483*, 136649. [CrossRef]
53. Noda, I. Two-dimensional correlation spectroscopy (2DCOS) analysis of polynomials. *J. Mol. Struct.* **2016**, *1124*, 53–60. [CrossRef]
54. Noda, I. Frontiers of Two-Dimensional Correlation Spectroscopy. Part 1. New concepts and noteworthy developments. *J. Mol. Struct.* **2014**, *1069*, 3–22. [CrossRef]
55. Zhu, G.; Wu, X.; Ge, J.P.; Liu, F.; Zhao, W.G.; Wu, C. Influence of mining activities on groundwater hydrochemistry and heavy metal migration using a self-organizing map (SOM). *J. Clean. Prod.* **2020**, *257*, 120664. [CrossRef]
56. Yan, C.X.; Liu, H.H.; Sheng, Y.R.; Huang, X.; Nie, M.H.; Huang, Q.; Baalousha, M. Fluorescence characterization of fractionated dissolved organic matter in the five tributaries of Poyang Lake, China. *Sci. Total Environ.* **2018**, *637–638*, 1311–1320. [CrossRef] [PubMed]
57. Cuss, C.W.; Guéguen, C. Analysis of dissolved organic matter fluorescence using self-organizing maps: Mini-review and tutorial. *Anal. Methods* **2016**, *8*, 716–725. [CrossRef]
58. Zhou, P.; Huang, J.L.; Pontius, R.G.; Hong, H.S. New insight into the correlations between land use and water quality in a coastal watershed of China: Does point source pollution weaken it? *Sci. Total Environ.* **2016**, *543*, 591–600. [CrossRef]
59. Deng, Y.D.; Lu, Y.; Du, X.Q.; Ye, X.Y.; Feng, J. Identifying spatial patterns and driving factors of anthropogenic impacts on the groundwater environment based on groundwater chemical kinetics. *J. Clean. Prod.* **2025**, *486*, 144436. [CrossRef]
60. Li, P.Y.; Wu, J.H.; Qian, H. Assessment of groundwater quality for irrigation purposes and identification of hydrogeochemical evolution mechanisms in Pengyang County, China. *Environ. Earth Sci.* **2012**, *69*, 2211–2225. [CrossRef]
61. Xing, S.P.; Zhang, C.R.; Guo, H.M.; Sheng, Y.Z.; Liu, X.Y. Hydrologic changes induced by groundwater abstraction lead to arsenic mobilization in shallow aquifers. *J. Hazard. Mater.* **2024**, *480*, 136133. [CrossRef] [PubMed]
62. Chen, K.; Liu, Q.M.; Peng, W.H.; Liu, X.H. Source apportionment and natural background levels of major ions in shallow groundwater using multivariate statistical method: A case study in Huaibei Plain, China. *J. Environ. Manag.* **2022**, *301*, 113806. [CrossRef] [PubMed]
63. Hagedorn, B.; Whittier, R.B. Solute sources and water mixing in a flashy mountainous stream (Pahsimeroi River, U.S. Rocky Mountains): Implications on chemical weathering rate and groundwater–surface water interaction. *Chem. Geol.* **2015**, *391*, 123–137. [CrossRef]

64. Yang, Y.J.; Yuan, X.F.; Deng, Y.M.; Xie, X.J.; Gan, Y.Q.; Wang, Y.X. Seasonal dynamics of dissolved organic matter in high arsenic shallow groundwater systems. *J. Hydrol.* **2020**, *589*, 125120. [CrossRef]
65. Tian, Y.; Zheng, Y.; Wu, B.; Wu, X.; Liu, J.; Zheng, C.M. Modeling surface water-groundwater interaction in arid and semi-arid regions with intensive agriculture. *Environ. Model. Softw.* **2015**, *63*, 170–184. [CrossRef]
66. Liu, J.R.; Song, X.F.; Yuan, G.F.; Sun, X.M.; Liu, X.; Wang, S.Q. Characteristics of $\delta^{18}O$ in precipitation over Eastern Monsoon China and the water vapor sources. *Chin. Sci. Bull.* **2009**, *55*, 200–211. [CrossRef]
67. Du, Y.; Deng, Y.M.; Ma, T.; Xu, Y.; Tao, Y.Q.; Huang, Y.W.; Liu, R.; Wang, Y.X. Enrichment of Geogenic Ammonium in Quaternary Alluvial–Lacustrine Aquifer Systems: Evidence from Carbon Isotopes and DOM Characteristics. *Environ. Sci. Technol.* **2020**, *54*, 6104–6114. [CrossRef]
68. Zeng, X.J.; Zheng, Y.X.; Chen, X.R.; Cao, X.; He, W.; Jiang, B.; Li, B.H.; Guo, H.M. Molecular Responses of Dissolved Organic Matter to Anthropogenic Groundwater Recharge: Characteristics, Transformations, and Sensitive Molecules. *Environ. Sci. Technol.* **2023**, *57*, 7789–7799. [CrossRef]
69. Singh, S.; Dutta, S.; Inamdar, S. Land application of poultry manure and its influence on spectrofluorometric characteristics of dissolved organic matter. *Agric. Ecosyst. Environ.* **2014**, *193*, 25–36. [CrossRef]
70. Harjung, A.; Schweichhart, J.; Rasch, G.; Griebler, C. Large-scale study on groundwater dissolved organic matter reveals a strong heterogeneity and a complex microbial footprint. *Sci. Total Environ.* **2023**, *854*, 158542. [CrossRef]
71. Wang, X.; Zhang, M.; Liu, L.L.; Wang, Z.P.; Lin, K.F. Using EEM-PARAFAC to identify and trace the pollution sources of surface water with receptor models in Taihu Lake Basin, China. *J. Environ. Manag.* **2022**, *321*, 115925. [CrossRef]
72. Du, C.; Chang, Z.B.; Tang, W.Z.; Hao, Z.N.; Zhang, M.Y.; Zhang, X.Y.; Liu, M.X.; Zhang, H. Molecular insights into the seasonal variation of fluorescent dissolved organic matter in a hydrologically isolated aquatic continuum. *J. Hydrol.* **2025**, *661*, 133620. [CrossRef]
73. Graham, P.W.; Baker, A.; Andersen, M.S. Dissolved Organic Carbon Mobilisation in a Groundwater System Stressed by Pumping. *Sci. Rep.* **2015**, *5*, 18487. [CrossRef] [PubMed]
74. Shi, J.H.; Zhao, Y.; Wei, D.; Zhang, D.Y.; Wei, Z.M.; Wu, J.Q. Insight into transformation of dissolved organic matter in the Heilongjiang River. *Environ. Sci. Pollut. Res.* **2018**, *26*, 3340–3349. [CrossRef] [PubMed]
75. Xue, C.H.; Yu, Y.; Huang, X. Comparison of Organic Matter Properties and Disinfection By-Product Formation between the Typical Groundwater and Surface Water. *Water* **2022**, *14*, 1418. [CrossRef]
76. Ding, H.Y.; Su, J.; Sun, Y.Y.; Yu, H.B.; Zheng, M.X.; Xi, B.D. Insight into spatial variations of DOM fractions and its interactions with microbial communities of shallow groundwater in a mesoscale lowland river watershed. *Water Res.* **2024**, *258*, 121797. [CrossRef]
77. Wilson, G.J.L.; Lu, C.; Lapworth, D.J.; Kumar, A.; Ghosh, A.; Niasar, V.J.; Krause, S.; Polya, D.A.; Gooddy, D.C.; Richards, L.A. Spatial and seasonal controls on dissolved organic matter composition in shallow aquifers under the rapidly developing city of Patna, India. *Sci. Total Environ.* **2023**, *903*, 166208. [CrossRef]
78. Ma, Y.; Liu, Z.-H.; Xi, B.-D.; Li, W.-T.; Xu, Y.-Q.; Zhao, H.-Z.; Chen, Z.-Q.; He, X.-S.; Xing, B. Molecular structure and evolution characteristics of dissolved organic matter in groundwater near landfill: Implications of the identification of leachate leakage. *Sci. Total Environ.* **2021**, *787*, 147649. [CrossRef]
79. Cuss, C.W.; Shi, Y.X.; McConnell, S.M.; Guéguen, C. Changes in the fluorescence composition of multiple DOM sources over pH gradients assessed by combining parallel factor analysis and self-organizing maps. *J. Geophys. Res. Biogeosci.* **2014**, *119*, 1850–1860. [CrossRef]
80. Jin, X.C.; Chen, X.Q.; Gao, L.M.; Wu, Y.F.; Lu, H.S.; Yuan, M.H.; Cui, J.H.; Wei, F.Y. Fluorescence Analysis of River DOM Spectra Using PARAFAC in Combination with a Self-Organizing Map to Distinguish Organic Matter Sources. *Int. J. Environ. Res.* **2024**, *18*, 20. [CrossRef]
81. Yang, X.L.; Yu, X.B.; Cheng, J.R.; Zheng, R.Y.; Wang, K.; Dai, Y.X.; Tong, N.J.; Chow, A.T. Impacts of land-use on surface waters at the watershed scale in southeastern China: Insight from fluorescence excitation-emission matrix and PARAFAC. *Sci. Total Environ.* **2018**, *627*, 647–657. [CrossRef]
82. Qin, X.; Wang, H.S.; Gong, J.S.; Ye, Y.H.; Zhou, K.; Xu, N.Z.; Li, L.; Li, J. Multivariate Statistics and Hydrochemistry Combined to Reveal the Factors Affecting Shallow Groundwater Evolution in a Typical Area of the Huaibei Plain, China. *Water* **2025**, *17*, 962. [CrossRef]
83. Kumar, A.; Mishra, S.; Taxak, A.K.; Pandey, R.; Yu, Z.-G. Nature rejuvenation: Long-term (1989–2016) vs short-term memory approach based appraisal of water quality of the upper part of Ganga River, India. *Environ. Technol. Innov.* **2020**, *20*, 101164. [CrossRef]

water

MDPI

Article

Study on Natural Background Levels and Mechanisms of Groundwater Contamination in an Overexploited Aquifer Region: A Case Study of Xingtai City, North China Plain

Qi Wang [1,2], Meili Wang [1,2], Yan Li [1,2], Binghao Guo [1,2], Hongchao Li [1,2], Yang Liu [1,2], Liya Zhao [1,2,*], Chunyang Ma [1,2] and Ziting Yuan [1,2,3,*]

1 Hebei Key Laboratory of Geological Resources and Environment Monitoring and Protection, Shijiazhuang 050021, China; qqwangqi@163.com (Q.W.); 15930055568m@sina.com (Y.L.); gbh1993716@163.com (B.G.); lihongchao@163.com (H.L.)
2 Hebei Geo-Environmental Monitoring, Shijiazhuang 050021, China
3 Institute of Hydrogeology and Environmental Geology, Chinese Academy of Geological Sciences, Shijiazhuang 050061, China
* Correspondence: 18101361909@163.com (L.Z.); yuanziting2025@163.com (Z.Y.)

Abstract: This study investigates the groundwater over-exploitation zone in Xingtai City, North China Plain, to address two critical gaps in the current understanding of groundwater chemistry: the lack of established natural background levels (NBLs) and the ambiguous mechanisms of groundwater contamination. Sixty shallow-groundwater samples were collected and analyzed using a combination of Piper diagrams, cumulative-probability statistics, contamination-index methods, and multivariate statistical techniques to determine NBLs and threshold values (TVs) for major chemical constituents and to clarify the contamination mechanisms. The results indicate that the groundwater is weakly alkaline, with the most prevalent water types being HCO_3–Na and SO_4·Cl–Na. The NBLs for Na^+, Ca^{2+}, Mg^{2+}, Cl^-, SO_4^{2-} and NO_3^- are 32.3 mg/L, 34.1 mg/L, 17.8 mg/L, 46.2 mg/L, 66.4 mg/L and 0.886 mg/L, respectively, and the corresponding TVs are 116 mg/L, 54.6 mg/L, 33.9 mg/L, 248 mg/L, 258 mg/L and 44.7 mg/L. Based on the TVs, 56.7% of the sampling sites are identified as anthropogenically contaminated. Principal component analysis reveals that groundwater over-extraction, industrial activities and water–rock interaction are the dominant drivers of groundwater contamination, whereas intensive abstraction, agricultural fertilization and domestic sewage discharge exert additional influence. The findings provide a scientific basis for pollution control and sustainable utilization of groundwater in over-exploited regions.

Keywords: water quality; hydrochemistry; multivariate statistical techniques; pollution sources

1. Introduction

Natural background levels (NBLs) of groundwater are fundamental for scientifically identifying contamination and establishing water-quality protection standards [1]. They not only reflect the intrinsic concentrations of chemical constituents under undisturbed hydrogeological conditions, but also serve as critical thresholds for distinguishing natural geochemical processes from anthropogenic impacts [2]. However, groundwater chemistry is influenced by a complex interplay of factors, including geological settings, climate, land use, and human activities [3]. Under the influence of intensive anthropogenic pressures, groundwater is currently threatened by diffuse agricultural pollution, as well as point sources such as industrial effluents and domestic sewage [4]. Accurate delineation of regional NBLs is essential for safeguarding drinking-water security and supporting the sustainable use of water resources.

In recent years, substantial advancements have been made globally in the determination of groundwater NBLs and the characterization of contamination. Urresti-Estala et al. [5] employed statistical, iterative and distribution-function methods to establish NBLs in Milan, Italy, and the Guadalhorce River basin, Spain, respectively. Their research revealed that industrial wastewater discharge and agricultural fertilizer application are primary factors causing concentrations to exceed NBLs and triggering contamination. Zhou et al. [6] demonstrated that agricultural activities are the primary driver of elevated NBLs and subsequent contamination in the Xi'an alluvial plain. Furthermore, numerous studies have investigated contamination mechanisms and controlling factors: in rapidly urbanizing areas, groundwater quality deterioration is mainly attributed to domestic sewage, industrial wastewater and chemical fertilizers [7,8], whereas in coastal zones, seawater intrusion, water–rock interactions, sewage discharge and agricultural pollution dominate [9,10]. Evidently, differing anthropogenic patterns and geological backgrounds among regions lead to distinct contamination mechanisms. With increasing groundwater abstraction, extensive cone-of-depression has formed in many areas, altering both hydraulic and hydrochemical regimes [11] and exacerbating groundwater quality degradation. Nevertheless, investigations into NBLs and contamination mechanisms in over-exploited zones remain scarce, constraining effective groundwater protection in such regions.

Consequently, this study selects Xingtai City in the North China Plain as the study area. Xingtai hosts the largest groundwater depression cone in China and represents a critical agricultural and industrial base. Groundwater over-extraction and quality deterioration are severe in this region. The aquifer system is subject to both naturally elevated geochemical backgrounds and anthropogenic contamination. The NBL characteristics and contamination mechanisms in this region may thus differ markedly from those reported elsewhere. This study employs cumulative probability methods, contamination index methods, and multivariate statistical techniques to determine NBLs, elucidates hydrochemical signatures driven jointly by natural and anthropogenic factors, and clarifies the mechanisms of water-quality deterioration. The results provide a scientific basis for rational groundwater development and contamination control in over-exploited regions.

2. Materials and Methods

2.1. Study Area

The study area is situated in the central-eastern part of Xingtai City, Hebei Province, China, covering a total area of 12,414 km^2, 70.8% of which is the plain region. Xingtai has a north-temperate continental semi-arid monsoon climate with four distinct seasons. The multi-year mean air temperature is 13.5 °C. Precipitation is concentrated mainly in July and August, with a multi-year mean of 519.7 mm, whereas the multi-year mean evaporation is 1953 mm [12].

2.2. Geological and Hydrogeological Setting

Xingtai city is located in the North China Platform in terms of regional geology. The western part is the Taihang Mountains, while the eastern part is the North China Plain. The aquifer system comprises primarily porous and karstic aquifer. Porous aquifer is dominant in the central and eastern plains, while karstic aquifer is primarily found in the western mountainous area; the present study focuses on the porous aquifer in the central and eastern plains. Carbonate rocks are widely distributed in the area. The regional groundwater flow direction is from west to east. Recharge sources include atmospheric precipitation, surface-water infiltration and irrigation return flow; discharge is dominated by artificial abstraction.

2.3. Sampling and Analytical Procedures

In May 2024, 60 shallow-groundwater samples were collected (Figure 1). Polyethylene bottles (500 mL) were used for sampling. All bottles were first rinsed with distilled water and then conditioned by rinsing three times with groundwater at the sampling point to ensure sample integrity. A submersible pump was employed for purging and sampling. A 1-inch submersible pump was used for well purging and sampling at a pumping rate of 1–3 L/min, with a purge volume of 100 L. Each well was purged until field pH and EC stabilized before sample collection. Each sample was split into two aliquots: one aliquot for cation analysis (Na^+, K^+, Ca^{2+} and Mg^{2+}) was acidified on site with ultrapure nitric acid to pH < 2; the second aliquot for anion analysis (Cl^-, SO_4^{2-}, F^- and NO_3^-) was left unacidified. All bottles were filled without headspace, stored in the dark at 4 °C and analyzed within one week of collection.

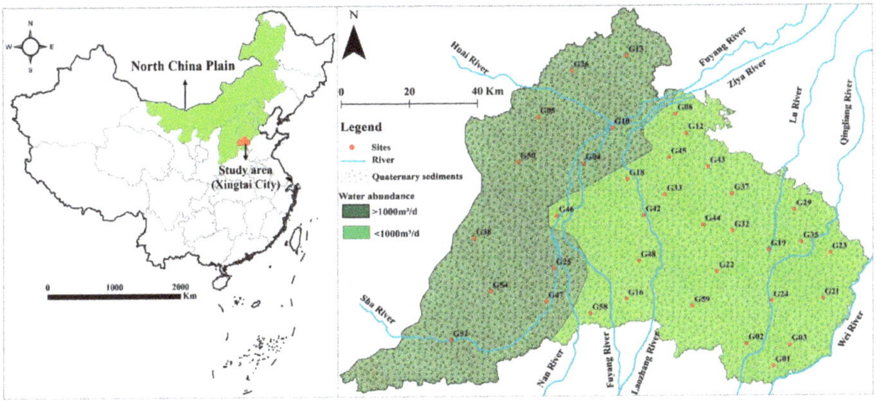

Figure 1. Distribution map of groundwater monitoring sites.

Field pH was measured with a portable multi-parameter analyzer (Hach-HQ40D, Hach Company, Shanghai). Cations (K^+, Na^+, Ca^{2+} and Mg^{2+}) were determined by atomic absorption spectrophotometry (AA240FS/GTA120, Agilent Technologies, Santa Clara, CA, USA). Anions (Cl^-, SO_4^{2-}, F^- and NO_3^-) were analyzed by ion chromatography (CIC-D180, Qingdao Shenghan Chromatography Technology Co., Ltd., Qingdao, China). HCO_3^- was measured by hydrochloric-acid titration. Total hardness (TH) was determined by EDTA titration, and total dissolved solids (TDS) by the drying-gravimetric method.

Following analysis, charge-balance errors for cations and anions were calculated and constrained to <5% to ensure data reliability.

2.4. Methodology

2.4.1. Determination of Hydrochemical Facies

The Piper trilinear diagram is commonly employed to portray the overall hydrochemical type of individual water samples; however, it does not directly provide the relative proportions of the different facies encountered within a dataset [13]. Consequently, it is used in conjunction with the Kurlov formula, in which anions and cations are ranked according to their milliequivalent percentages, and only those ions exceeding 25% of the total anionic or cationic charge participate in the classification [14]. This combined approach enables both the identification of the prevailing hydrochemical facies and the quantification of their areal frequencies.

2.4.2. Calculation of NBLs

Cumulative-probability plots, widely accepted for partitioning hydrochemical datasets into distinct populations [15], allow an intuitive delineation of the NBLs for single or multiple constituents.

Let the cumulative distribution function be F(X), where $F(X) = P_r\ (X \leq x)$. For a discrete random variable X, F(x) is expressed as

$$F(X) = \sum_{k=1}^{n} P_r(X = x_k) \tag{1}$$

in which x_k denotes the largest value in X that does not exceed x, thereby defining the upper bound for cumulative-probability evaluation.

The first inflection point on the cumulative-probability curve separates the first, geogenic population from subsequent populations [16]; the abscissa of this inflection is adopted as the NBLs for the constituent under consideration.

2.4.3. Calculation of TVs

Given that hydrogeological settings and the intensity of anthropogenic pressures vary across regions, the groundwater-quality criteria required for each area also differ accordingly. Consequently, the natural background level (NBL) of each hydrochemical constituent is compared with the Grade-III groundwater quality standard of China or the corresponding guideline value recommended by the World Health Organization, which are adopted as the reference value (REF). The TVs is then calculated using the BRIDGE methodology [17], as expressed by the following equations:

$$TV = (REF + NBL)/2 \quad \text{If} \quad NBL < REF \tag{2}$$

$$TV = NBL \quad \text{If} \quad NBL > REF \tag{3}$$

Establishing the TVs for groundwater hydrochemistry enables a more accurate assessment of the contamination status. Any constituent whose concentration exceeds its respective TV can be defined as water-quality pollution resulting from anthropogenic activities.

2.4.4. Determination of Pollution Load

The contamination percentage (%) for each well was calculated using the pollution index proposed by Pacheco and van der Weijden [18] and subsequently refined by Soumya et al. [19].

$$\text{Percentage of Pollution Index (PPI)} = \frac{[Cl^-] + [SO_4^{2-}] + [NO_3^-]}{[Cl^-] + [SO_4^{2-}] + [NO_3^-] + [HCO_3^-]} \times 100 \tag{4}$$

In the equation, all ionic concentrations are expressed in $meq \cdot L^{-1}$. By examining the relationship between the pollution load and the relevant ionic ratios, the provenance of contaminants can be inferred.

2.4.5. Identification of Controlling Factors Using Principal Component Analysis (PCA)

PCA is a multivariate statistical technique that employs an orthogonal linear transformation to reduce the dimensionality of a complex, correlated dataset into a smaller set of uncorrelated principal components. These components capture the dominant structures and patterns within the data and thereby facilitate the identification of the principal factors governing groundwater contamination [20]. In recent years, PCA has been widely applied to the identification of pollution sources in soil and water environmental pollution [21].

3. Results and Discussion

3.1. Hydrochemical Characteristics of Groundwater

Table 1 summarizes the descriptive statistics for twelve parameters: pH, K^+, Na^+, Ca^{2+}, Mg^{2+}, HCO_3^-, Cl^-, SO_4^{2-}, NO_3^-, F^-, TH and TDS. The pH values range from 6.78 to 8.47, with a mean of 7.84, indicating weakly alkaline conditions; none of the samples exceed the permissible limit. The mean TH and TDS are 533 mg/L and 1271 mg/L, respectively. Notably, 40.0% of the samples exceed the standard for TH, and 33.3% exceed the standard for TDS.

Table 1. Statistical table of groundwater chemical indicators.

Parameters	Range (mg/L)	Average (mg/L)	CV (%)	Standard *	Over Standard Rate (%)
pH	6.78–8.47	7.84	5.46	6.5–8.5	0
K^+	0.420–15.4	2.21	137	-	3.3
Na^+	6.26–1275	285	112	200	38.3
Ca^{2+}	4.01–297	86.7	78.8	-	46.7
Mg^{2+}	2.92–402	74.7	127	-	33.3
HCO_3^-	40.2–1008	405	58.2	-	-
Cl^-	15.4–2017	301	132	250	33.3
SO_4^{2-}	20.3–2355	344	149	250	31.7
NO_3^-	0–106	16.0	183	88.6	8.3
F^-	0.11–3.91	0.84	83.1	1	23.3
TH	24.0–2202	533	97.1	450	40.0
TDS	169–5659	1271	99.6	1000	33.3

Note: * refers to the Class III water quality standard in the "Groundwater Quality Standard" [22]. The dash (-) indicates the absence of a standard. CV = coefficient of variation.

Among cations, the mean concentrations decrease in the order Na^+ (285 mg/L) > Ca^{2+} (86.7 mg/L) > Mg^{2+} (74.7 mg/L) > K^+ (2.21 mg/L); with exceedance rate for Na^+ is 38.3%. For anions, the order is HCO_3^- (405 mg/L) > SO_4^{2-} (344 mg/L) > Cl^- (301 mg/L) > NO_3^- (16.0 mg/L), with exceedance rates of 33.3% for Cl^-, 31.7% for SO_4^{2-} and 8.33% for NO_3^-. The elevated concentrations and high coefficients of variation for both anions and cations reflect strong anthropogenic impacts [23] and pronounced spatial heterogeneity in geological background and human activities.

3.2. Hydrochemical Facies

A Piper diagram was constructed to classify the hydrochemical types of the study area. The results show that Na^+ dominates among cations (field D), and HCO_3^- dominates among anions (field F) (Figure 2).

Application of the Kurlov classification further indicates that the 60 samples represent 20 distinct hydrochemical facies, attesting to the complexity of the groundwater chemistry. The predominant types—HCO_3–Na, SO_4·Cl–Na·Ca, SO_4·Cl–Na, HCO_3·Cl–Na and HCO_3·Cl–Ca·Mg—account for 15.0%, 10.0%, 8.3%, 8.3% and 8.3% of the samples, respectively. The predominance of Na^+ and HCO_3^- is generally associated with aquifers rich in carbonate and silicate minerals [24]. Conversely, the occurrence of the SO_4·Cl–Na·Ca facies suggests that the influence of industrial effluents or agricultural activities [25].

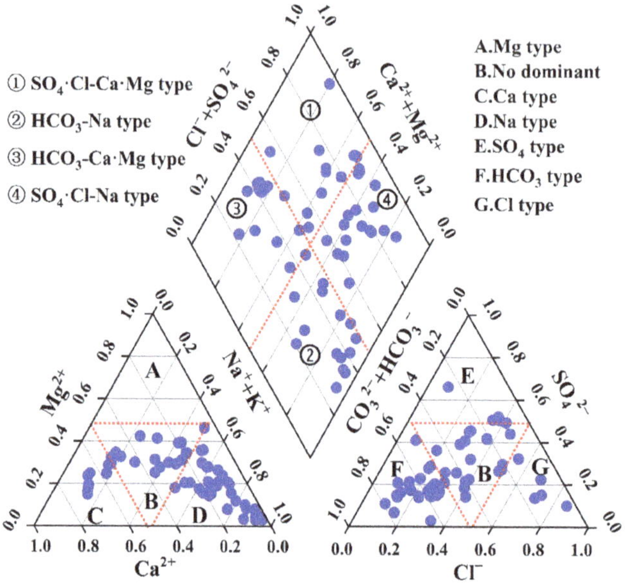

Figure 2. Groundwater hydrochemical type map.

3.3. Controlling Factors and Sources of Major Ions

3.3.1. Controlling Factors

The evolution of groundwater chemistry is primarily governed by rock weathering, precipitation, and evaporation processes [26]. Gibbs diagrams were employed to identify the dominant processes in the study area. Figure 3a,b shows that most samples cluster in the rock-weathering and evaporation–concentration fields, indicating that these two factors are the primary controls on regional groundwater chemistry. A subset of samples plots outside the Gibbs envelope, suggesting additional influence from anthropogenic activities such as agriculture or industry [27].

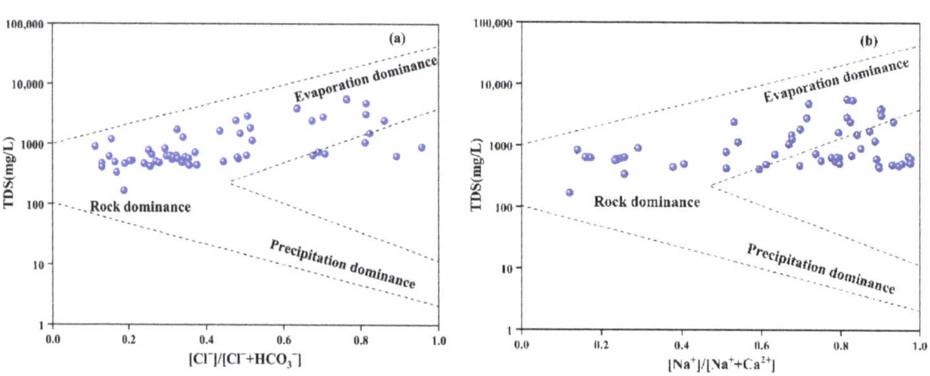

Figure 3. Gibbs diagrams for groundwater in this region (**a**) TDS VS $[Cl^-]/[Cl^- + HCO_3^-]$ and (**b**) TDS VS $[Na^+]/[Na^+ + Ca^{2+}]$.

3.3.2. Ion Sources

The ion ratio method is commonly used to identify the sources of ionic components in the aqueous environment. Na-normalized molar ratios are used to constrain the lithological end-members governing groundwater composition [28]. Published values indicate that silicate weathering yields Ca^{2+}/Na^{+} and Mg^{2+}/Na^{+} molar ratios of 0.35 ± 0.15 and 0.24 ± 0.12, respectively; carbonate weathering yields 50 ± 20 and 20 ± 8; and evaporite dissolution yields 0.17 ± 0.09 and 0.02 ± 0.01 [29]. Figure 4a shows that the majority of samples fall within the silicate and carbonate end-member ranges, with only a few approaching the evaporite field, implying that groundwater chemistry is mainly controlled by silicate and carbonate weathering, whereas evaporite dissolution plays a minor role.

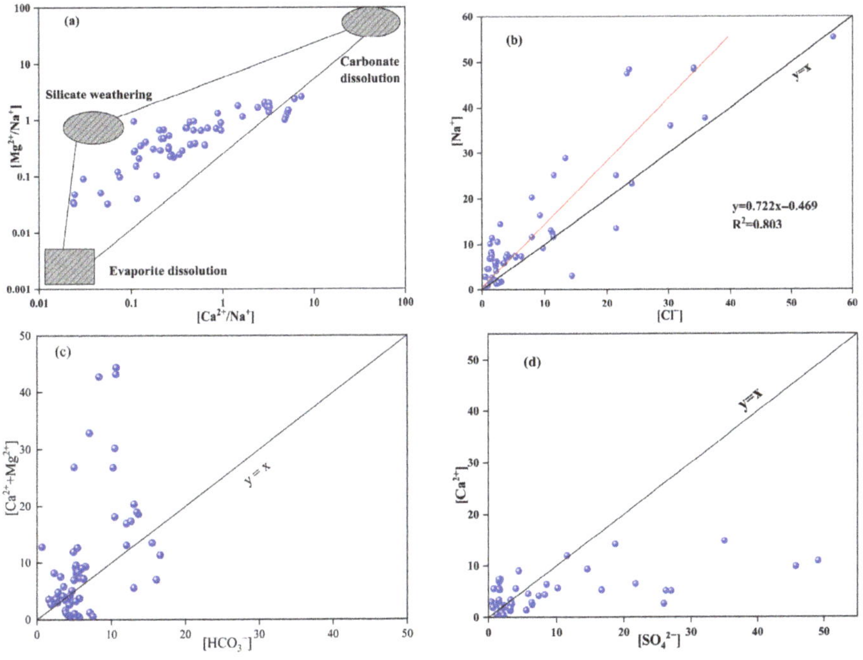

Figure 4. Relationships between major ions for groundwater samples in the study (**a**) $[Ca^{2+}]/[Na^{+}]$ VS $[Mg^{2+}]/[Na^{+}]$, (**b**) $[Na^{+}]$ VS $[Cl^{-}]$, (**c**) $[HCO_3^{-}]$ VS $[Ca^{2+} + Mg^{2+}]$, (**d**) $[Ca^{2+}]$ VS $[SO_4^{2-}]$.

The Na^{+}/Cl^{-} molar ratio was calculated to further constrain the origin of Na^{+} and Cl^{-}. In Figure 4b, Na^{+} and Cl^{-} exhibit a strong positive correlation (R^2 = 0.803), and 86.7% of the samples plot above the 1:1 line. These findings suggest a common source for Na^{+} and Cl^{-}, most likely halite dissolution, while the excess Na^{+} (points above the 1:1 line) may also originate from silicate weathering or domestic sewage [30].

In natural waters, Ca^{2+} and Mg^{2+} are derived mainly from the dissolution of carbonate rocks, silicate minerals and evaporites, whereas HCO_3^{-} predominantly originates from carbonate dissolution and is less affected by anthropogenic inputs [31]. If carbonate dissolution is dominant, the molar ratio of $[Ca^{2+} + Mg^{2+}]/[HCO_3^{-}]$ should approximate 1:1; when evaporites participate, $[Ca^{2+}]/[SO_4^{2-}]$ tends toward 1:1 [32]. Figure 4c indicates that 80.0% of the samples plot above the 1:1 line for $[Ca^{2+} + Mg^{2+}]/[HCO_3^{-}]$, implying that silicate weathering also contributes to Ca^{2+} and Mg^{2+} in addition to carbonate dissolution [33]. Figure 4d shows that most samples cluster around the 1:1 line for $[Ca^{2+}]/[SO_4^{2-}]$, confirming the involvement of evaporite (e.g., gypsum) dissolution; however, 78.3% of the samples plot below the 1:1 line, suggesting that the excess SO_4^{2-} is derived from other sources, such as industrial effluent discharge [34].

3.4. NBLs and TVs of Groundwater Chemical Components in Overexploited Areas

3.4.1. Analysis of NBLs and TVs

Cumulative-probability diagrams were constructed for Na^+, Ca^{2+}, Mg^{2+}, Cl^-, SO_4^{2-}, F^- and NO_3^- using the regional hydrochemical dataset. As shown in Figure 5, the abscissa of the first inflection point on each curve was adopted as the NBLs. The derived NBLs are 32.3 mg/L for Na^+, 34.1 mg/L for Ca^{2+}, 17.8 mg/L for Mg^{2+}, 46.2 mg/L for Cl^-, 66.4 mg/L for SO_4^{2-}, 0.89 mg/L for NO_3^- and 0.33 mg/L for F^-. Based on these, NBLs and TVs were subsequently calculated using Equations (2) and (3) in Section 2.4.3.

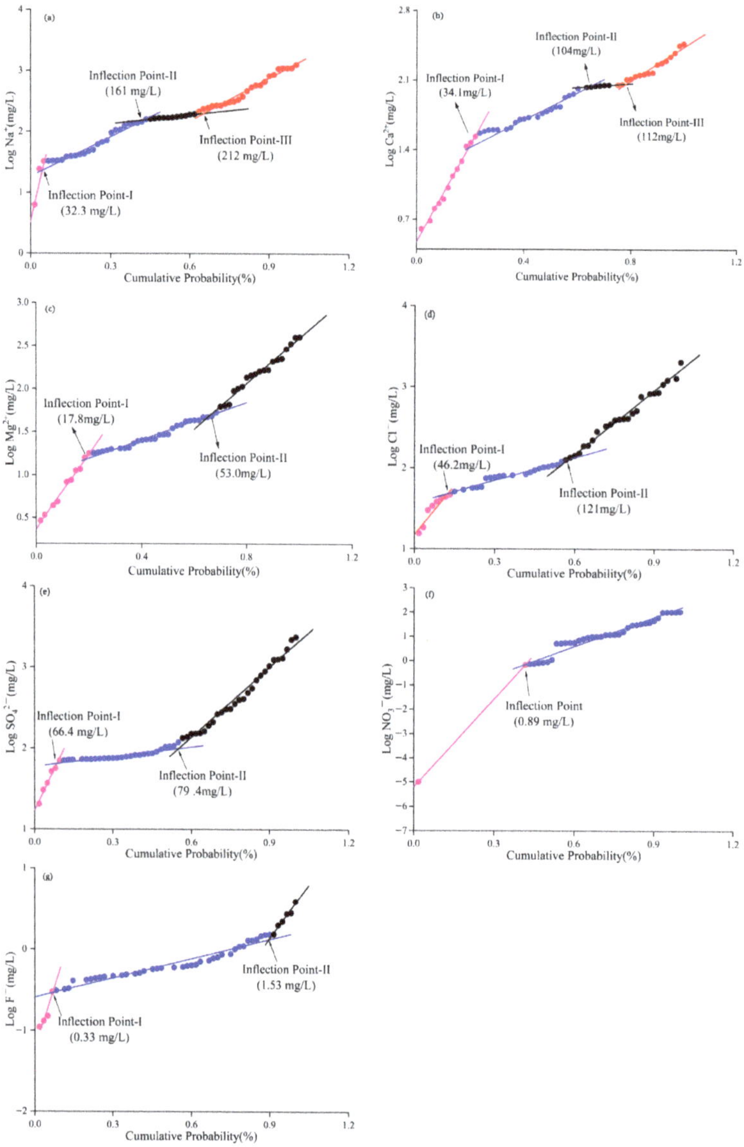

Figure 5. Cumulative probability plot of Na^+ (**a**), Ca^{2+} (**b**), Mg^{2+} (**c**), Cl^- (**d**), SO_4^{2-} (**e**), NO_3^- (**f**) and F^- (**g**) concentrations in groundwater.

Table 2 indicates that all constituents exhibit concentrations above their respective TVs. Over 50% of sampling sites exceed the TVs for Na^+, Ca^{2+} and Mg^{2+}, demonstrating pronounced anthropogenic impacts on groundwater quality.

Table 2. Statistics of natural background values and threshold characteristics of groundwater hydrochemical components.

Parameters	Standard (mg/L)	NBLs	TVs	Exceeding the TVs Points	Percentage Occupied (%)
Na^+	200	32.3	116	40	66.7
Ca^{2+}	75	34.1	54.6	33	55.0
Mg^{2+}	50	17.8	33.9	30	50.0
Cl^-	450	46.2	248	20	33.3
SO_4^{2-}	450	66.4	258	19	31.7
NO_3^-	88.6	0.89	44.7	7	11.7
F^-	1	0.33	0.67	23	38.3

The cumulative-probability curves further reveal that Mg^{2+}, Cl^-, SO_4^{2-} and F^- display two inflection points, whereas Na^+ and Ca^{2+} display three. The varying numbers of inflection points suggest that multiple, distinct anthropogenic sources, including groundwater over-extraction, agricultural fertilization, and industrial or domestic sewage discharge, are superimposed on the natural background [35].

3.4.2. Comparative Analysis of Groundwater Chemical NBLs in Different Regions

The NBLs of Ca^{2+} and Mg^{2+} in the study area (34.1 mg/L and 17.8 mg/L, respectively) are obviously lower than those reported in the Xi'an alluvial plain (76–80 mg/L and 43–184 mg/L) and Slovenia (110 mg/L and 25.9 mg/L) (Table 3). This difference can be attributed to the predominant carbonate rock composition of the unconfined aquifers in the latter two regions, which typically results in elevated concentrations of Ca^{2+} and Mg^{2+}. Additionally, the NBLs of NO_3^- in the study area (0.89 mg/L) is substantially lower than that observed in Milan, Italy (12 mg/L) and Rajasthan, India (26 mg/L). This lower NBLs of NO_3^- is likely due to the limited application of fertilizers in the study area compared to the extensive agricultural activities in the latter regions, which have been shown to increase NO_3^- background levels. Furthermore, the NBLs of SO_4^{2-} in the study area and the Xi'an alluvial plain are higher than those in other regions, likely due to the high degree of industrialization in these areas.

Table 3. Statistical Analysis of NBLs in Different Regions (mg/L).

Region	Na^+	Ca^{2+}	Mg^{2+}	Cl^-	SO_4^{2-}	NO_3^-	F^-
Study area	32.3	34.1	17.8	46.2	66.4	0.89	0.33
Xi'an alluvial plain [6]	101–121	76.1–80.0	43–184	40.0–73.6	340–696	-	-
Milan area, Italy [2]	25.2	-	-	29.3	44	12	-
Rajasthan, India [1]	158	-	-	242	28	26	0.35
Slovenia [36]	12	110	25.9	30.8	26	6.3	-

3.5. Contamination Characteristics and Controlling Factors in the Over-Exploited Zone

3.5.1. Contamination Characteristics

The contamination percentage at each sampling site was calculated using Equation (4) and interpreted in conjunction with the Na^+/Cl^- molar ratio to distinguish between chemical weathering and anthropogenic impacts. Sites with contamination percentages < 50% are considered to be controlled primarily by chemical weathering, whereas those >50% are deemed anthropogenically contaminated. Figure 6 shows that 56.7% of the samples exhibit contamination percentages above 50%, signifying substantial anthropogenic influence.

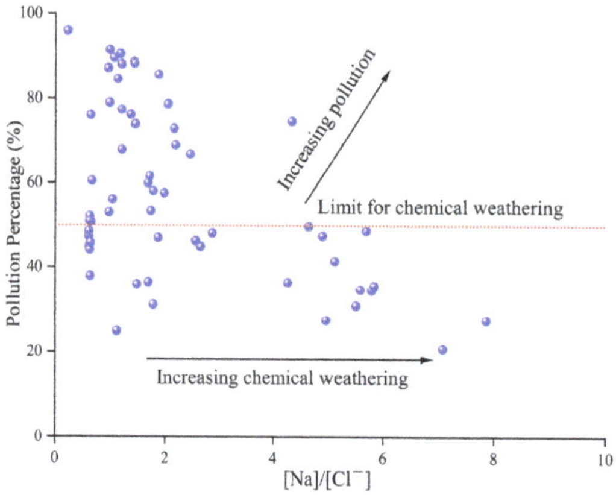

Figure 6. Relationship between Pollution Percentage and [Na^+/Cl^-] Ratio.

Combined with the evidence presented in Section 3.3.2, it is concluded that groundwater in the study area is severely affected by intensive abstraction and other human activities. Owing to persistent over-extraction, the regional flow field has been altered, lengthening flow paths and residence times and enhancing mineral dissolution. Consequently, carbonates and silicates within the aquifer are extensively leached, leading to concentrations of Na^+, Ca^{2+}, Mg^{2+}, $Cl^-$and F^- that exceed the calculated TVs [37]. As demonstrated in Section 3.3.2, SO_4^{2-} originates from both evaporite dissolution and industrial effluent discharge; the latter is the principal anthropogenic contributor to SO_4^{2-} concentrations above the threshold. Furthermore, because abstracted groundwater is predominantly used for agricultural irrigation and domestic supply [38], irrigation return flows enriched in nitrogen fertilizers [39] and direct discharge of domestic sewage are responsible for NO_3^- contamination, with 11.7% of the sampling sites exceeding the NO_3^- threshold.

3.5.2. Controlling Factors of Groundwater Contamination

To elucidate the principal drivers governing groundwater quality deterioration, eleven hydrochemical variables (K^+, Na^+, Ca^{2+}, Mg^{2+}, Cl^-, $SO_4{}^{2-}$, HCO_3^-, NO_3^-, TDS, TH and pH) were subjected to PCA. Prior to modeling, the Kaiser–Meyer–Olkin (KMO) measure and Bartlett's test of sphericity were applied to verify the suitability of the dataset. The KMO value was 0.632 and Bartlett's test value was 1 110 ($p < 0.001$), confirming the adequacy of the data for PCA. Three principal components with eigenvalues > 1 were extracted, together accounting for 51.8%, 18.9% and 14.6% of the total variance (Table 4).

Principal component 1 (PC1) exhibits an eigenvalue of 6.65 and explains 51.8% of the variance. It displays strong positive loadings on SO_4^{2-}, Mg^{2+}, Cl^-, TDS, TH, K^+, Na^+ and Ca^{2+}. As discussed in Section 3.3.2, SO_4^{2-} is derived primarily from evaporite dissolution and industrial effluents, whereas Na^+, Ca^{2+}, Mg^{2+}, Cl^-, TH and TDS are closely linked to rock weathering and water–rock interaction. Consequently, PC1 is interpreted as the combined influence of water–rock interaction and industrial wastewater discharge.

Principal component 2 (PC2) has an eigenvalue of 1.61 and contributes 18.9% of the variance. It exhibits a strong association with pH and HCO_3^-, and a moderate correlation with Mg^{2+}, Ca^{2+}, TDS, and TH. Previous studies have shown that carbonate dissolution tends to depress pH [40]. In the present dataset, pH decreases with increasing concentrations of HCO_3^-, Mg^{2+}, Ca^{2+}, TDS and TH, a pattern attributable to intensive groundwater abstraction. Over the past decades, persistent over-exploitation has created a pronounced cone of depression in the study area [41]. The consequent thickening of the unsaturated zone has shifted the hydrochemical regime toward more oxidizing conditions, promoting the oxidation of sulfide minerals, thereby elevating sulfate and TH concentrations while lowering pH [42]. Thus, PC2 is ascribed to the impacts of severe groundwater over-extraction.

Table 4. Loadings of 11 selected variables on VARIMAX rotated factors in this area.

Parameters	PC1	PC2	PC3
SO_4^{2-}	0.941	0.205	0.108
Mg^{2+}	0.912	0.308	0.091
TDS	0.860	0.396	−0.006
Cl^-	0.850	0.073	0.496
K^+	0.836	−0.306	−0.026
Ca^{2+}	0.828	0.395	−0.165
TH	0.817	0.369	0.261
Na^+	0.669	0.291	0.579
HCO_3^-	0.241	0.855	−0.133
pH	−0.223	−0.712	−0.467
NO_3^-	−0.165	−0.084	0.897
Eigenvalue	6.65	1.61	1.14
Contribution rate (%)	51.8	18.9	14.6
Cumulative contribution rate (%)	51.8	70.7	85.3

Principal component 3 (PC3) possesses an eigenvalue of 1.14 and explains 14.6% of the variance. It is strongly positively loaded on NO_3^- and moderately correlated with Na^+ and Cl^-. Nitrate in groundwater typically originates from agricultural fertilizers, domestic sewage, industrial effluents, manure and atmospheric deposition [43]. Given that the majority of samples were collected from farmland and villages, agricultural fertilization and domestic wastewater are identified as the principal sources of nitrate. Moreover, domestic sewage is characterized by elevated Na^+ and Cl^- concentrations. Accordingly, PC3 represents the influence of agricultural fertilization and domestic sewage.

4. Limitations and Future Directions

Although this study determined the NBLs and TVs of hydrochemical constituents in an over-exploited aquifer and identified the dominant contamination processes, several limitations remain. First, the temporal variability of groundwater contamination was not addressed, and a systematic analysis of the dynamic evolution of pollution is lacking. Second, source apportionment of dissolved ions relied solely on hydrochemical approaches, yielding only qualitative insights; the specific contributions of individual sources remain unquantified. Furthermore, while PCA elucidated the controlling factors, the relative importance of each factor was not quantitatively assessed.

In view of the above research deficiencies, future research could be carried out in the following directions: (1) implement multi-temporal, high-density sampling campaigns, employ time-series analyses to capture pollution dynamics, and integrate geostatistical interpolation with GIS to refine spatial sampling networks and delineate contamination patterns at finer scales; (2) combine stable-isotope tracers with multivariate statistical techniques to quantitatively apportion ion provenance and to discriminate the contributions of natural versus anthropogenic sources, thereby providing a dynamic and comprehensive scientific basis for groundwater protection.

5. Conclusions

Based on 60 shallow-groundwater samples and the combined application of hydrochemical and multivariate statistical techniques, this study determined the NBLs and TVs of major chemical constituents in a groundwater over-exploitation zone and identified the principal drivers of contamination. The dominant hydrochemical facies are HCO_3–Na and SO_4·Cl–Na, indicating pronounced anthropogenic influence. Using the established NBLs and TVs, 56.7% of the samples were identified as anthropogenically contaminated, underscoring the necessity of defining NBLs prior to contamination assessment.

The study demonstrates that groundwater over-extraction, industrial activities, water–rock interaction, agricultural fertilization and domestic sewage are the key factors controlling groundwater quality in the over-exploited region. These findings provide a scientific basis for pollution control in such areas. Nevertheless, the temporal dynamics of contamination and the quantitative contributions of individual sources remain inadequately resolved. Future investigations should adopt multi-temporal, high-resolution sampling strategies and integrate isotope tracers to elucidate spatio-temporal pollution evolution and to quantify source-specific contributions, thereby furnishing more comprehensive support for groundwater management and protection.

Author Contributions: Conceptualization, M.W., B.G., H.L., L.Z. and Z.Y.; Methodology, Q.W., Y.L. (Yan Li), B.G. and C.M.; Software, Q.W., Y.L. (Yang Liu) and L.Z.; Validation, Y.L. (Yan Li), L.Z. and Z.Y.; Formal analysis, Y.L. (Yang Liu); Investigation, Q.W., M.W., Y.L. (Yan Li), B.G., H.L., Y.L. (Yang Liu), L.Z. and C.M.; Resources, L.Z.; Data curation, M.W.; Writing—original draft, Q.W.; Writing—review & editing, Z.Y.; Supervision, L.Z. and Z.Y.; Project administration, Z.Y.; Funding acquisition, Z.Y. All authors have read and agreed to the published version of the manuscript.

Funding: This work was supported by the Hebei Key Laboratory of Geological Resources and Environment Monitoring and Protection (Grant No. JCYKT202305), Hebei Provincial Water Resources Science and Technology Program Project (Grant No. 2024-23), Hebei Provincial Project (Grant No. 13000024P00329410255J) and the Report on Investigation and Assessment of Groundwater Environment Status of Prefecture-level centralized Drinking water source in Hebei Province.

Data Availability Statement: The data presented in this study are available on request from the corresponding author. (the data are not publicly available due to privacy restrictions.)

Acknowledgments: The authors gratefully acknowledge the editor and anonymous reviewers for their valuable comments on this manuscript. The authors also appreciate the financial support from the different organizations.

Conflicts of Interest: The authors declare no conflicts of interest.

References

1. Rahman, A.; Tiwari, K.K.; Mondal, N.C. Assessment of hydrochemical backgrounds and threshold values of groundwater in a part of desert area, Rajasthan, India. *Environ. Pollut.* **2020**, *266*, 115150. [CrossRef] [PubMed]
2. De, C.M.; Crosta, G.B.; Frattini, P. Hydrogeochemical characterization and natural background levels in urbanized areas: Milan Metropolitan area (Northern Italy). *J. Hydrol.* **2017**, *547*, 455–473. [CrossRef]
3. Kuang, X.; Liu, J.; Scanlon, B.R.; Jiao, J.J.; Jasechko, S.; Lancia, M.; Zheng, C. The changing nature of groundwater in the global water cycle. *Science* **2024**, *383*, eadf0630. [CrossRef] [PubMed]
4. Davis, G.B.; Salama, R.B. Modeling point sources of groundwater contamination: A review and case study. In *Contamination of Groundwaters*; CRC Press: Boca Raton, FL, USA, 2020.
5. Urresti-Estala, B.; Carrasco-Cantos, F.; Vadillo-Pérez, I.; Jiménez-Gavilán, P. Determination of background levels on water quality of groundwater bodies: A methodological proposal applied to a Mediterranean river basin (Guadalhorce River, Málaga, southern Spain). *J. Environ. Manag.* **2013**, *117*, 121–130. [CrossRef]

6. Zhou, Y.; Wu, J.; Gao, X.; Guo, W.; Chen, W. Hydrochemical background levels and threshold values of phreatic groundwater in the greater Xi'an region, China: Spatiotemporal distribution, influencing factors and implication to water quality management. *Expo. Health* **2022**, *15*, 757–771. [CrossRef]
7. Ijioma, U.D. Delineating the impact of urbanization on the hydrochemistry and quality of groundwater wells in Aba, Nigeria. *J. Contam. Hydrol.* **2021**, *240*, 103792. [CrossRef]
8. Ren, C.; Zhang, Q. Groundwater chemical characteristics and controlling factors in a region of Northern China with intensive human activity. *Int. J. Environ. Res. Public Health* **2020**, *17*, 9126. [CrossRef]
9. Huang, G.; Sun, J.; Zhang, Y.; Chen, Z.; Liu, F. Impact of anthropogenic and natural processes on the evolution of groundwater chemistry in a rapidly urbanized coastal area, South China. *Sci. Total Environ.* **2013**, *463–464*, 209–221. [CrossRef]
10. Qin, R.; Wu, Y.; Xu, Z.; Xie, D.; Zhang, C. Assessing the impact of natural and anthropogenic activities on groundwater quality in coastal alluvial aquifers of the lower Liaohe River Plain, NE China. *Appl. Geochem.* **2013**, *31*, 142–158. [CrossRef]
11. Wang, L.; Zhang, Q.; Wang, H. Rapid urbanization has changed the driving factors of groundwater chemical evolution in the large groundwater depression funnel area of Northern China. *Water* **2023**, *15*, 2917. [CrossRef]
12. Zhinan, Z.; Lirong, W.; Congmei, W.; Xiaoqing, H. Risk Distribution Characteristics of Rainstorm and Flood Disaster Based on Flood Area Model in the Xiaoma River Basin of Xingtai. *J. Arid Meteorol.* **2021**, *39*, 486.
13. Amadi, A.N.; Olasehinde, P.I.; Nwadioha, I.J.; Okunlola, I.A.; Shaibu, I.; Nwakife, C.N. Comparative assessment of groundwater suitability for irrigation and drinking purposes in Agbara and Ota area of Southwestern Nigeria. *Int. J. Sci. Glob. Sustain.* **2016**, *2*, 1–12. [CrossRef]
14. Bu, J.; Liu, W.; Pan, Z.; Ling, K. Comparative study of hydrochemical classification based on different hierarchical cluster analysis methods. *Int. J. Environ. Res. Public Health* **2020**, *17*, 9515. [CrossRef] [PubMed]
15. Runnells, D.D.; Dupon, D.P.; Jones, R.L.; Cline, D.J. Determination of background chemistry of water at mining and milling sites, Salt Lake Valley, Utah, USA. In Proceedings of the 9th International Symposium on Water-Rock Interaction, Taupo, New Zealand, 30 March–3 April 1998; A A Balkema: Rotterdam, The Netherlands, 1998.
16. Reimann, C.; Filzmoser, P.; Garrett, R.G. Background and threshold: Critical comparison of methods of determination. *Sci. Total Environ.* **2005**, *346*, 1–16. [CrossRef] [PubMed]
17. Nayak, A.; Matta, G.; Uniyal, D.P.; Kumar, A.; Kumar, P.; Pant, G. Assessment of potentially toxic elements in groundwater through interpolation, pollution indices, and chemometric techniques in dehradun in uttarakhand state. *Environ. Sci. Pollut. Res.* **2024**, *31*, 36241–36263. [CrossRef] [PubMed]
18. Pacheco, F.; Weijden, C.H. Contributions of water-rock interactions to the composition of groundwater in areas with a sizeable anthropogenic input: A case study of the waters of the Fundo area, Central Portugal. *Water Resour. Res.* **1996**, *32*, 3553–3570. [CrossRef]
19. Soumya, B.S.; Sekhar, M.; Riotte, J.; Banerjee, A.; Braun, J.J. Characterization of groundwater chemistry under the influence of lithologic and anthropogenic factors along a climatic gradient in upper Cauvery basin, South India. *Environ. Earth Sci.* **2013**, *69*, 2311–2335. [CrossRef]
20. Yeater, K.M.; Duke, S.E.; Riedell, W.E. Multivariate analysis: Greater insights into complex systems. *Agron. J.* **2015**, *107*, 799–810. [CrossRef]
21. Onjia, A.; Huang, X.; Trujillo González, J.M.; Egbueri, J.C. Chemometric approach to distribution, source apportionment, ecological and health risk of trace pollutants. *Front. Environ. Sci.* **2022**, *10*, 1107465. [CrossRef]
22. *GB/T 14848-2017*; Standard for Groundwater Quality. Ministry of Natural Resources of the People's Republic of China: Beijing, China, 2017.
23. Nganje, T.N.; Hursthouse, A.S.; Edet, A.; Stirling, D.; Adamu, C.I. Hydrochemistry of surface water and groundwater in the shale bedrock, cross river basin and niger delta region, nigeria. *Appl. Water Sci.* **2017**, *7*, 961–985. [CrossRef]
24. Liang, Z.; Chen, J.; Jiang, T.; Li, K.; Gao, L.; Wang, Z.; Xie, Z. Identification of the dominant hydrogeochemical processes and characterization of potential contaminants in groundwater in Qingyuan, China, by multivariate statistical analysis. *RSC Adv.* **2018**, *8*, 33243–33255. [CrossRef]
25. Chen, J.; Huang, Q.; Lin, Y.; Fang, Y.; Qian, H.; Liu, R.; Ma, H. Hydrogeochemical characteristics and quality assessment of groundwater in an irrigated region, Northwest China. *Water* **2019**, *11*, 96. [CrossRef]
26. Herojeet, R.; Dewangan, R.K.; Naik, P.K.; Verma, J.R. Probabilistic modelling is superior to deterministic approaches in the human health risk assessment: An example from a tribal stretch in Central India. *Sci. Rep.* **2023**, *13*, 19351. [CrossRef] [PubMed]
27. Liu, F.; Song, X.; Yang, L.; Han, D.; Zhang, Y.; Ma, Y.; Bu, H. The role of anthropogenic and natural factors in shaping the geochemical evolution of groundwater in the Subei Lake basin, Ordos energy base, Northwestern China. *Sci. Total Environ.* **2015**, *538*, 327–340. [CrossRef] [PubMed]

28. Gaillardet, J.; Dupre, B.; Louvat, P.; Allegre, C.J. Global silicate weathering and CO_2 consumption rates deduced from the chemistry of large rivers. *Chem. Geol.* **1999**, *159*, 3–30. [CrossRef]
29. Liu, S.T.; Zhang, D.; Li, Y.H.; Yang, J.M.; Jia, B.J. Water sources and factors controlling on hydro-chemical compositions in the Yiluo River basin. *Environ. Sci.* **2020**, *41*, 1184–1196.
30. Yusuf, M.A.; Abiye, T.A.; Ibrahim, K.O.; Abubakar, H.O. Hydrogeochemical and salinity appraisal of surficial lens of freshwater aquifer along Lagos coastal belt, South West, Nigeria. *Heliyon* **2021**, *7*, e08231. [CrossRef]
31. Sheikhy, N.T.; Ramli, M.F.; Aris, A.Z.; Sulaiman, W.N.A.; Juahir, H.; Fakharian, K. Identification of the hydrogeochemical processes in groundwater using classic integrated geochemical methods and geostatistical techniques, in Amol-Babol Plain, Iran. *Sci. World J.* **2014**, *2014*, 419058. [CrossRef]
32. Zhan, H.; Wu, Q.; Liu, B.; Zhou, G. Exploration of hydrogeochemical characterization and assessment of organic pollution characteristics of shallow groundwater near a chemical plant that discharged sewage illegally. *Sustainability* **2022**, *14*, 660. [CrossRef]
33. Wang, L.; Yang, N.; Zhao, Y.; Zhang, Q. Research on the features and driving factors of shallow groundwater quality in arid areas, Northwest China. *Water* **2025**, *17*, 934. [CrossRef]
34. Yuan, B.; Hao, Y.; Guo, Z.; Hu, Y.; Li, J.; Meng, J.; Wang, F. Hydrogeochemical disparities and constraints of water produced from various coal seams in the Baode Block, Ordos Basin, China. *ACS Omega* **2024**, *9*, 4905–4919. [CrossRef] [PubMed]
35. Edmunds, W.M.; Doherty, P.; Griffiths, K.; Shand, P.; Peach, D. *Baseline Report Series: The Chalk of Dorset*; The Environment Agency: London, UK, 2002.
36. Serianz, L.; Cerar, S.; Šraj, M. Hydrogeochemical characterization and determination of natural background levels (NBL) in groundwater within the main lithological units in Slovenia. *Environ. Earth Sci.* **2020**, *79*, 373. [CrossRef]
37. Morris, B.L.; Lawrence, A.R.; Chilton, P.J.; Adams, B.; Calow, R.C.; Klinck, B.A. *Groundwater and Its Susceptibility to Degradation: A Global Assessment of the Problem and Options for Management*; The United Nations Environment Programme (UNEP): Nairobi, Kenya, 2003.
38. Cao, R.J.; Jiang, R.G.; Zhao, Y.; Xie, J.C.; Yu, X. Spatiotemporal characteristics of groundwater depth in Xingtai City on the North China Plain: Changing patterns, causes and prediction. *Appl. Ecol. Environ. Res.* **2020**, *18*, 6605–6622. [CrossRef]
39. Zhang, L.; Dong, D.; Lv, S.; Zhang, J.; Yan, M.; Han, G.; Li, H. Hydrochemical characterization and quality assessment of groundwater in the southern plain of Hebei Province, China. *Water* **2023**, *15*, 3791. [CrossRef]
40. Wang, Z.; Shimabukuro, M.; Kishida, R.; Yokoi, T.; Kawashita, M. Effects of pH on the microarchitecture of carbonate apatite granules fabricated through a dissolution–precipitation reaction. *Front. Bioeng. Biotechnol.* **2024**, *12*, 1396275. [CrossRef]
41. Zhang, Q.; Wang, H.; Wang, Y.; Yang, M.; Zhu, L. Groundwater quality assessment and pollution source apportionment in an intensely exploited region of northern China. *Environ. Sci. Pollut. Res.* **2017**, *24*, 16639–16650. [CrossRef]
42. Zhang, Q.; Liu, W.; Sun, H.; Wang, H.; Wang, L.; Zhang, J.; Xu, Z. Cascade reservoir regulations on nitrogen source and transformation in the Tibetan Plateau river: Constraints from high frequency data of Lancang river. *J. Hydrol.* **2025**, *650*, 132563. [CrossRef]
43. Gibert, O.; Assal, A.; Devlin, J.F. Performance of a field-scale biological permeable reactive barrier for in-situ remediation of nitrate-contaminated groundwater. *Sci. Total Environ.* **2018**, *643*, 1295–1306. [CrossRef]

water

MDPI

Article

A Study on the Hydrochemical Evolution Property and Pollution Source Attribution of Groundwater in Highly Urbanized Areas: A Case Study of Shenzhen City

Yue Wei [1], Yang Li [2], Lihong Zhang [1], Chuane Liu [1], Qingzhai Meng [1], Jianbo Yin [1,*] and Long Wang [3]

[1] 801 Institute of Hydrogeology and Engineering Geology, Shandong Provincial Bureau of Geology and Mineral Resources (SPBGM), Jinan 250014, China
[2] Land and Resources Exploration Center of Hebei Bureau of Geology and Mineral Resources Exploration (Hebei Mining and Geological Disaster Emergency Rescue Center), Shijiazhuang 050061, China
[3] Chinese Academy of Geological Sciences, Beijing 100037, China
* Correspondence: jianbo2025524@163.com

Abstract: Accurate identification of groundwater pollution sources is crucial for the socio-economic development of a region. In highly urbanized areas, where human activities have a pronounced impact on groundwater, however, the hydrochemical evolution patterns and sources of pollutants remain unclear. Taking Shenzhen, a highly urbanized city in China, as a case study, this research employed a combination of multivariate statistical techniques and the Positive Matrix Factorization (PMF) model to elucidate the hydrochemical evolution and quantitatively parse the pollution sources of groundwater in such regions. The results revealed that the pH of groundwater in the study area ranged from 4.24 to 7.31, indicating weak acidity to neutrality. The exceedance rates for pH, NH_4^+, COD, Mn, and Fe were as high as 67.1%, 44.3%, 44.3%, 34.3%, and 31.4%, respectively. The Water Quality Index assessment revealed that 32.9% of the groundwater samples were classified as poor, highlighting the significant impact of human activities. Land-use types significantly affected groundwater quality, with urban areas exhibiting higher concentrations of the COD, NO_3^-, Mn, and Fe compared to agricultural and forested areas. The predominant hydrochemical type of groundwater in the study area was $HCO_3{\cdot}Cl$—$Ca{\cdot}Na$, with rock weathering (primarily silicate weathering) being the dominant process controlling groundwater chemistry. The PMF model identified three major pollution sources in the highly urbanized region: domestic and industrial wastewater, enhanced water-rock interactions leading to the release of hydrochemical components, and agricultural fertilizers, contributing 43.9%, 37.0%, and 19.1% to groundwater pollution, respectively. Geostatistical spatial interpolation techniques demonstrated that in urban areas, groundwater quality was primarily controlled by the discharge of domestic and industrial wastewater, while in agricultural areas, excessive fertilizer application was the main driver of groundwater degradation. These findings provide a scientific basis for groundwater pollution prevention and sustainable utilization in highly urbanized regions.

Keywords: urbanization; groundwater; pollution source apportionment; PMF model; land use

1. Introduction

As one of the world's most vital freshwater resources, groundwater accounts for approximately 96% of the planet's total freshwater reserves. It serves not only as a crucial source of drinking water for numerous regions but also plays an essential role in agricultural irrigation and industrial water use [1–3]. In arid and semi-arid areas, groundwater often represents the sole reliable water supply [4]. However, the increasing global population and rapid urbanization have led to a continuous rise in groundwater extraction and utilization. Concurrently, human activities have intensified the impact on groundwater quality [5,6]. Groundwater contamination poses a threat not only to the safety of drinking water but also to human health and ecosystems through its use in irrigation and industry [7–9]. Thus, accurately identifying the sources of groundwater pollution has become of paramount importance.

In recent years, the issue of groundwater contamination has garnered increasing attention, with notable advancements in the identification of pollution sources both domestically and internationally. Traditionally, the identification of groundwater pollution sources has primarily relied on chemical analysis methods, which involve measuring the concentrations of various pollutants in groundwater and utilizing ratios between different water chemical components to identify pollution sources [10,11]. However, this approach faces limitations when dealing with complex pollution sources, as it struggles to accurately differentiate pollutants from various origins. Some scholars have employed periodic sampling and detection in conjunction with multivariate statistical techniques, such as correlation analysis and principal component analysis, to qualitatively identify the sources of pollutants in groundwater [12]. Nevertheless, these multivariate statistical techniques fail to accurately quantify the contribution rates of individual pollution sources [13]. In recent years, the Positive Matrix Factorization (PMF) model, developed by the U.S. Environmental Protection Agency, has been widely applied in source apportionment studies of atmospheric, soil, and surface water environments [14,15]. This model not only accurately identifies the sources of pollutants in the environment but also quantifies the contribution rates of each pollution source to the entire study area and the monitoring sites [16,17], thereby enhancing the efficiency of source tracing. Although the PMF model is widely used in other environmental media, its application to identify groundwater pollution sources, especially in highly urbanized areas with multiple pollution sources, remains limited.

The groundwater systems in urbanized regions are influenced by both geological structures and human activities, such as industrial production, agricultural practices, urban construction, and domestic wastewater discharge. These influences lead to significant variations in the recharge, flow, and discharge mechanisms across different areas, resulting in complex and dynamically changing pollution sources [18]. Therefore, investigating the sources of groundwater pollution in highly urbanizing areas holds great scientific significance and practical application value. Shenzhen, located in the Pearl River Delta, is a highly developed and densely populated modern city. With the acceleration of urbanization, the city's groundwater system is under multiple pressures. On one hand, urbanization has increased impervious surfaces, altering the recharge conditions of groundwater [19]. On the other hand, human activities like industrial production, agricultural practices, and domestic wastewater discharge have significantly heightened the risk of groundwater contamination [20]. At present, systematic studies on the sources of groundwater pollution in Shenzhen are still lacking. Most research has focused on the current status and extent of groundwater quality, with insufficient in-depth analysis of the dynamic changes in pollution sources [21,22]. This limitation makes it difficult to fully understand the complex pollution sources. Thus, conducting a comprehensive study on the sources of groundwater pollution in Shenzhen is of great scientific and practical significance.

This study systematically investigates the hydrochemical evolution and pollution sources of groundwater in Shenzhen by integrating a suite of techniques, including chemical analysis, multivariate statistical techniques, the PMF model, and GIS spatial interpolation.

The research focuses on three main aspects: (1) the characteristics and assessment of groundwater quality in highly urbanized areas; (2) the hydrochemical evolution patterns of groundwater in these regions; and (3) the quantitative identification of groundwater pollution sources. The findings are expected to provide crucial insights for groundwater protection and pollution control in rapidly urbanizing areas.

2. Materials and Methods

2.1. Study Area

2.1.1. Overview of the Study Area

Shenzhen is a modern international metropolis situated in the southern part of Guangdong Province, China. It lies on the eastern shore of the Pearl River estuary, borders Daya Bay and Dapeng Bay to the east, fronts the Pearl River estuary and Lingdingyang to the west, adjoins Hong Kong along the Shenzhen River to the south, and abuts Dongguan and Huizhou to the north. The city is located south of the Tropic of Cancer between longitudes 113°43′–114°38′ E and latitudes 22°24′–22°52′ N, covering a total area of 1997.47 km^2. The multi-year mean annual air temperature is 23 °C, and the multi-year mean annual precipitation is 1935.8 mm, with an average of 131 rainy days per year. Among them, farmland accounts for 7.65%, urban and rural land accounts for 24.3%, forest and grassland account for 64.1%, and water area accounts for 3.95%.

Rapid socio-economic development in recent decades has made Shenzhen one of China's most urbanized cities, with an urbanization rate approaching 100%. Its industrial activities include high-tech industries, advanced manufacturing, traditional manufacturing, electroplating, printed circuit board manufacturing, chemical industry, textile and dyeing industry, papermaking industry, and food processing, among others. Concomitant growth in industrial and agricultural water demand has exerted substantial pressure on groundwater quality.

2.1.2. Geological and Hydrogeological Conditions

Shenzhen lies within the hydrogeological sub-region of the Pearl River Delta. Exposed strata in the area comprise mainly the Lower Paleozoic, Middle and Upper Devonian, Carboniferous, Jurassic, Cretaceous, Tertiary and Quaternary systems. Structural features are well developed; NE-trending structures predominate, followed by NW-trending and E–W-trending structures, with minor N–S-trending faults sporadically exposed. Quaternary deposits are widespread, consisting predominantly of alluvial layers along riverbanks, fluvio-marine deltaic sequences at estuaries, and marine deposits along the coastal zone.

Groundwater in Shenzhen can be classified into three types: pore water in unconsolidated Quaternary sediments, fissure water in bedrock, and karst water. The primary recharge sources are atmospheric precipitation and lateral inflow from rivers. In addition, groundwater can also be replenished through human-made structures such as reservoirs, agricultural irrigation systems, and ponded water in pits. Groundwater flow is controlled by topography, generally moving from mountainous or hilly areas toward valley basins, ultimately discharging southward into the sea or northward into Dongguan and Huizhou. Discharge mechanisms include spring outflow, baseflow to surface water bodies, evaporation and artificial abstraction.

2.2. Sample Collection and Analysis

2.2.1. Acquisition of Groundwater Samples

Groundwater samples were collected during June–July 2021. A total of 70 samples were obtained exclusively from domestic wells; their spatial distribution is shown in Figure 1. Prior to sampling, bottles were rinsed with deionized water; immediately before collection, each bottle was conditioned by rinsing three times with the groundwater to be sampled to ensure cleanliness. After collection, samples were placed in portable refrigerated coolers, transported to the laboratory, filtered through 0.45 μm membranes, and stored in 500 mL polyethylene bottles. Aliquots designated for cation analysis were acidified to pH < 2 using 1:1 HNO_3.

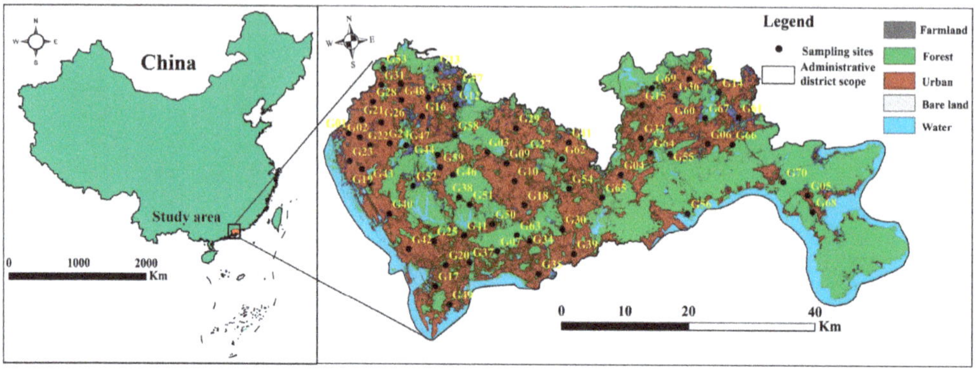

Figure 1. Distribution Map of Sampling Sites.

2.2.2. Hydrochemical Analysis

Measured parameters comprised field-determined pH and dissolved oxygen (DO), together with laboratory analyses for K^+, Na^+, Ca^{2+}, Mg^{2+}, Cl^-, SO_4^{2-}, HCO_3^-, NO_3^-, NO_2^-, NH_4^+, chemical oxygen demand (COD), total hardness (TH), total dissolved solids (TDS), Fe, and Mn. Analytical methods, instrumentation, and method detection limits for each hydrochemical variable are summarized in Supplementary Table S1.

2.2.3. Analysis of Data

Water Quality Index (WQI) Approach

The Water Quality Index is a widely recognized tool for assessing groundwater quality and its suitability for drinking purposes. In the WQI calculation, each hydrochemical parameter is assigned a weight (W_i) that reflects its relative importance to overall groundwater quality. The present study adopts the Class III thresholds prescribed by the Chinese "Standard for Groundwater Quality" (GB/T 14848-2017) [23] as the primary reference; where a parameter is not included in that standard, the guideline value given by the World Health Organization (WHO, 2011) [24] is used instead.

The procedural steps for computing *WQI* are summarized below:

$$W_i = \frac{W_i}{\sum_{i=1}^{n} W_i}$$

$$Q_i = C_i/S_i \times 100$$

$$SI_j = W_i \times Q_i$$

$$WQI = \sum SI_j$$

where Q_i denotes the quality rating; C_i and S_i denote the concentration (mg/L) and the water quality standard of each indicator, respectively; and SI_j is the sub-index of the i-th indicator. The specific values of the weights assigned to each water chemistry index are detailed in Supplementary Table S2.

Positive Matrix Factorization (PMF) Model

The PMF model, recommended by the United States Environmental Protection Agency, is a receptor-oriented source-apportionment technique. By decomposing the sample-variable matrix into factor contributions and profiles through non-negativity constraints on both matrices, PMF reduces high-dimensional hydrochemical data to a small number of latent factors without requiring prior source-composition spectra. Owing to its operational simplicity and robustness, the method has been extensively applied to atmospheric, soil, and aquatic environmental systems.

The model computation is expressed as Equation:

$$X_{ij} = \sum_{k=1}^{p} g_{ik} f_{ki} + e_{ij}$$

where X_{ij}: Concentration of species j in sample i; g_{ik}: Contribution of source k to sample i; f_{kj}: Concentration of species j in source k; and e_{ij}: Residual, i.e., the portion of X<sub> not explained by the PMF model. PMF iteratively decomposes the original data matrix to obtain the optimal matrices G and F by minimizing the objective function Q.

Spatial Interpolation

Spatial interpolation is a fundamental technique in geographic information systems (GIS) and spatial analysis that estimates values at unsampled locations from point observations. In water-quality studies, it is routinely employed to map pollutant concentrations, water temperature, bathymetry, and other variables. Kriging-an optimal geostatistical interpolator-exploits spatial autocorrelation by fitting a semivariogram model to the data; it minimizes the estimation variance while preserving the spatial structure of the variable. Beyond yielding accurate, smooth prediction surfaces, Kriging also quantifies the uncertainty of each estimate. In the present work, Kriging was applied to project the source contributions obtained from the PMF model onto the study-area basemap, thereby visualizing the spatial distribution of contaminant sources across Shenzhen.

Multivariate Statistical Techniques

To assess the variations in groundwater quality among different land-use types, the nonparametric Kruskal–Wallis test was utilized, with a significance level set at $p < 0.05$. Subsequently, pairwise comparisons between the three groups were carried out using the Mann–Whitney U test. The R computing platform was employed to perform all statistical analyses (version 4.3.2 (New Zealand)).

Chloro-Alkaline Indices (CAI)

To elucidate the chemical evolution process of groundwater, we utilized *CAI* to evaluate the cation exchange among its chemical constituents.

$$CAI-1 = \frac{[\gamma(Na^{+}) + \gamma(K^{+})] - \gamma(Cl^{-})}{\gamma(Cl^{-})}$$

$$CAI-2 = \frac{[\gamma(Na^{+}) + \gamma(K^{+})] - \gamma(Cl^{-})}{\gamma\left(SO_4^{2-}\right) + \gamma(HCO_3^{-}) + \gamma\left(CO_3^{2-}\right)}$$

3. Results

3.1. Properties of Groundwater Hydrochemistry

To thoroughly assess the groundwater quality within the study area, a detailed statistical analysis was performed on the hydrochemical parameters obtained from all sampling points. The compiled data for each chemical parameter can be found in Table 1. Groundwater pH across the study area ranges from 4.24 to 7.31, with a mean of 6.08; 67.1% of the samples fall below the national Class III threshold (pH < 6.5), indicating a predominantly weakly acidic to neutral signature. Previous studies have found that the frequency of acid rain in the Shenzhen area is as high as 70%, with an average pH value of 4.25. Therefore, acidic precipitation could be an important reason for the low pH of groundwater [25]. Dissolved-oxygen (DO) concentrations vary between 0.950 and 8.19 mg/L (mean value is 2.85 mg/L); The low concentration of DO is closely related to the shallow depth of groundwater and its susceptibility to human activities [26]. Cation abundances follow the order Ca^{2+} (46.5 mg/L) > Na^{+} (46.0 mg/L) > K^{+} (21.9 mg/L) > Mg^{2+} (5.22 mg/L) > NH_4^{+} (4.78 mg/L). NH_4^{+} exceeds the Class III limit in 44.3% of the samples, evidencing pronounced anthropogenic inputs. Anion abundances decrease as HCO_3^{-} (158 mg/L) > Cl^{-} (66.7 mg/L) > SO_4^{2-} (39.5 mg/L) > NO_3^{-} (31.3 mg/L) > NO_2^{-} (1.94 mg/L); NO_2^{-} and NO_3^{-} surpass the respective thresholds in 14.3% and 10.0% of the samples, further confirming intensive human impact. Chemical oxygen demand (COD) ranges from 0.530 to 47.3 mg/L (mean 5.37 mg/L), with an exceedance rate of 44.3%. The wide concentration span and high non-compliance rate stem from domestic and industrial effluents, agricultural non-point sources, and heterogeneities in hydrogeological settings. Mean Fe and Mn concentrations are 0.992 and 0.278 mg/L, respectively, with exceedance rates of 31.4% and 34.3%. Overall, the exceedance rate of groundwater chemical components in highly urbanized areas is higher than that in areas with lower levels of urbanization, such as riparian zones and agricultural areas [27,28]. This further indicates that the groundwater quality in highly urbanized areas has been strongly affected by human activities.

Table 1. Statistical table of hydrochemical components of groundwater in the study area.

Parameters	Units	Range	Mean	SD	Standard	Exceed Standards for All Sites (%)
pH	—	4.24–7.31	6.08	0.678	6.5~8.5	67.1
DO	mg/L	0.950–8.19	2.85	1.43	—	—
TDS	mg/L	44.8–1906	465	424	≤1000	12.9
K^{+}	mg/L	0.26–104	21.3	25.5	—	—
Na^{+}	mg/L	2.91–342	46.0	51.1	≤200	1.43
Ca^{2+}	mg/L	3.19–247	46.5	46.6	—	—
Mg^{2+}	mg/L	0–23.2	5.22	4.57	—	—
NH_4^{+}	mg/L	0–45.0	4.78	10.2	≤0.643	44.3
Cl^{-}	mg/L	3.57–496	66.7	74.3	≤250	2.86
SO_4^{2-}	mg/L	0.142–295	39.5	47.0	≤250	1.43
HCO_3^{-}	mg/L	6.04–673	158	173	—	—
NO_3^{-}	mg/L	0.662–185	31.3	37.5	≤88.6	10.0
NO_2^{-}	mg/L	0.003–33.2	1.94	5.71	≤3.29	14.3
TH	mg/L	13.8–617	142	125	≤450	1.43
Mn	mg/L	0.003–2.96	0.278	4.75	≤0.1	34.3
Fe	mg/L	0–11.5	0.992	36.5	≤0.3	31.4
COD	mg/L	0.530–47.3	5.37	23.2	≤3	44.3

Note: The Standard Refers to the Class III Standard of the Chinese Groundwater Quality Standard (GB/T 14848-2017) [23].

3.2. *Types of Groundwater Hydrochemistry*

Piper diagrams were employed to delineate the hydrochemical facies, dominant ion assemblages, and evolutionary trends of groundwater within the study area. As illustrated in Figure 2, samples were classified according to the Shukalev scheme. Carbonate alkalinity (HCO_3^-) predominates among anions, whereas sodium (Na^+) constitutes the principal cation. Thirteen distinct hydrochemical types were identified; the four most prevalent are HCO_3·Cl–Ca·Na (50.0%), HCO_3·SO_4·Cl–Ca·Na (12.9%), HCO_3–Ca·Na (10.0%), and HCO_3·Cl–Ca (7.14%). Notably, Cl^-, Na^+, and SO_4^{2-} appear in 82.9%, 85.7%, and 20.0% of the facies designations, respectively—an unambiguous signal of anthropogenic modification of the groundwater system.

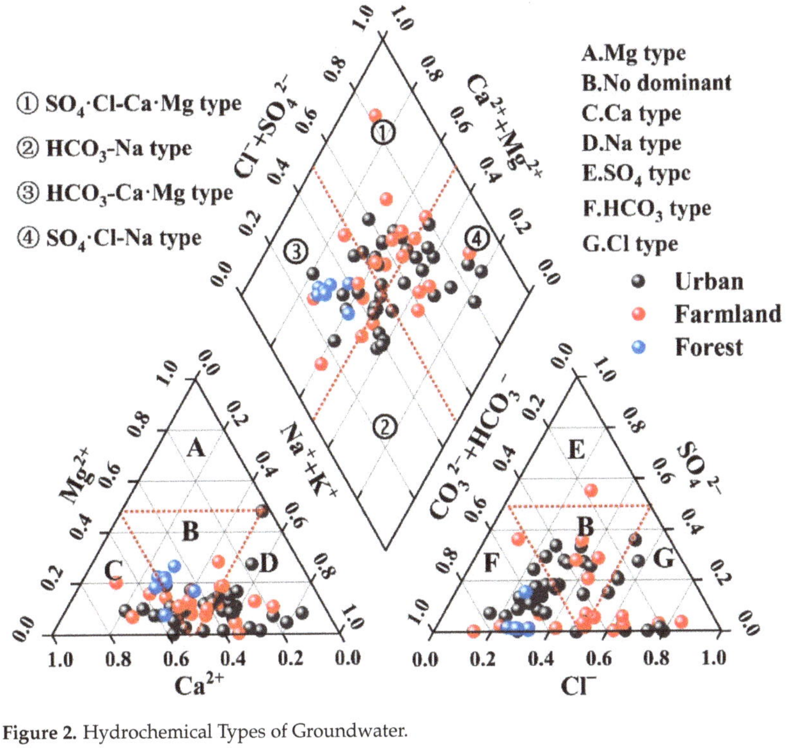

Figure 2. Hydrochemical Types of Groundwater.

3.3. *Water Quality Assessment by Using WQI Methods*

The WQI was applied to assess groundwater potability across the study area (Table 2). Of the sampled sites, 60.0% were classified as "excellent", 7.1% as "good", whereas 4.3% were rated "poor" and a further 4.3% as "very poor". Approximately 24.3% of the stations yielded water deemed "unsuitable for drinking". Significantly, all non-potable sites are located within urban land-use zones, highlighting that rapid urbanization and the associated anthropogenic pressures are the primary drivers of groundwater quality degradation [29].

Table 2. Statistics of water quality assessment results.

Water Quality Level	Comprehensive Water Quality Score	Monitoring Stations	Proportion (%)	Urban (%)	Forest (%)	Farmland (%)
Excellent water	<50	42	60.0%	24.29%	14.3%	21.4%
Good water	50–100	5	7.14%	4.29%	—	2.86%
Poor water	100.1–200	3	4.29%	4.29%	—	—
Very poor water	200.1–300	3	4.29%	2.86%	—	1.43%
Water unsuitable for drinking purposes	>300	17	24.3%	24.3%	—	—

4. Discussion

4.1. Influence of Land-Use Type on Groundwater Hydrochemistry

Land-use type serves as a direct proxy for the intensity and nature of anthropogenic disturbance, and pollutant sources differ markedly among urban, agricultural, and forested areas. Previous studies have documented that urban zones usually exhibit elevated concentrations of heavy metals (e.g., Fe and Mn) and organic contaminants such as polycyclic aromatic hydrocarbons, primarily because of untreated or partially treated industrial effluents and domestic sewage. Agricultural regions, in contrast, are characterized by high NO_3^- and NH_4^+ levels derived from intensive fertilizer and pesticide application. Forested catchments, benefiting from natural attenuation and minimal human interference, generally display the lowest contaminant burdens [30].

In this study, six hydrochemical indicators (pH, DO, COD, NO_3^-, Cl^-, and Mn) were selected to quantify the impact of land-use on groundwater quality (Figure 3). As depicted in Figure 3a, the mean pH of groundwater in urban areas (6.24) is significantly higher than that in agricultural areas (5.77) ($p < 0.05$). Previous studies have demonstrated that the majority of rainfall in this region is acidic, with a mean pH value of 4.25 [25]. This difference is attributed to the extensive impervious surfaces in urban areas, which prevent acidic precipitation from directly infiltrating the aquifer [5], whereas in agricultural fields, acid rain can percolate immediately, depressing groundwater pH.

COD concentrations in urban groundwater (6.13 mg/L) are significantly elevated relative to forest sites (1.97 mg/L) (Figure 3b). Likewise, NO_3^- in urban groundwater (43.8 mg/L) exceeds that in agricultural (19.1 mg/L, $p < 0.01$) and forest (8.87 mg/L, $p < 0.001$) groundwater; the difference between agricultural and forest samples is not statistically significant ($p > 0.05$) (Figure 3c). These patterns confirm that domestic and industrial effluents are the dominant nitrate source in urban aquifers, consistent with the findings of Zhang et al. [31].

Cl^- and Mn concentrations are also significantly higher in urban (79.5 and 0.288 mg/L) and agricultural (68.4 and 0.380 mg/L) groundwater than in forest groundwater (13.1 and 0.026 mg/L) (Figure 3d,e). Conversely, DO is markedly higher in forest groundwater (4.49 mg/L) than in urban or agricultural groundwater (both 2.58 mg/L) ($p < 0.001$) (Figure 3f). This is primarily because forestry land is mostly situated in mountainous areas, distant from human settlements, which results in weaker human impacts on the local groundwater.

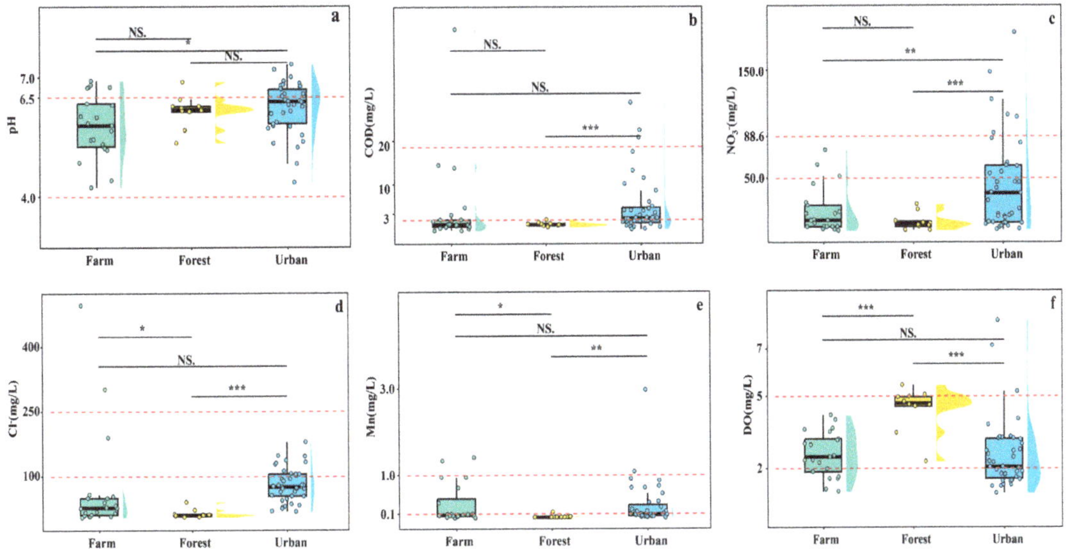

Figure 3. Comparison of Significant Differences in Hydrochemical Component Concentrations (**a**) pH; (**b**) COD; (**c**) NO_3^-; (**d**) Cl^-; (**e**) Mn; (**f**) DO. Note: *** denotes a p-value < 0.001; ** denotes a p-value < 0.01; * denotes a p-value < 0.1; NS: not significant. The red dash line represents the groundwater quality standard limit in China (GB/T 14848-2017) [23].

4.2. Mechanisms of Groundwater Hydrochemical Evolution in Rapidly Urbanizing Areas

4.2.1. Factors Controlling Groundwater Hydrochemistry

Gibbs diagrams are widely employed to qualitatively evaluate the relative influence of atmospheric precipitation, rock weathering, and evaporation–crystallization on the origin of dissolved ions in groundwater [32]. As shown in Figure 4, the majority of samples plot within the central region of the TDS versus $Na^+/(Na^+ + Ca^{2+})$ and $Cl^-/(Cl^- + HCO_3^-)$ fields, i.e., within the "rock-weathering" end-member domain. This distribution indicates that rock weathering reactions constitute the primary control on the major-ion composition of local groundwater. Nevertheless, a subset of samples collected beneath agricultural and forested land-use types falls into the precipitation-dominated sector, whereas several urban and agricultural samples shift toward the evaporation–crystallization quadrant, demonstrating that both rainfall inputs and evaporative concentration also modulate hydrochemical signatures. In addition, approximately 20% of the sampling sites lie outside the three conventional Gibbs sectors, implying an additional overprint of anthropogenic activities on the natural ionic assemblages. Zhou et al. also found that some sampling points were located outside the end-member range, indicating that human activities have had an impact on the groundwater chemistry [33].

4.2.2. Analysis of Hydrochemical Evolution Processes

To further elucidate the hydrochemical evolution of groundwater in the study area, ion-ratio analysis was employed to identify the dominant lithological weathering signatures and associated hydrogeochemical processes. Na-normalized molar ratios were first used to assign samples to specific weathering endmembers. Typical molar ratios for silicate weathering are $Ca^{2+}/Na^+ = 0.35 \pm 0.15$ and $Mg^{2+}/Na^+ = 0.24 \pm 0.12$; for carbonate weathering, 50 ± 20 and 20 ± 8; and for evaporite dissolution, 0.17 ± 0.09 and 0.02 ± 0.013. As shown in Figure 5a, most urban and agricultural plots cluster close to the silicate-evaporite join, whereas forested (and a minority of agricultural) samples shift toward the carbonate field. This distribution reflects the regional aquifer lithology: shallow horizons are dominated by silicate rocks, with only minor carbonate outcrops whose dissolution imprint remains limited.

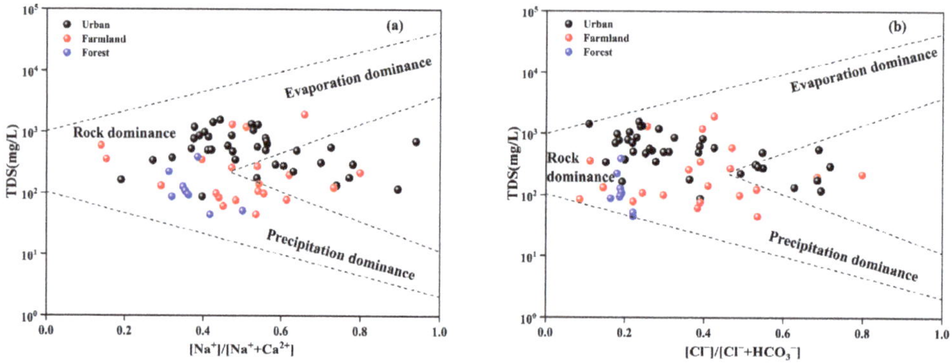

Figure 4. Identification plot of hydrochemical controls (**a**) $[Na^+]/[Na^+ + Ca^{2+}]$ vs. TDS; (**b**) $[Cl^-]/[Cl^- + HCO_3^-]$ vs. TDS.

When evaporite dissolution controls water chemistry, the molar ratio $Cl^-/(Na^+ + K^+)$ should equal unity (i.e., fall on the y = x line) [34]. In Figure 5b, 91.4% of the samples plot above this line, with $Na^+ + K^+$ significantly exceeding Cl^- and exhibiting a strong linear relationship ($R^2 = 0.861$). This indicates that silicate weathering constitutes the principal source of Na^+ and K^+, supplemented by minor evaporite dissolution. Moreover, urban samples display markedly higher $(Na^+ + K^+)$ relative to Cl^-, implying an additional anthropogenic contribution from industrial and domestic effluents [35].

The ratio of $[Ca^{2+} + Mg^{2+}]$ to $[HCO_3^- + SO_4^{2-}]$ can be used to determine the sources of Ca^{2+} and Mg^{2+}. A ratio greater than 1 indicates that Ca^{2+} and Mg^{2+} primarily originate from the dissolution of carbonate minerals. Conversely, a ratio of less than 1 suggests that these ions derive from the dissolution of sulfate or silicate minerals. When the ratio equals 1, it signifies that both carbonate and silicate rock weathering and dissolution are occurring simultaneously [36]. As depicted in Figure 5c, sample points from agricultural and forestry land primarily cluster around the y = x line, whereas those from urban land are concentrated below this line. The concentrations of $[Ca^{2+} + Mg^{2+}]$ and $[HCO_3^- + SO_4^{2-}]$ in urban land samples exceed those in agricultural and forestry land samples. This suggests that the weathering and dissolution of silicate rocks supply substantial amounts of $Ca^{2+} + Mg^{2+}$ to urban land water samples. In contrast, $Ca^{2+} + Mg^{2+}$ in agricultural and forestry land samples originate from the weathering and dissolution of both carbonate and silicate rocks. Additionally, acid rain (characterized by low pH values) is prevalent in the study area. This acid rain erodes surface carbonate rocks, which then recharge the shallow groundwater (with an average burial depth of 2.17 m). Consequently, the erosion and dissolution of carbonate rocks serve as the primary source of HCO_3^-.

The ratio plot of $[Ca^{2+}]/[SO_4^{2-}]$ can be used to assess the impact of evaporite weathering and dissolution on water chemistry [37]. As shown in Figure 5d, Ca^{2+} and SO_4^{2-} exhibit a strong linear relationship, indicating that evaporite weathering and dissolution are one of the sources of these ions. The sample points from all three types of land use are generally located above the y = x line, suggesting that there are additional sources contributing to the excess Ca^{2+}. As previously discussed, the weathering and dissolution of silicate rocks and the erosion of carbonate rocks also play significant roles in supplying Ca^{2+}.

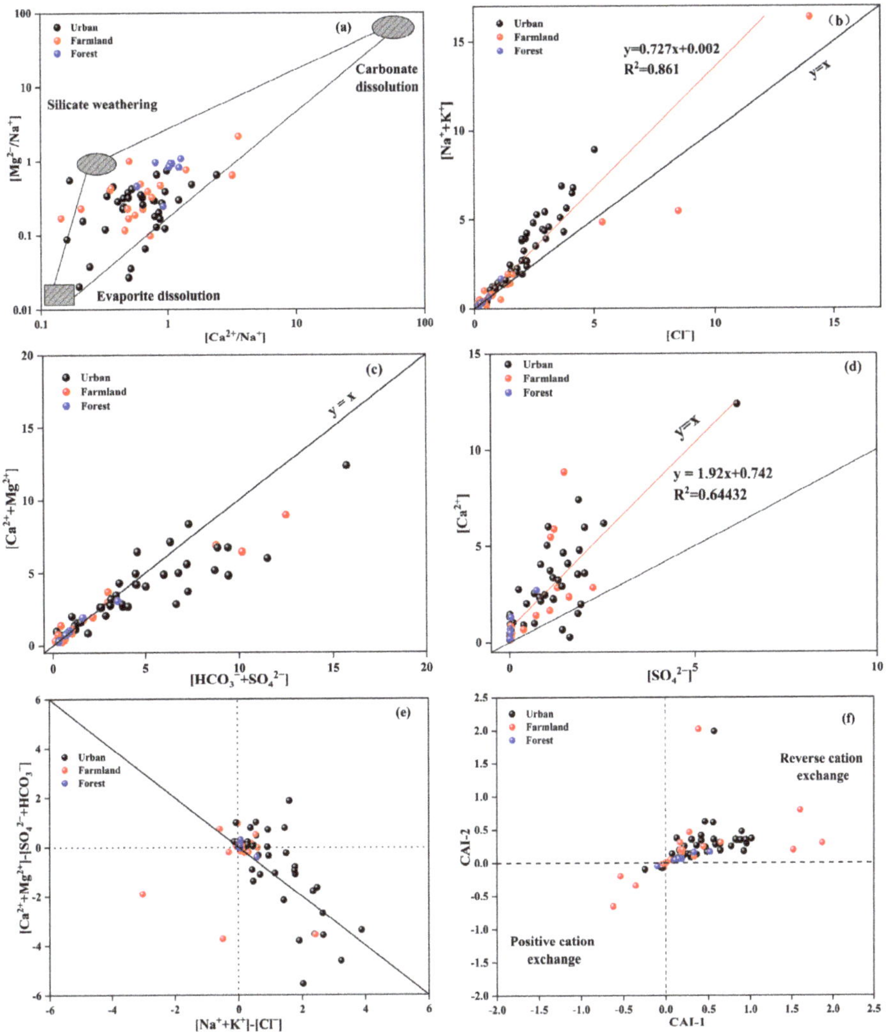

Figure 5. Ion ratio diagram: (**a**) $[Ca^{2+}/Na^{+}]$ vs. $[Mg^{2+}/Na^{+}]$; (**b**) $[Cl^{-}]$ vs. $[Na^{+}]$; (**c**) $[HCO_3^{-} + SO_4^{2-}]$ vs. $[Ca^{2+} + Mg^{2+}]$; (**d**) $[SO_4^{2-}]$ vs. $[Ca^{2+}]$; (**e**) $[Na^{+} + K^{+}]$-$[Cl^{-}]$ vs. $[Ca^{2+} + Mg^{2+}]$-$[HCO_3^{-} + SO_4^{2-}]$; (**f**) CAI-1 vs. CAI-2.

To further illustrate the influence of hydrochemical processes on water chemistry, we examined cation exchange [38,39]. Figure 5e shows that all sample points cluster around the $y = -x$ line, indicating that cation exchange is occurring. According to Figure 5f, more than 80% of the samples from each of the three land-use types fall in the CAI > 0 zone, demonstrating that reverse cation exchange dominates [40]. Combined with the preceding discussion, this indicates that the concentrations of Ca^{2+} and Mg^{2+} in the groundwater are jointly controlled by silicate weathering, carbonate dissolution, and reverse cation exchange.

4.2.3. Impact of Human Activities on Groundwater Hydrochemical Evolution

Human activities are also one of the significant factors influencing the hydrochemical evolution of groundwater. Typically, in regions with intense human activities, the concentrations of NO_3^-, Cl^-, SO_4^{2-}, and Na^+ in groundwater tend to increase markedly. Among these, SO_4^{2-} is mainly influenced by industrial activities, whereas NO_3^-, Cl^-, and Na^+ are affected by agricultural fertilization, manure, and domestic wastewater [41].

Based on the milliequivalent ratios of $[NO_3^-]/[Ca^{2+}]$ vs. $[SO_4^{2-}]/[Ca^{2+}]$, we can identify which type of human activity has impacted the groundwater [42]. When $[NO_3^-]/[Ca^{2+}] > [SO_4^{2-}]/[Ca^{2+}]$, groundwater is strongly influenced by agricultural activities or rainfall. Conversely, when $[NO_3^-]/[Ca^{2+}] < [SO_4^{2-}]/[Ca^{2+}]$, the impact of industrial activities is more pronounced. As shown in Figure 6, 61.5% of the sampling points are located in the mixed impact zone, indicating that the groundwater in the study area is affected by a combination of industrial wastewater, domestic sewage, and fertilizers. Zhang et al. also found that in rapidly urbanizing areas, groundwater is subjected to mixed contamination from industrial wastewater, domestic sewage, and landfill leachate [18]. The study area is a highly urbanized metropolis with a high population density and highly developed industry, resulting in a large volume of wastewater. The shallow groundwater level means that the discharge of domestic and industrial wastewater can directly affect the chemical composition of groundwater. Additionally, the frequent rainfall (with an annual precipitation of about 1900 mm) causes agricultural fertilizers to enter the groundwater through leaching and surface runoff, thereby causing groundwater pollution.

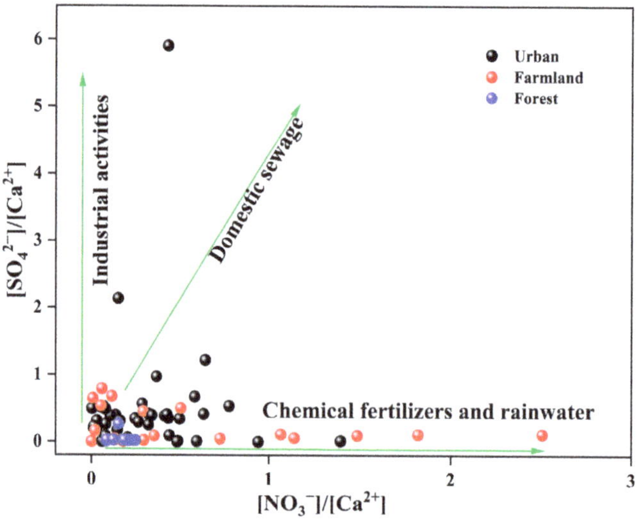

Figure 6. Identification plot of human activity impacts $[NO_3^-]/[Ca^{2+}]$ vs. $[SO_4^{2-}]/[Ca^{2+}]$.

4.3. Source Apportionment of Groundwater Quality Pollution in Rapidly Urbanizing Areas

4.3.1. Pollution Source Apportionment Based on the PMF Model

In this study, the PMF model was employed to quantitatively identify the sources of groundwater pollution in the study area. Concentration data and uncertainty data for 17 hydrochemical indicators (K^+, Na^+, Ca^{2+}, Mg^{2+}, Cl^-, SO_4^{2-}, HCO_3^-, NO_3^-, NO_2^-, NH_4^+, COD, TH, TDS, Fe, and Mn) from 70 groundwater samples were input into the PMF model. After running the model, we iteratively adjusted the "strong" and "weak" settings for elements and the factor format to optimize the fit. The goal was to achieve a fitting coefficient greater than 0.5 for the measured versus simulated element concentrations and a difference of less than 10% between the calculated Q value and the theoretical Q value. Ultimately, a three-factor solution was determined to be the optimal configuration.

As shown in Table 3, factor 1 accounts for 43.9% of the total variance and is heavily weighted by NH_4^+, NO_2^-, Cl^-, SO_4^{2-}, HCO_3^-, TH, COD, Ca^{2+}, Fe, and Mn; NO_3^- also displays moderately high loading. NH_4^+ and NO_3^- in shallow aquifers are typically derived from domestic sewage, chemical fertilizers, soil organic nitrogen, and atmospheric deposition [43], whereas Cl^- originates from halite dissolution, wastewater, seawater intrusion, or road-de-icing salts [44]. The mean water-table depth (2.17 m) is shallow, and most sampling wells are large-diameter domestic wells located within villages; thus, untreated domestic effluent can readily infiltrate. COD, a proxy for organic and reduced-matter contamination, exhibits an exceedance rate of 44.3%, indicating prominent organic pollution. Previous study indicates that COD is supplied by domestic and industrial effluents, agricultural runoff, urban stormwater, natural organic matter, and landfill leachate [45]. Fe and Mn are additionally introduced by industrial discharges (mining, steel making, electroplating) and corroded municipal pipelines. Shenzhen's leading industries—electronics, high-end equipment manufacturing, metal products, general machinery, and automotive production—generate large volumes of processed wastewater that, if inadequately treated, can readily contaminate the shallow aquifer. In addition, the concentrations of Fe and Mn in groundwater range from 0 to 11.5 mg/L and from 0.003 to 2.96 mg/L, respectively, with high spatial variability and high exceedance rates, indicating that they are affected by human activities (such as industrial wastewater). Therefore, Factor 1 is designated "sewage and industrial pollution". Source-contribution analysis indicates that this factor supplies 51.0% of NH_4^+, 65.1% of NO_2^-, 53.6% of Cl^-, 59.0% of SO_4^{2-}, 60.6% of HCO_3^-, 58.6% of TH, 55.3% of COD, 49.9% of Ca^{2+}, 75.6% of Fe, and 56.3% of Mn (Figure 7).

Table 3. Source profiles obtained from the PMF model.

Parameters	Factor 1	Factor 2	Factor 3
pH	2.04	2.60	1.36
DO	0.773	1.55	0.005
NO_3^-	0.0009	0	0.003
NH_4^+	0.00004	0.00001	0.00003
NO_2^-	0.010	0.003	0.002
Cl^-	7.50	4.66	1.82
SO_4^{2-}	0.0008	0.0006	0
COD	1.11	0.596	0.3
TDS	12.8	46.2	30.8
TH	21.4	10.7	4.41
HCO_3^-	11.4	1.55	5.85
K^+	0.006	1.05	0.207
Na^+	0.692	3.35	1.64
Ca^{2+}	4.00	2.93	1.08
Mg^{2+}	0	0.681	0.375
Fe	0.00006	0	0.00002
Mn	0.005	0.004	0.001
Possible sources	Sewage and and industrial pollution	Water–rock interaction	Agricultural fertilizer
Contribution (%)	43.9	37.0	19.1

Factor 2 explains 37.0% of the variance and is dominated by pH, DO, TDS, K^+, Na^+ and Mg^{2+}. Na^+, K^+, Mg^{2+}, and TDS are released primarily by water–rock interaction [46], whereas pH and DO are bulk physicochemical parameters that reflect the integrated outcome of aqueous chemical and biological processes rather than specific pollutant sources [47]. As analyzed in Section 4.2.2, the K^+, Na^+, and Mg^{2+} in the groundwater of the study area mainly originate from the weathering and dissolution of rocks. Thus, Factor 2 is consequently interpreted as "water–rock interaction". According to Figure 7, this factor contributes 43.6% of pH, 66.6% of DO, 51.3% of TDS, 83.1% of K^+, and 64.5% of Na^+.

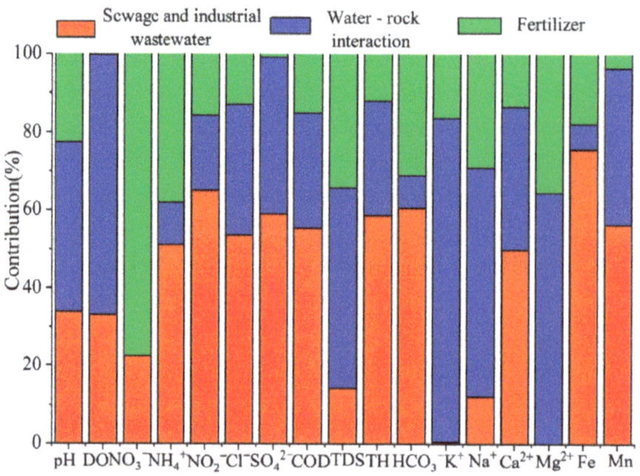

Figure 7. Contribution rates of groundwater pollution sources.

Factor 3 accounts for 19.1% of the variance and is heavily loaded on NO_3^-, with secondary contributions from NH_4^+, TDS, and Na^+. As discussed earlier, NO_3^- and NH_4^+ are introduced by chemical fertilizers, sewage, rain, etc. Because the concentration of NO_3^- in rain is low (1.47 mg/L) [25], and PC1 and PC3 are independent of each other. Due to PC1 represents sewage, therefore, Factor 3 is assigned to "agricultural fertilizer". Figure 7 shows that this source contributes 77.5% of NO_3^-, 38.1% of NH_4^+, 34.3% of TDS, and 28.9% of Na^+.

4.3.2. Spatial Distribution of Pollution Sources

Spatial interpolation of the factor scores for the three PMF-derived sources (PC1–PC3) was conducted to generate distribution maps of groundwater contamination across Shenzhen (Figure 8).

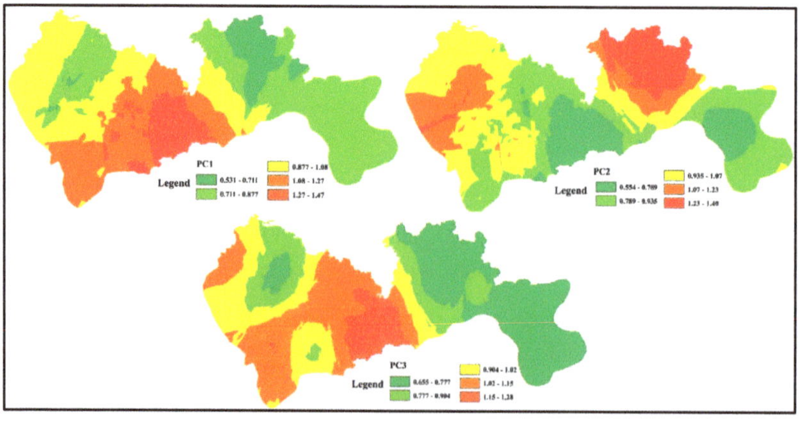

Figure 8. Spatial Distribution of Groundwater Pollution Sources.

High PC1 scores are concentrated around the peripheral districts of the central metropolitan area, indicating severe impacts from domestic sewage and industrial effluence. These peri-urban zones are densely populated and host numerous small-scale factories that often lack wastewater-treatment facilities; consequently, uncontrolled discharge constitutes the principal threat to groundwater quality. Elevated PC2 scores occur mainly in the north-eastern and north-western parts of the study area, reflecting regions where water–rock interaction governs the hydrochemical signature. High PC3 scores are observed southeast of the city core, demonstrating that intensive agricultural activity—and associated fertilizer use—is the dominant driver of nitrate contamination in that sector.

Overall, the spatial pattern of pollution sources is tightly coupled to land–use type. In highly urbanized districts, groundwater quality is controlled primarily by municipal and industrial wastewater; therefore, comprehensive collection and treatment of all sewage streams is the critical measure for preventing further deterioration. In agricultural areas, excessive fertilizer application is the key factor, so strict control of diffuse agricultural inputs is essential for groundwater protection.

5. Limitations of the Study

This study combined hydrochemical characterization with the PMF model to quantify and visualize groundwater pollution sources in a rapidly urbanizing area, thereby providing environmental managers with a clear basis for targeted control strategies. Nevertheless, three main limitations should be acknowledged:

(1) Sampling frequency was low; samples were collected only during the wet season. Consequently, temporal variability in groundwater quality and potential seasonal shifts in contaminant sources could not be resolved. In addition, due to confidentiality issues regarding the maps, we were unable to obtain high-precision land-use type maps for the study area. Consequently, we could not acquire the distances between sampling wells and specific human activity sites such as industrial parks, wastewater treatment plants, drainage networks, and landfills. This limitation prevented us from accurately analyzing the relationship between specific human activities and groundwater pollution.

(2) Although PMF successfully apportioned overall contaminant loads, the accuracy of source attribution could be further improved. In recent years, dual-isotope techniques (e.g., δ^{15} N and δ^{18} O of NO_3^-; δ^{34} S and δ^{18} O of SO_4^{2-}) coupled with receptor models have markedly refined source quantification for specific solutes [48,49]. Owing to budget constraints, such isotopic analyses were not conducted in the present work.

(3) The investigation generated an extensive hydrochemical data set, but background hydrogeochemical information remains incomplete. As a result, advanced machine learning algorithms (e.g., Random Forest) that could identify key drivers of water-quality degradation and predict future trends were not employed.

Future research should therefore: (i) implement high-frequency sampling campaigns covering both wet and dry seasons, (ii) integrate multi-isotope tracers with PMF or other receptor models, and (iii) apply machine learning approaches to achieve more robust source apportionment and predictive capability for groundwater quality in rapidly urbanizing regions.

6. Conclusions

This study has provided valuable insights into the hydrochemical evolution and pollution sources of groundwater in highly urbanized regions, using Shenzhen, China, as a case study. By employing a combination of multivariate statistical techniques and the Positive Matrix Factorization (PMF) model, we have elucidated the complex interactions between human activities and groundwater quality. The main conclusions of this study are as follows:

(1) The groundwater quality in highly urbanized areas shows a high exceedance rate of water quality parameters, with the exceedance rates of pH, NH_4^+, COD, Mn, and Fe being 67.1%, 44.3%, 44.3%, 34.3% and 31.4%, respectively. The Water Quality Index assessment revealed that 32.9% of the groundwater samples were classified as poor, underscoring the substantial impact of human activities on groundwater quality. Land-use types significantly affected groundwater quality; the concentration of COD, NO_3^-, Mn, and Fe in the urban areas were significantly higher than in the agricultural and forested areas.

(2) Rock weathering is the primary control on the groundwater chemistry in the region, and human activities also have an important impact on the formation of groundwater chemistry. Among them, silicate weathering constitutes the principal source of Na^+ and K^+, while Ca^{2+}, Mg^{2+} and HCO_3^- primarily originate from the dissolution of carbonate minerals. Moreover, the weathering and dissolution of evaporites are one of the main sources of SO_4^{2-}. In addition, domestic sewage, industrial wastewater, and fertilizers are the main sources of NO_3^- in the groundwater of the region.

(3) The PMF model identified three major pollution sources: domestic and industrial wastewater (43.9%), enhanced water–rock interactions (37.0%), and agricultural fertilizers (19.1%). Geostatistical spatial interpolation techniques further demonstrated that groundwater quality in urban areas is primarily controlled by the discharge of domestic and industrial wastewater, while in agricultural areas, excessive fertilizer application is the main driver of groundwater degradation. Therefore, future research should advocate for strict control and compliance policies for municipal sewage discharge, as well as optimize agricultural fertilization strategies, such as formula fertilization and the use of slow-release fertilizers, to prevent further deterioration of groundwater quality.

Although this study identified groundwater pollution sources in highly urbanized regions, it faced several limitations: low sampling frequency restricted to the wet season, lack of isotopic analyses due to budget constraints, and incomplete background data that hindered the application of advanced machine learning techniques. Future work should include year-round high-frequency sampling, the integration of multi-isotope tracers with PMF, and the application of machine learning methods to enhance source apportionment and predictive capabilities, thereby supporting improved groundwater management in urban areas.

Supplementary Materials: The following supporting information can be downloaded at: https://www.mdpi.com/article/10.3390/w17202945/s1. Table S1: Hydrochemical parameters, analytical method, equipment and detection limits; Table S2: Table of weight values for groundwater chemistry indicators.

Author Contributions: Y.W.: Investigation, methodology, software, data curation, and writing—original draft preparation. Y.L.: Investigation, software, and methodology. L.Z.: Investigation, methodology, and software. C.L.: Investigation, methodology, and data curation. Q.M.: Investigation and methodology. J.Y.: Supervision, methodology, and writing—reviewing and editing. L.W.: Investigation and methodology. All authors have read and agreed to the published version of the manuscript.

Funding: This research received no external funding.

Data Availability Statement: The original contributions presented in this study are included in the article/Supplementary Material. Further inquiries can be directed to the corresponding author.

Acknowledgments: The authors gratefully acknowledge the editor and anonymous reviewers for their valuable comments on this manuscript. The authors also appreciate financial support from the different organizations.

Conflicts of Interest: The authors declare no conflicts of interest.

References

1. Peng, H.Q.; Yang, W.; Ferrer, A.; Xiong, S.; Li, X.; Niu, G.; Lu, T. Hydrochemical characteristics and health risk assessment of groundwater in karst areas of southwest China: A case study of Bama, Guangxi. *J. Clean. Prod.* **2022**, *341*, 130872. [CrossRef]
2. Zhang, Q.Q.; Wang, H.W.; Xu, Z.; Li, G.; Yang, M.N.; Liu, J. Quantitative identification of groundwater contamination sources by combining isotope tracer technique with PMF model in an arid area of northwestern China. *J. Environ. Manag.* **2023**, *325*, 116588. [CrossRef]
3. Panneerselvam, B.; Muniraj, K.; Duraisamy, K.; Pande, C.; Karuppannan, S.; Thomas, M. An integrated approach to explore the suitability of nitrate-contaminated groundwater for drinking purposes in a semiarid region of India. *Environ. Geochem. Health* **2023**, *45*, 647–663. [CrossRef]
4. Hasan, M.S.U.; Rai, A.K. Groundwater quality assessment in the Lower Ganga Basin using entropy information theory and GIS. *J. Clean. Prod.* **2020**, *274*, 123077. [CrossRef]
5. Wang, L.; Zhang, Q.Q.; Wang, H.W. Rapid urbanization has changed the driving factors of groundwater chemical evolution in the large groundwater depression funnel area of northern China. *Water* **2023**, *15*, 2917. [CrossRef]
6. Torres-Martínez, J.A.; Mora, A.; Knappett, P.S.K.; Ornelas-Soto, N.; Mahlknecht, J. Tracking nitrate and sulfate sources in groundwater of an urbanized valley using a multi-tracer approach combined with a Bayesian isotope mixing model. *Water Res.* **2020**, *182*, 115962. [CrossRef]
7. Wang, F.; Liu, L.; Xu, W.; Li, Y.; Ruan, Q.; Cao, W. Multiple stable isotopic approaches for tracing nitrate contamination sources: Implications for nitrogen management in complex watersheds. *Ecotoxicol. Environ. Safe* **2024**, *269*, 115822. [CrossRef] [PubMed]
8. Li, Y.; Tu, Y.; Sun, T.; Duan, Y.; Kou, J.; Li, W.; Gao, J. Source apportionment of organic carbon and nitrogen in sediments from river and lake in the highly urbanized Changjiang Delta. *J. Hazard. Mater.* **2024**, *478*, 135590. [CrossRef] [PubMed]
9. Li, X.; Liang, G.; He, B.; Ning, Y.; Yang, Y.; Wang, L.; Wang, G. Recent advances in groundwater pollution research using machine learning from 2000 to 2023: A bibliometric analysis. *Environ. Res.* **2025**, *267*, 120683. [CrossRef]
10. Wang, L.; Wang, Q.; Zheng, D. Study on the pollution mechanism and driving factors of groundwater quality in typical industrial areas of China. *Water* **2025**, *17*, 1420. [CrossRef]
11. Marazuela, M.Á.; Jiménez, J.; Baquedano, C.; Martínez-León, J.; Gasco-Cavero, S.; Cruz-Pérez, N.; Santamarta, J.C.; Garcia-Gil, A. Hydrogeological and hydrochemical processes affecting groundwater quality on volcanic islands: Insights from El Hierro (Canary Islands, Spain). *J. Hydrol.* **2025**, *654*, 132874. [CrossRef]
12. Khaskheli, A.; Khuhawar, T.M.J.; Khuhawar, M.Y.; Khuhawar, F.Y.; Lanjwani, M.F.; Landar, A.Q. Evaluation of groundwater quality of Mirpurkhas District, Sindh, Pakistan, for drinking, irrigation and multivariate analysis. *Arab. J. Geosci.* **2025**, *18*, 136. [CrossRef]
13. Deng, Y.; Lu, Y.; Du, X.; Ye, X.; Feng, J. Identifying spatial patterns and driving factors of anthropogenic impacts on the groundwater environment based on groundwater chemical kinetics. *J. Clean. Prod.* **2025**, *486*, 144436. [CrossRef]
14. Wang, Q.; Ou, X.; Wu, C.; Yao, S.; Huang, S.; Zhu, C.; Liao, Z.; Wang, K.; Zhang, S.; Chen, J. Machine learning-driven PMF modeling for accurate and objective source identification of VOCs. *Atmos. Pollut. Res.* **2025**, 102721. [CrossRef]
15. Lv, X.; Li, C.; Wang, H.; Yang, S.; Wang, S.; Yang, C.; Teng, F.; Zhao, J.; Xu, X.; Zhang, H.; et al. Source apportionment and risk quantification of soil heavy metals using SOM-PMF model: Implications for ecological and human health management. *Environ. Monit. Assess.* **2025**, *197*, 868. [CrossRef]
16. Liu, Z.; Wang, L.; Yan, M.; Ma, B.; Cao, R. Source apportionment of soil heavy metals based on multivariate statistical analysis and the PMF model: A case study of the Nanyang Basin, China. *Environ. Technol. Innov.* **2024**, *33*, 103537. [CrossRef]
17. Wang, A.; Wang, J.; Luan, B.; Wang, S.; Yang, D.; Wei, Z. Classification of pollution sources and their contributions to surface water quality using APCS-MLR and PMF model in a drinking water source area in southeastern China. *Water* **2024**, *16*, 1356. [CrossRef]
18. Zhang, Q.Q.; Miao, L.; Wang, H.W.; Hou, J.; Li, Y. How rapid urbanization drives deteriorating groundwater quality in a provincial capital of China. *Pol. J. Environ. Stud.* **2020**, *29*, 441–450. [CrossRef]
19. Attard, G.; Winiarski, T.; Rossier, Y.; Eisenlohr, L. Impact of underground structures on the flow of urban groundwater. *J. Hydrol.* **2016**, *24*, 5–19. [CrossRef]

20. Ruan, D.; Bian, J.; Wang, Y.; Wu, J.; Gu, Z. Identification of groundwater pollution sources and health risk assessment in the Songnen Plain based on PCA-APCS-MLR and trapezoidal fuzzy number-Monte Carlo stochastic simulation model. *J. Hydrol.* **2024**, *632*, 130897. [CrossRef]
21. Jiang, Z.; Wang, C.; Liu, Y.; Feng, Z.; Ji, C.; Zhang, H. Study on the raw water allocation and optimization in Shenzhen city, China. *Water* **2019**, *11*, 1426. [CrossRef]
22. Zhang, Q.Q.; Sun, J.; Liu, J.; Huang, G.; Lu, C.; Zhang, Y. Driving mechanism and sources of groundwater nitrate contamination in the rapidly urbanized region of south China. *J. Contam. Hydrol.* **2015**, *182*, 221–230. [CrossRef] [PubMed]
23. *GB/T 14848-2017*; Standard for Groundwater Quality. Ministry of Natural Resources of the People's Republic of China: Beijing, China, 2017.
24. World Health Organization. *Guidelines for Drinking-Water Quality*, 4th ed.; World Health Organization: Geneva, Switzerland, 2011.
25. Zhou, X.; Xu, Z.; Liu, W.; Wu, Y.; Zhao, T.; Jiang, H.; Zhang, X.; Zhang, J.; Zhou, L.; Wang, Y. Chemical composition of precipitation in Shenzhen, a coastal mega-city in South China: Influence of urbanization and anthropogenic activities on acidity and ionic composition. *Sci. Total Environ.* **2019**, *662*, 218–226. [CrossRef] [PubMed]
26. Zhang, Q.Q.; Wang, H.W.; Wang, Y.C.; Yang, M.N.; Zhu, L. Groundwater quality assessment and pollution source apportionment in an intensely exploited region of northern China. *Environ. Sci. Pollut. Res.* **2017**, *24*, 16639–16650. [CrossRef]
27. Wang, Q.; Wang, M.; Li, Y.; Guo, B.; Li, H.; Liu, Y.; Zhao, L.; Ma, C.; Yuan, Z. Study on Natural Background Levels and Mechanisms of Groundwater Contamination in an Overexploited Aquifer Region: A Case Study of Xingtai City, North China Plain. *Water* **2025**, *17*, 2836. [CrossRef]
28. Wang, L.; Yang, N.; Zhao, Y.; Zhang, Q. Research on the Features and Driving Factors of Shallow Groundwater Quality in Arid Areas, Northwest China. *Water* **2025**, *17*, 934. [CrossRef]
29. Rai, S.C.; Saha, A.K. Impact of urban sprawl on groundwater quality: A case study of Faridabad city, National Capital Region of Delhi. *Arab. J. Geosci.* **2015**, *8*, 8039–8045. [CrossRef]
30. Lockhart, K.; King, A.; Harter, T. Identifying sources of groundwater nitrate contamination in a large alluvial groundwater basin with highly diversified intensive agricultural production. *J. Contam. Hydrol.* **2013**, *151*, 140–154. [CrossRef]
31. Zhang, Q.Q.; Wang, H.W. Assessment of sources and transformation of nitrate in the alluvial–pluvial fan region of north China using a multi-isotope approach. *J. Environ. Sci.* **2020**, *89*, 9–22. [CrossRef]
32. Shi, X.; Gao, Y.; Qian, H.; Chen, J.; Li, W.; Li, S.; Liu, Y. Elucidating the hydrochemistry and REE evolution of surface water and groundwater affected by acid mine drainage. *Environ. Pollut.* **2025**, *366*, 125495. [CrossRef]
33. Zhou, B.; Wang, H.W.; Zhang, Q.Q. Assessment of the evolution of groundwater chemistry and its controlling factors in the Huangshui River Basin of northwestern China, using hydrochemistry and multivariate statistical techniques. *Int. J. Environ. Res. Public Health* **2021**, *18*, 7551. [CrossRef] [PubMed]
34. Sun, R.; Zhang, X.Q.; Wu, Y.H. Major ion chemistry of water and its controlling factors in the Yamzhog Yumco basin, south Tibet. *J. Lake Sci.* **2012**, *24*, 600–608. [CrossRef]
35. Sun, H.; Huffine, M.; Husch, J.; Sinpatanasakul, L. Na/Cl molar ratio changes during a salting cycle and its application to the estimation of sodium retention in salted watersheds. *J. Contam. Hydrol.* **2012**, *136*, 96–105. [CrossRef] [PubMed]
36. Yang, J.; Zhao, W.; Liang, X.; Xu, F. The hydrochemical characteristics and formation mechanism of highly mineralized coal mine water in semi-arid regions in northwest China. *Water* **2024**, *16*, 2244. [CrossRef]
37. Shen, B.; Wu, J.; Zhan, S.; Jin, M.; Saparov, A.S.; Abuduwaili, J. Spatial variations and controls on the hydrochemistry of surface waters across the Ili–Balkhash Basin, arid Central Asia. *J. Hydrol.* **2021**, *600*, 126565. [CrossRef]
38. Lakhdari, A.S.; Bouselsal, B.; Saibi, H.; Ouarekh, M. Assessment of groundwater quality and hydrogeochemical properties in the adrar continental intercalaire aquifer of the Algerian Sahara. *Appl Water Sci.* **2025**, *15*, 175. [CrossRef]
39. Boualem, B.; Egbueri, J.C. Graphical, statistical and index-based techniques integrated for identifying the hydrochemical fingerprints and groundwater quality of in Salah, Algerian Sahara. *Environ. Geochem. Health* **2024**, *46*, 158. [CrossRef]
40. Bouselsal, B.; Satouh, A.; Egbueri, J.C. Evaluating water quality, mineralization mechanisms, and potential health risks of nitrate contamination in the Continental Intercalaire aquifer of Reggane, Algeria. *Environ. Earth Sci.* **2024**, *83*, 539. [CrossRef]
41. Saccon, P.; Leis, A.; Marca, A.; Kaiser, J.; Campisi, L.; Böttcher, M.E.; Savarino, J.; Escher, P.; Eisenhauer, A.; Erbland, J. Multi-isotope approach for the identification and characterisation of nitrate pollution sources in the Marano lagoon (Italy) and parts of its catchment area. *Appl. Geochem.* **2013**, *34*, 75–89. [CrossRef]
42. Jiang, W.; Li, B.; Zhang, Z.; Zhang, Y. Hydrochemical characteristics, controlling factors and groundwater sources of Zaozigou gold mine. *Sustainability* **2024**, *16*, 7989. [CrossRef]
43. Wu, Y.; Ju, H.; Jiang, H.; Zhang, G.; Qi, P.; Li, Z. Identifying nitrate sources and transformations in an agricultural watershed in Northeast China: Insights from multiple isotopes. *J. Environ. Manag.* **2023**, *340*, 118023. [CrossRef] [PubMed]

44. Jin, Z.; Xue, Q.; Chen, L.; Jin, M.; Li, F. Using dual isotopes to evaluate sources and transformations of nitrate in the West Lake watershed, eastern China. *J. Contam. Hydrol.* **2015**, *177–178*, 64–75. [CrossRef]
45. Lee, J.; Lee, S.; Yu, S.; Rhew, D. Relationships between water quality parameters in rivers and lakes: BOD5, COD, NBOPs, and TOC. *Environ. Monit. Assess.* **2016**, *188*, 252. [CrossRef]
46. Fan, W.; Zhou, J.; Zheng, J.; Guo, Y.; Hu, L.; Shan, R. Hydrochemical characteristics, control factors and health risk assessment of groundwater in typical arid region Hotan Area, Chinese Xinjiang. *Environ. Pollut.* **2024**, *363*, 125301. [CrossRef]
47. Xiao, J.; Jin, Z.D.; Wang, J.; Zhang, F. Hydrochemical characteristics, controlling factors and solute sources of groundwater within the Tarim River Basin in the extreme arid region, NW Tibetan Plateau. *Quat. Int.* **2015**, *380*, 237–246. [CrossRef]
48. Ji, X.; Shu, L.; Chen, W.; Chen, Z.; Shang, X.; Yang, Y.; Dahlgren, R.A.; Zhang, M. Nitrate pollution source apportionment, uncertainty and sensitivity analysis across a rural–urban river network based on $\delta^{15}N/\delta^{18}O\text{–}NO_3^-$ isotopes and SIAR modeling. *J. Hazard. Mater.* **2022**, *438*, 129480. [CrossRef]
49. Jung, H.; Kim, J.; Lee, J. Seasonal and spatial contributions of sulfate and trace elements in river water in mining districts: Insights from hydrogeochemical and isotopic analysis based on statistical models. *J. Hazard. Mater.* **2025**, *488*, 137246. [CrossRef] [PubMed]

water

MDPI

Article

Key Controlling Factors and Sources of Water Quality in Agricultural Rivers: A Study on the Water Source Area for the South-to-North Water Transfer Project

Congcong Yang [1,2], Zeliang Qu [3], Xiaoyu Shi [4], Li Yang [1], Nan Yang [3], Fan Yang [5,*] and Qianqian Zhang [6,*]

[1] Shandong Provincial Geo-Mineral Engineering Exploration Institute (801 Institute of Hydrogeology and Engineering Geology, Shandong Provincial Bureau of Geology and Mineral Resources), Jinan 250014, China; congyang2024@163.com (C.Y.); yangli_202509@163.com (L.Y.)
[2] Shandong Engineering Research Center for Environmental Protectionand Remediation on Groundwater, Jinan 250014, China
[3] Shandong Coalfield Geological Planning Research Institute, Jinan 250000, China; 18660806081@163.com (Z.Q.)
[4] Shandong Provincial NO. 4 Institute of Geological and Mineral Survey, Weifang 261021, China; m18865365053@163.com
[5] Huaxin College, HeBei GEO University, Shijiazhuang 050700, China
[6] Hebei and China Geological Survey Key Laboratory of Groundwater Remediation, Institute of Hydrogeology and Environmental Geology, Chinese Academy of Geological Sciences, Shijiazhuang 050061, China
* Correspondence: yangfan130184@163.com (F.Y.); zhangqianqian@mail.cgs.gov.cn (Q.Z.)

Abstract: River water quality is a direct determinant of both drinking water security and regional economic vitality. However, the hydrochemical trajectories and solute provenance of agricultural streams remain only fragmentarily understood. Here, we examine the Jinqian River—a representative agricultural tributary of the Danjiangkou Reservoir source area for the South-to-North Water Diversion Project—by coupling hydrochemistry with multivariate statistics techniques. The results revealed that the pH values of the river water ranged from 7.55 to 8.30, indicating a weakly alkaline condition. During all three hydrological periods, the concentrations of total nitrogen (TN) exceeded the limits set by the Class III surface water quality standards in China, suggesting that the agricultural river has been significantly impacted by human activities. Solute dynamics followed three rainfall-modulated patterns: (i) dilution-driven decreases in the flood season (e.g., Na^+), (ii) concentration via flushing or evaporative concentration (e.g., SO_4^{2-}), and (iii) reservoir-induced damping of seasonal contrasts (e.g., TN), the latter attributable to nitrogen retention behind upstream dams. Geochemical fingerprints reveal that Cl^- and Na^+ originate predominantly from halite dissolution; Ca^{2+}, Mg^{2+} and HCO_3^- from coupled carbonate–silicate weathering; and SO_4^{2-} from evaporite dissolution. Principal component analysis distills four dominant quality controlling factors: agricultural fertilizers, halite weathering, evaporite dissolution, and domestic effluent. These findings provide a quantitative basis for managing nutrient and salt fluxes in agricultural rivers and for safeguarding water sustainability within water-diversion source regions.

Keywords: water quality; sources; human activities; river; controlling factors

1. Introduction

Hydrochemical characterization of river systems underpins both rational water resource management and effective environmental protection [1]. Agricultural tributaries are of particular concern because they must simultaneously sustain ecological functions and support intensive human use. These channels are frequently confronted with multiple challenges, such as agricultural runoff, industrial emissions, and domestic sewage. These factors can alter the chemical composition of the water, thereby leading to water quality degradation [2–4]. Accurate and comprehensive investigation of river water chemistry is crucial for assessing the health of aquatic ecosystems, ensuring the safety of drinking water, and guiding sustainable agricultural practices [5,6]. Within the context of large-scale water-diversion schemes such as China's South-to-North project, maintaining pristine source-water quality is a prerequisite for long-term engineering success and sustainability [7,8].

The composition of river water reflects a dynamic interplay between natural templates and anthropogenic forcings. Natural factors, including geological composition, hydrological conditions, and climatic variability, determine the baseline water quality and natural variations of water bodies [9]. Anthropogenic factors, such as agricultural activities, industrial processes, and urban development, introduce additional substances and pollutants into river systems. For instance, agricultural runoff can introduce nutrients, pesticides, and sediments, while industrial emissions may contain heavy metals and organic compounds [10]. Previous studies have identified a variety of factors that influence river water chemistry. However, the relative importance of these factors may vary significantly across different regions and river systems.

Preventing hydrochemical degradation hinges on the unambiguous identification of solute provenance. Accurate identification and quantification of these sources are essential steps in developing effective pollution control strategies [11,12]. Globally, researchers have exploited chemical fingerprints, stable-isotope tracers, and statistical unmixing models to reconstruct the origin of dissolved and suspended constituents [13–15]. However, the current understanding of the patterns of water quality variation and the sources of substances in complex agricultural river systems remains inadequate.

To fill these knowledge gaps, this study focuses on the Jinqian River, a typical agricultural river in the water source area of the South-to-North Water Diversion Project. This study investigates the characteristics and controlling factors of water chemistry in agricultural rivers by conducting periodic sampling and applying hydrochemical and multivariate statistical techniques. The primary objectives of this study were to systematically investigate the water quality characteristics and seasonal variations in agricultural rivers, identify the controlling factors influencing these variations, and elucidate the sources of substances affecting agricultural rivers. The findings of this study provide a scientific basis for understanding and managing water quality in agricultural river systems and can be applied to similar regions worldwide.

2. Materials and Methods

2.1. Description of the Study Region

The Jinqian River, a left-bank tributary of the Hanjiang, drains the southern slopes of the Qinling Mountains (109°09′–110°19′ E, 33°11′–33°52′ N). Originating at Jinjing River in Zhashui County, Shaanxi Province, the river traces a 261 km southeasterly course through Zhashui and Shanyang counties before crossing into Hubei and joining the Han River at Yunxi County (Figure 1). Its 5610 km^2 catchment experiences a mean annual air temperature of 13.1 °C, 752 mm of precipitation, and 889 mm of evaporation. Roughly four-fifths of the rainfall is delivered between May and October, imparting a pronounced monsoon signature to the hydrological regime.

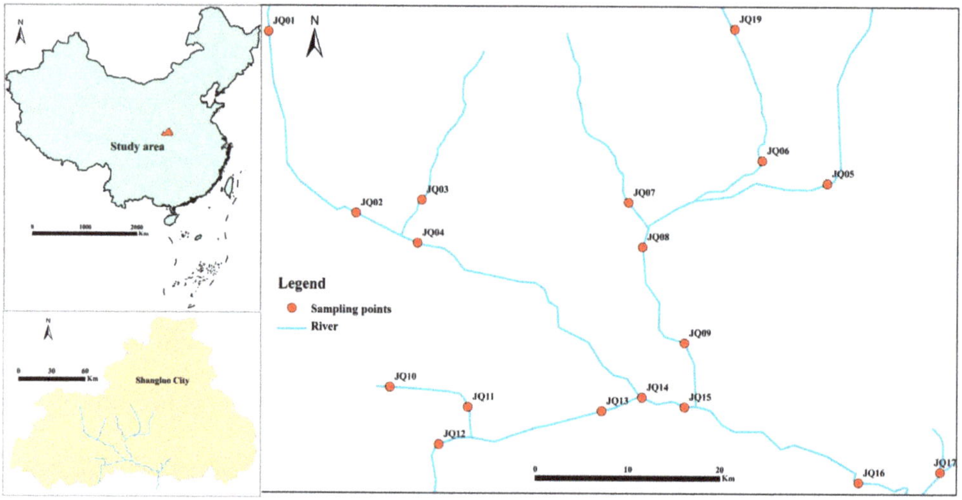

Figure 1. Location of the study area and distribution of sampling sites.

2.2. Specimen Collection and Testing

2.2.1. Specimen Collection Strategy

Field campaigns were conducted during three hydrologic windows in 2022—dry season (January), normal season (May), and flood season (August)—with one sampling event per season. Along the Jinqian main stem and its tributaries, 19 stations were established (12 on the trunk, 7 on feeders) and sampled sequentially from headwaters to confluence. Ancillary materials comprised four bulk-rainfall collectors and eight municipal effluent grabs. At each river site, water was withdrawn 0.5 m below the mid-channel surface with a 5-L polyethylene sampler, then transferred to pre-rinsed 2-L HDPE bottles that had been triple-washed with de-ionized water and conditioned three times with on-site river water. Samples were kept in portable coolers at 4 °C, transported to the laboratory, and analyzed within seven days of collection.

2.2.2. Analysis of Water Chemistry

The hydrochemical analysis included pH, total dissolved solids (TDS), major ions (K^+, Na^+, Ca^{2+}, Mg^{2+}, NH_4^+, Cl^-, SO_4^{2-}, NO_3^-, HCO_3^-, F^-), total nitrogen (TN), and dissolved organic carbon (DOC).

In the field, pH was assessed with a portable multi-parameter analyzer (Hach-HQ40D, Hach Company, Shanghai, China). The concentrations of cations, including K^+, Na^+, NH_4^+, Ca^{2+}, and Mg^{2+}, were quantified using atomic absorption spectrophotometry (Agilent 7500ce ICP-MS, Agilent Technologies, Tokyo, Japan). Meanwhile, anions such as Cl^-, SO_4^{2-}, F^-, and NO_3^- were evaluated via ion chromatography (CIC-D180, Qingdao Shenghan Chromatography Technology Co., Ltd., Qingdao, China). HCO_3^- levels were determined through hydrochloric acid titration. The TDS were calculated using the drying-gravimetric technique. For TN analysis, the alkaline potassium persulfate digestion combined with UV spectrophotometry was employed (Spectrophotometer, Jingke L9, Shanghai, China). Lastly, DOC was measured by high-temperature catalytic oxidation with a TOC-VCPH analyzer (Shimadzu Corporation, Kyoto, Japan). The detection limits of the analytical methods are summarized in Supplementary Table S1.

2.3. Data Analytics

2.3.1. Water Quality Index (WQI)

The WQI condenses a suite of hydrochemical variables into a single metric that reflects the suitability of surface water for potable use. Each parameter is assigned a weight (Wi) proportional to its perceived health relevance. Thresholds follow the Chinese Grade-III surface-water standard (GB 3838-2002) [16]; analytes not regulated therein adopt guideline values from World Health Organization published in 2017 were utilized [17].

The procedural steps are as follows:

$$RW_i = \frac{W_i}{\sum_{i=1}^{n} W_i}$$

$$Q_i = C_i/S_i \times 100$$

$$SI_i = W_i \times Q_i$$

$$WQI = \sum SI_i$$

2.3.2. Statistical Analysis

In this study, we employed the Kolmogorov–Smirnov (KS) test and one-way analysis of variance (ANOVA) to examine the seasonal variations in the concentration values of water quality parameters, including pH, SO_4^{2-}, Na^+, HCO_3^-, NO_3^-, and TN, in river water. Initially, the KS test was utilized to evaluate the normality of the data. The results revealed that the concentration values of all six water chemistry parameters conformed to a normal distribution ($p > 0.05$). Thereafter, one-way ANOVA was conducted to assess the seasonal differences in these water chemistry parameters. Principal component analysis was used to identify the main controlling factors affecting the water quality of the Jinqian River. All the statistical analyses were performed using R software (version 4.4.1, New Zealand).

3. Result

3.1. The Hydrochemical Characteristics of Jinqian River

Table 1 summarizes the hydrochemical signatures of the Jinqian River across the three hydrologic seasons. pH drifted within narrow, slightly alkaline bands: 7.55–7.99 (dry season), 7.96–8.25 (normal season), and 7.74–8.30 (flood season), with respective means of 7.81, 8.10, and 8.13, respectively. The concentrations of DOC during the three periods ranged from 0.765 to 2.38 mg/L, 0.740 to 4.00 mg/L, and 0.700 to 3.23 mg/L, with mean values of 1.14 mg/L, 1.24 mg/L, and 1.31 mg/L, respectively, exhibiting a pattern of flood season > normal season > dry season. The mean concentrations of TN in the three hydrological periods were 3.32 mg/L, 3.22 mg/L, and 3.23 mg/L, respectively, showing a trend of dry season > flood season > normal season. The concentrations of TN in all three hydrological periods exceeded the limits of Class III surface water quality standards in China, indicating a substantial impact of human activities on the water body. Cationic abundance followed the sequence $Ca^{2+} > Mg^{2+} > Na^+ > K^+ > NH_4^+$ in all campaigns, while anions ranked $HCO_3^- > SO_4^{2-} > NO_3 > Cl^- > F^-$. Cl^- exhibited substantial spatial variability in all three periods (CV > 50%), suggesting that the river water may have been continuously influenced by localized pollution sources, which caused strong spatial heterogeneity. The mean concentrations of TDS followed the order of normal season (278 mg/L) > dry season (275 mg/L) > flood season (249 mg/L), which further indicates that precipitation has a certain diluting effect on the hydrochemical components of the water body.

Table 1. Statistical table of characteristics of water quality variation in Jinqian River.

Parameters	Units	Standard	Dry Season			Normal Season			Flood Season		
			Range	Mean	CV	Range	Mean	CV	Range	Mean	CV
pH	—	6~9	7.55–7.99	7.81	1.61	7.96–8.25	8.10	1.00	7.74–8.3	8.13	1.51
K^+	mg/L	12	0.680–3.87	2.00	45.7	0.650–3.20	1.63	46.5	0.720–6.44	2.01	63.9
Na^+	mg/L	200	4.47–15.9	7.93	30.9	2.93–10.6	6.24	33.1	2.68–18.0	6.02	51.9
Ca^{2+}	mg/L	75	36.3–58.7	48.9	9.89	41.3–55.1	49.3	6.56	41.0–51.9	47.6	6.59
Mg^{2+}	mg/L	50	8.44–16.2	11.6	18.3	7.36–13.7	10.9	14.0	5.27–14.1	10.4	26.0
Cl^-	mg/L	≤250	3.86–24.9	7.63	67.0	2.52–14.6	6.51	50.1	3.86–21.7	7.45	52.5
SO_4^{2-}	mg/L	≤250	18.3–61.6	39.4	26.0	24.0–70.2	50.0	21.6	20.9–80.6	56.0	25.7
HCO_3^-	mg/L	—	146–180	165	5.78	147–176	157	6.43	131–159	145	6.37
NO_3^-	mg/L	≤44.3	6.07–15.4	8.98	27.3	2.27–15.2	7.54	38.7	6.56–17.7	9.45	29.3
NH_4^+	mg/L	≤1.29	0.352–1.33	0.710	38.7	0.303–1.20	0.720	36.7	0.280–1.36	0.620	49.8
TN	mg/L	≤1	2.56–3.84	3.32	10.6	2.79–4.18	3.22	9.09	2.56–4.01	3.23	11.6
F^-	mg/L	≤1	0.893–1.12	0.990	6.59	0.399–1.34	0.810	33.6	0.599–2.26	1.02	38.2
DOC	mg/L	—	0.765–2.38	1.14	32.0	0.740–4.00	1.24	57.5	0.700–3.23	1.31	44.9
TDS	mg/L	≤1000	160–357	275	16.6	169–349	278	15.1	160–357	249	16.6

Note: CV: Coefficient of variation; Standard: Class III standard of the national surface water quality standard, Chinese (GB 3838−2002) [16].

3.2. *Evaluation of Water Quality in the Jinqian River*

Water Quality Index (WQI) screening shows that "good" sites predominate in all campaigns, but their proportion declines temporally from 100% (dry season) to 94.7% (normal season) and 89.5% (flood season). Stations classed as "poor" emerge only during normal and flood season phases (5.3% and 10.5%, respectively) (Table 2), underscoring the dual control of hydrologic dilution and episodic external inputs on river water quality.

Table 2. Water quality evaluation results of the Jinqian River in different seasons.

Water Quality Level	Comprehensive Water Quality Score	Proportion of Sites in 3 Periods (%)		
		Dry Season	Normal Season	Flood Season
Excellent water	<50	0	0	0
Good water	50–100	100	94.7	89.5
Poor water	100.1–200	0	5.26	10.5
Very poor water	200.1–300	0	0	0
Water unsuitable for drinking purposes	>300	0	0	0

3.3. *The Hydrochemical Type of the Jinqian River*

The Piper trilinear diagram classifies the hydrochemical types of river water based on the ionic composition and concentration ratios of the water. As shown in Figure 2, the sample points from all three hydrological periods predominantly fall within the region dominated by HCO_3–Ca·Mg type water, indicating that this type is the primary hydrochemical type in the study area. Further analysis of the hydrochemical types using the Schoeller classification method revealed that during the dry season, the water type is mainly HCO_3–Ca (accounting for 42.1%), while in the normal and flood seasons, the dominant type is $HCO_3 \cdot SO_4$–Ca·Mg, with respective proportions of 36.8% and 52.6%. The seasonal variation in hydrochemical types may be related to the increased dissolution of gypsum in the strata due to higher rainfall, as well as the higher evaporation intensity caused by higher temperatures during the flood and normal seasons [18].

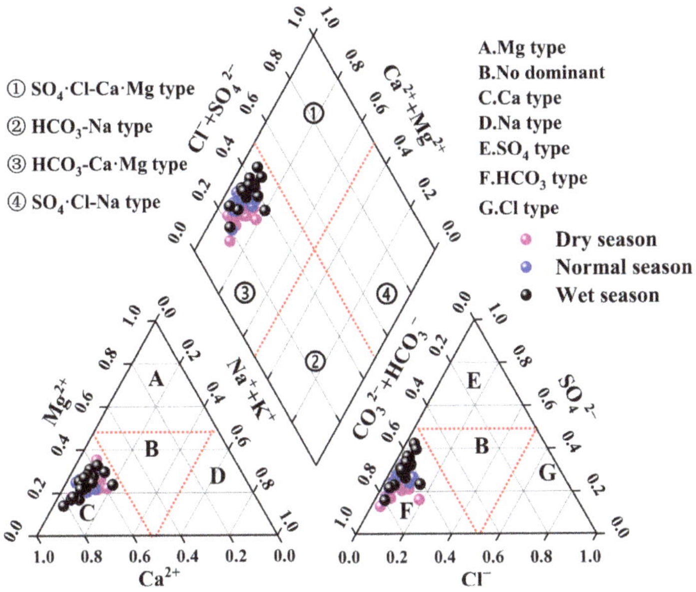

Figure 2. Hydrochemical type of the Jinqian River.

4. Discussion

4.1. The Temporal and Spatial Variations of Water Quality in the Jinqian River

The temporal variation of river water quality is an important aspect in water environment research, as it reflects the impact of natural processes and human activities on river water quality. In this study, we examined the temporal variations of 14 water quality parameters and summarized 4 seasonal variation characteristics of agricultural rivers. The first characteristic is that the dry season has significantly lower values than the flood ($p < 0.001$) and normal seasons ($p < 0.01$), as evidenced by the representative indicators of pH and SO_4^{2-} (Figure 3a,b). This result is consistent with that reported by Esquivel-Hernández et al., who found that during the flood season, storm runoff transports a large number of alkaline substances from roads and soils into the river, thereby causing an increase in the pH of the river [19]. The sulfate exhibited similar variations to pH, mainly due to abundant precipitation and increased surface runoff during the flood and normal seasons. These runoffs carried a large amount of sulfate from the soil strata into the river, leading to an increase in sulfate concentration. As detailed in Section 4.2, the study found that sulfate in the Jinqian River primarily originates from sulfate minerals in the soil. The second variation is that the concentration of water quality parameters is significantly higher only in the dry season than in the flood season, with the representative indicator being Na^+ ($p < 0.05$) (Figure 3c). This is mainly related to the discharge of domestic sewage from surrounding villages during the dry season and the dilution of river water by rainwater during the flood season [20]. The third variation characteristic is that the concentration of water quality parameters in the dry season is significantly higher than that in the flood season ($p < 0.001$) and normal season ($p < 0.05$), and the normal season is significantly higher than the flood season ($p < 0.001$), with the representative indicator being HCO_3^- (Figure 3d). This is primarily attributed to the dilution effect of rainfall and the transport of residual acidic substances (such as sulfates and nitrates) from agricultural fields into the river by storm runoff during the flood season, which further reduces the concentration of

HCO_3^- in the river water. The fourth seasonal variation is that the concentrations of water quality parameters show no difference in the three hydrological periods ($p > 0.05$), with representative indicators being NO_3^- and TN (Figure 3e,f). This indicates that the nitrogen input and transformation processes in the region are relatively stable. In addition, there are multiple hydropower stations built on the Jinqian River, and the reservoirs have a strong buffering and retention capacity for nitrogen, which further weakens the seasonal variation of nitrogen [21].

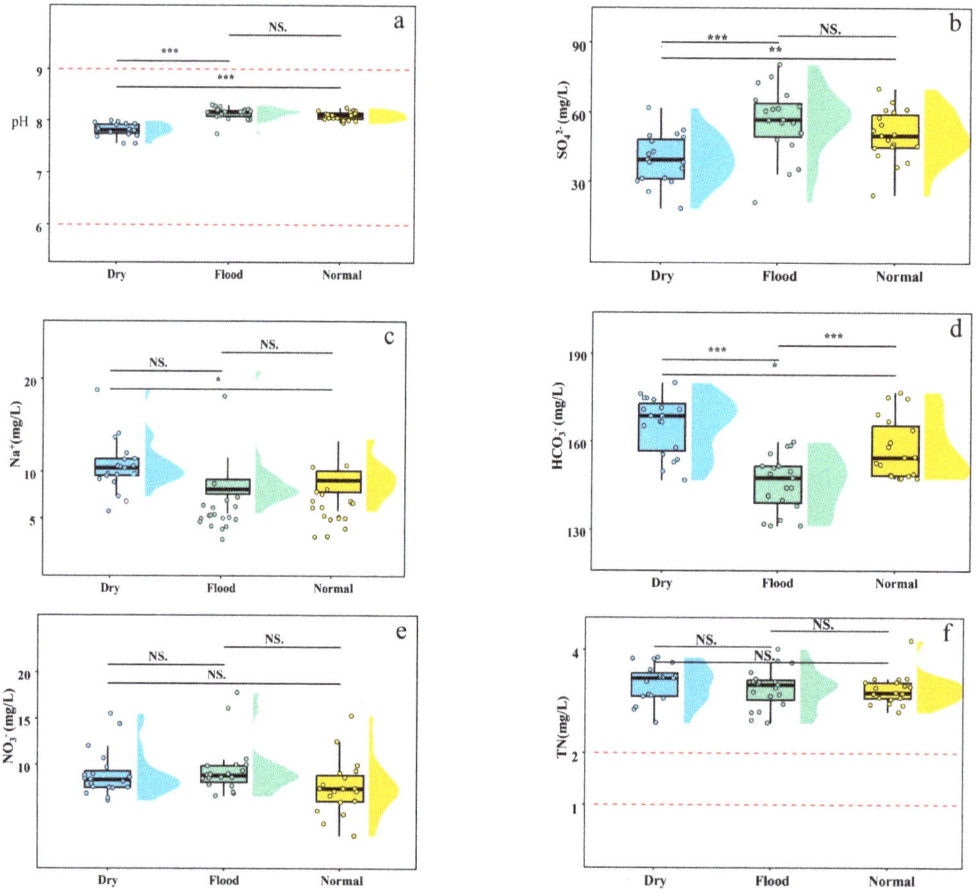

Figure 3. Seasonal variations of hydrochemical parameters in the Jinqian River (**a**) pH; (**b**) SO_4^{2-}; (**c**) Na^+; (**d**) HCO_3^-; (**e**) NO_3^-; (**f**) TN. Note: *** denotes a p-value < 0.001, ** denotes a p-value < 0.01, * denotes a p-value < 0.05, NS: Not Significant. In (**a**,**f**), the red lines represent the limits set by the Chinese Surface Water Quality Standards (GB3838-2002) [16].

4.2. Identification of Hydrochemical Component Sources in Agricultural Rivers

The ion ratio coefficient method is frequently employed to investigate the hydrogeochemical processes of water bodies and the origins of their hydrochemical constituents [22]. The molar ratios of (Ca^{2+}/Na^+) and (Mg^{2+}/Na^+) are commonly used to distinguish the control of water chemical composition by the weathering and dissolution of carbonate rocks, silicate rocks, and evaporite rocks. Previous studies have shown that the molar concentration ratios of Ca^{2+}/Na^+ and Mg^{2+}/Na^+ produced by the weathering of silicate rocks are 0.35 ± 0.15 and 0.24 ± 0.12, respectively. For carbonate rock weathering, these ratios are 50 ± 20 and 20 ± 8, and for evaporite rocks, they are 0.17 ± 0.09 and 0.02 ± 0.013 [23]. As shown in Figure 4a, the sample points from the three periods are generally located in the intermediate zone between the weathering dissolution of silicate rocks and carbonate rocks. In combination with the analysis of hydrochemical types mentioned earlier, it is evident that the weathering dissolution of carbonate rocks is the primary source of hydrochemical components in the river water, followed by that of silicate rocks.

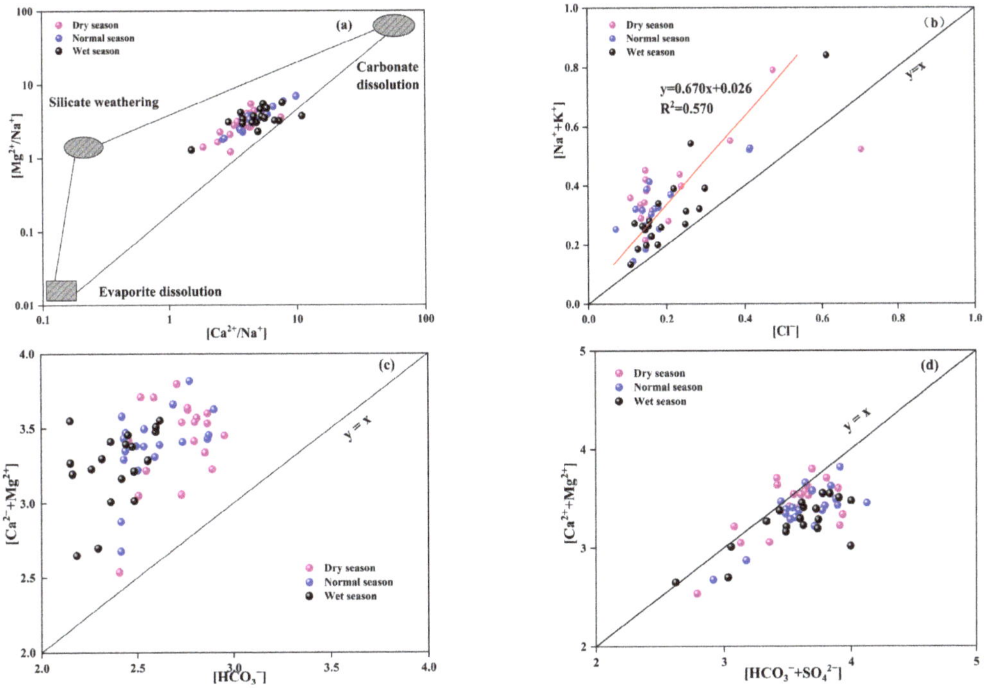

Figure 4. Source identification of hydrochemical components (**a**) $[Ca^{2+}/Na^+]$ vs. $[Mg^{2+}/Na^+]$; (**b**) $[Cl^-]$ vs. $[Na^+ + K^+]$; (**c**) $[HCO_3^-]$ vs. $[Ca^{2+} + Mg^{2+}]$; (**d**) $[HCO_3^- + SO_4^{2-}]$ vs. $[Ca^{2+} + Mg^{2+}]$.

The origin of Na^+, K^+, and Cl^- was evaluated via the $[Cl^-]$ versus $[Na^+ + K^+]$ ratio. Previous studies have found that when rock salt is the primary source, the ratio of $[Cl^-]$ to $[Na^+ + K^+]$ should be approximately equal to 1 [24]. Figure 4b shows that all sample points from the three periods are situated above the y = x line and display a significant correlation ($R^2 = 0.570$). This suggests that in addition to the Na^+ and K^+ contributed by the weathering dissolution of rock salt, the weathering dissolution of silicate rocks also provides a significant portion of Na^+ and K^+ to the water. This result is consistent with the conclusions of Ren et al., who also found that in the Yellow River, Na^+, K^+, and Cl^- are not only derived from the dissolution of rock salt but are also influenced by the dissolution of silicate rocks [20].

Previous work has established that Ca^{2+} and Mg^{2+} in natural waters originate predominantly from the dissolution of carbonate, silicate, and evaporite lithologies [25], whereas HCO_3^- is derived mainly from carbonate and silicate weathering with limited anthropogenic influence. When carbonate dissolution prevails, the equivalent ratio $[Ca^{2+} + Mg^{2+}]/[HCO_3^-]$ should approximate 1:1; involvement of silicate weathering shifts the expected ratio toward $[Ca^{2+} + Mg^{2+}]/[HCO_3^- + SO_4^{2-}] \approx 1:1$ [26]. In Figure 4c, all flood and dry season samples plot above the y = x line for $[Ca^{2+} + Mg^{2+}]/[HCO_3^-]$, evidencing an additional silicate contribution. After incorporating SO_4^{2-} (Figure 4d), more than 75.0% of the sample points in all three periods fall below the y = x line, indicating that the weathering-dissolution of evaporite rocks (such as gypsum dissolution) is the primary source of SO_4^{2-} in the water.

4.3. The Impact of Human Activities on Agricultural Rivers

The concentrations of Cl^- and NO_3^- in the aquatic environment are closely related to human activities. Therefore, the Cl^-/NO_3^- ratio serves as a diagnostic tracer for discriminating different human impacts on river chemistry [27]. Chloride is effectively conservative in surface waters because it is unaffected by common physical, chemical, or microbial transformations. Therefore, the NO_3^-/Cl^- ratio could as a robust proxy for nitrate provenance [28]. Under normal circumstances, the primary sources of Cl^- contamination include domestic sewage, feces, industrial wastewater, road salts, and agricultural chemicals (such as potassium chloride or KCl) [29]. Manure and sewage are characterized by elevated Cl^- concentrations and low NO_3^-/Cl^- ratios, whereas synthetic fertilizers and atmospheric deposition exhibit the opposite signature [30].

As shown in Figure 5, the rainwater samples (n = 4) exhibited low Cl^- concentrations (2.01 mg/L) and high NO_3^-/Cl^-ratios (4.44), whereas the sewage samples (n = 8) had high Cl^- concentrations (7.5 mg/L) and low NO_3^-/Cl^- ratios (1.16), which is consistent with previous studies [30]. In the study area, NO_3^- in the river is likely primarily derived from a mixture of sewage and fertilizer pollution, as evidenced by the relatively high Cl^- concentration and moderate to low NO_3^-/Cl^- ratio. The impact of rainfall on nitrogen in the Jinqian River is minimal, given the low NO_3^- concentration in precipitation (4.60 mg/L). During the flood season, approximately 58% of the samples are close to the fertilizer pollution endmember. This is a typical characteristic of agricultural rivers [31]. While, during the dry season, 50% of the samples are located at the sewage endmember, with the remaining samples positioned at the mixed endmember.

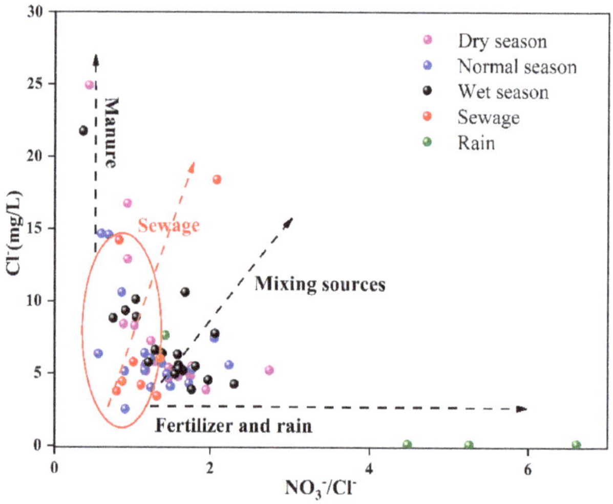

Figure 5. Plot of source identification of nitrate.

4.4. Source Apportionment of Pollution in Agricultural Rivers

Effective river-basin management requires unambiguous identification of the factors driving water quality degradation. Multivariate statistics have been extensively applied to source apportionment in aquatic systems. In this study, principal component analysis (PCA) was employed to differentiate seasonal pollution sources in the Jinqian River. Thirteen routinely monitored physicochemical parameters (TDS, TH, Na^+, K^+, Ca^{2+}, Mg^{2+}, HCO_3^-, SO_4^{2-}, NO_3^-, Cl^-, COD, Fe, and Mn) were selected for analysis. Prior to extraction, the data matrix was subjected to Kaiser–Meyer–Olkin (KMO) and Bartlett's sphericity tests. The resultant KMO value of 0.604 and Bartlett's $\chi^2 = 364$ ($p < 0.001$) confirm the suitability of the hydrochemical data set for PCA. Four principal components (PCs) with eigenvalues > 1 were retained, collectively explaining 79.9% of the total variance and thereby preserving the information embedded in the original thirteen variables.

Principal component 1 (PC1) accounts for 27.6% of the variance and exhibits strong positive loadings on Na^+, Cl^-, F^-, K^+, and NO_3^-, with moderate positive loadings on NH_4^+ and TN (Table 3). Na^+ and Cl^- exhibit the strongest correlation with PC1. As demonstrated in Section 4.2, the primary sources of Na^+ and Cl^- in the Jinqian River are the weathering and dissolution of rock salt (NaCl). Therefore, PC1 primarily represents the influence of rock salt dissolution on the river water. Under normal circumstances, NO_3^-, NH_4^+, and TN in the water environment are unequivocally linked to anthropogenic inputs such as agricultural runoff, domestic effluent, and atmospheric deposition [32,33]. Because elevated NH_4^+ and TN are robust indicators of sewage, and these variables are most strongly correlated with PC4. In addition, as presented in Supplementary Table S2, a notably significant positive correlation ($p < 0.01$) is observed between K^+ and NO_3^-. This finding suggests that these two ions likely stem from a common origin. Moreover, previous scholars have found that the co-enrichment of K^+ and NO_3^- in the aquatic environment can indicate that the water body is affected by agricultural fertilizer pollution [34]. Since different principal components represent independent sources of pollution [35], thus, PC1 mainly represents the influence of agricultural fertilization and rock salt weathering on the water chemistry of the Jinqian River.

Table 3. The results of principal component analysis in the Jinqian River.

Parameters	PC1	PC2	PC3	PC4
Na^+	0.877	0.031	0.156	0.233
Cl^-	0.843	−0.204	0.091	0.020
F^-	0.809	−0.151	0.021	−0.234
K^+	0.808	0.370	−0.149	−0.053
NO_3^-	0.803	0.434	−0.085	0.245
SO_4^{2-}	0.031	0.838	0.185	−0.109
Ca^{2+}	0.101	0.823	0.015	0.044
Mg^{2+}	−0.053	−0.002	0.830	−0.213
TDS	0.055	0.594	0.704	0.056
HCO_3^-	−0.157	0.120	0.672	0.467
DOC	0.206	−0.593	−0.635	−0.003
TN	0.364	−0.136	−0.109	0.883
NH_4^+	0.214	−0.104	−0.180	0.692
Eigenvalue	3.59	2.52	2.18	1.50
Contribution Rate (%)	27.6	19.4	17.8	15.1
Cumulative Contribution Rate (%)	27.6	47.0	64.8	79.9

Principal component 2 (PC2) explains 19.4% of the variance and is dominated by SO_4^{2-} and Ca^{2+}. These ions are diagnostic of gypsum ($CaSO_4{\cdot}2H_2O$) dissolution. And in Section 4.2, it has been confirmed that the SO_4^{2-} in the Jinqian River water originates from the dissolution of gypsum. Consequently, PC2 represents the control of gypsum weathering and dissolution on the water chemistry of the Jinqian River.

Principal component 3 (PC3) accounts for 17.8% of the variance and is characterized by high positive loadings on Mg^{2+}, TDS, and HCO_3^-, together with a moderate negative loading on DOC. Mg^{2+} and HCO_3^- are typical products of both silicate and carbonate weathering [36]; earlier analyses demonstrated that carbonate dissolution dominates their supply in this basin. Therefore, PC3 represents the influence of carbonate weathering and dissolution on the water chemistry of the Jinqian River.

Principal component 4 (PC4) explains 15.1% of the variance and is defined by strong positive loadings on TN and NH_4^+. As shown in Supplementary Table S2, the two parameters also exhibit a significant positive correlation ($p < 0.01$), indicating that they originate from a common source [37]. Riverine nitrogen is derived from domestic sewage, synthetic fertilizers, and manure [38]. Because PC1 already encapsulates agricultural fertilization, the segregation of TN and NH_4^+ on PC4 points to domestic wastewater. Field observations confirm that untreated sewage from riverside villages is discharged directly into the Jinqian River, representing a critical cause of TN exceedances.

5. Research Limitations and Future Prospects

This study provides an in-depth investigation into the hydrochemical evolution and pollution sources of agricultural rivers, offering valuable insights for the protection of water environments in such regions. However, several limitations should be acknowledged and addressed in future research.

Firstly, this study did not include analyses of trace metals, which are often contributed by fertilizers and may have significant environmental impacts. Future research should incorporate assessments of trace metals to provide a more comprehensive understanding of pollution sources and their potential risks to river ecosystems.

Secondly, the study failed to consider emerging contaminants arising from human activities within the catchment area. These contaminants, such as pharmaceuticals, personal care products, and microplastics, can exert complex and long-term effects on water quality and aquatic life. Future investigations should focus on identifying and quantifying these emerging contaminants to better assess the overall health of the river.

Additionally, this study primarily utilized principal component analysis to qualitatively identify pollution sources in the Jinqian River. While the source apportionment results offer some reference value, their accuracy needs further improvement. In recent years, isotope tracing techniques have been widely applied and developed. For instance, using the dual $\delta^{15}N$ and $\delta^{18}O\text{-}NO_3^-$ in combination with source apportionment models can not only accurately identify the sources of nitrogen in water bodies but also quantify the contribution rates of each source to nitrogen pollution [39,40]. Future research should pay more attention to applying this technique in this region.

6. Research Recommendations

To address the deterioration of water quality in agricultural rivers, we propose the following evidence-based measures:

(1) Optimize Fertilizer Management

Excessive application of synthetic fertilizers is a primary driver of nutrient pollution. Current local farming practices often involve the broadcast application of chemical fertilizers at excessively high rates. To mitigate agricultural non-point source pollution caused by over-application, it is essential to improve local farmers' fertilization methods by promoting precision agriculture technologies. These technologies can match nutrient inputs to crop demands and soil fertility, thereby reducing surplus nutrient application. We recommend the application of well-composted organic amendments or slow-release fertilizers to enhance nutrient use efficiency and support sustainable agricultural practices.

(2) Upgrade Domestic Wastewater Treatment

At present, the domestic wastewater treatment capacity of villages along the Jinqian River is evidently insufficient. Untreated or partially treated sewage from riverside settlements significantly increases the river's nitrogen and other pollution loads. It is crucial to extend centralized wastewater treatment infrastructure to all villages within the riparian zone. Where connection to a centralized system is not yet feasible, compact package plants or constructed wetlands should be installed to ensure that effluents meet discharge standards before being released into the environment.

(3) Strengthen long-term monitoring and assessment

A permanent water quality monitoring network should be established and operated across the catchment, with sampling conducted at biweekly to monthly intervals. Key parameters, including TN, NO_3^-, NH_4^+, chemical oxygen demand, total phosphorus, metal ions, emerging contaminants, and others, must be analyzed under strict quality assurance and quality control (QA/QC) protocols. In addition, more accurate assessments of water quality and pollution sources within the catchment are essential to ensure the sustainable use and scientific protection of water resources in the region.

7. Conclusions

This study systematically analyzed the characteristics of water quality changes, material sources, and main controlling factors of agricultural rivers and conducted a comprehensive assessment using hydrochemical ratio methods and multivariate statistical techniques. The results show that agricultural rivers are significantly affected by agricultural fertilizers and domestic sewage. Based on principal component analysis, the main factors identified that control the water quality of agricultural rivers include the use of agricultural fertilizers, the dissolution of rock salt and evaporites, and the discharge of domestic sewage. Future research should focus more on the issues of heavy metal and emerging contaminant pollution in rivers, as well as the application of more precise source apportionment techniques, such as isotope tracing technology, to accurately identify pollution sources in agricultural rivers. This will provide more accurate technical support for the prevention and control of pollution in agricultural rivers.

Supplementary Materials: The following supporting information can be downloaded at: https://www.mdpi.com/article/10.3390/w17213111/s1, Table S1. Hydrochemical parameters, analytical method, equipment and detection limits; Table S2. Correlation analysis of hydrochemical indicators in the Jinqian River.

Author Contributions: C.Y., investigation, methodology, writing—original draft preparation and editing, Z.Q., investigation, methodology, data curation, supervision. X.S., investigation, methodology, data curation. L.Y., investigation, methodology, software, data curation. N.Y., investigation, methodology, software. F.Y., investigation, methodology, writing—review and editing. Q.Z., supervision, methodology, writing—review and editing. All authors have read and agreed to the published version of the manuscript.

Funding: This work was funded by the National Natural Science Foundation of China (Grant No. 42377232), Natural Science Foundation of Hebei Province of China (Grant No. D2022504015).

Data Availability Statement: The data presented in this study are available from the corresponding author upon request due to privacy restrictions.

Acknowledgments: The authors gratefully acknowledge the editor and anonymous reviewers for their valuable comments on this manuscript. The authors also appreciate the financial support from the different organizations.

Conflicts of Interest: The authors declare no conflicts of interest.

References

1. Zhang, Q. Water quality assessment of river basins: New insights and practical solutions. *Water* **2025**, *17*, 1207. [CrossRef]
2. Li, S.; Gu, S.; Liu, W.; Han, H.; Zhang, Q. Water quality in relation to the landuse and land cover in the upper Han River basin, China. *Catena* **2008**, *75*, 216–222. [CrossRef]
3. Zhang, Q.; Liu, W.; Sun, H.; Wang, H.; Wang, L.; Zhang, J.; Xu, Z. Cascade reservoir regulations on nitrogen source and transformation in the Tibetan Plateau river: Constraints from high frequency data of Lancang River. *J. Hydrol.* **2025**, *650*, 132563. [CrossRef]
4. Liu, W.; Jiang, H.; Zhang, J.; Xu, Z. Driving forces of nitrogen cycling and the climate feedback loops in the Yarlung Tsangpo River Basin, the highest-altitude large river basin in the world. *J. Hydrol.* **2022**, *610*, 127974. [CrossRef]
5. Gandhimathi, G.; Chellaswamy, C.; Thiruvalar, S.P. Comprehensive river water quality monitoring using convolutional neural networks and gated recurrent units: A case study along the Vaigai River. *J. Environ. Manag.* **2024**, *365*, 121567. [CrossRef]
6. Mustapha, A.; Aris, A.Z.; Juahir, H.; Ramli, M.F.; Kura, N.U. River water quality assessment using environmetric techniques: Case study of Jakara River Basin. *Environ. Sci. Pollut. Res.* **2013**, *20*, 5630–5644. [CrossRef]
7. Zhao, Z.Y.; Zuo, J.; Zillante, G. Transformation of water resource management: A case study of the South-to-North Water Diversion project. *J. Clean. Prod.* **2017**, *163*, 136–145. [CrossRef]
8. Zhang, Q. The South-to-North Water Transfer Project of China: Environmental implications and monitoring strategy. *J. Am. Water Resour. Assoc.* **2009**, *45*, 1238–1247. [CrossRef]
9. Liu, J.; Han, G.; Liu, M.; Zeng, J.; Liang, B.; Qu, R. Distribution, sources and water quality evaluation of the riverine solutes: A case study in the Lancangjiang River Basin, Tibetan Plateau. *Int. J. Environ. Res. Public Health* **2019**, *16*, 4670. [CrossRef] [PubMed]
10. Mbaye, M.L.; Gaye, A.T.; Spitzy, A.; Dähnke, K.; Gaye, B. Seasonal and spatial variation in suspended matter, organic carbon, nitrogen, and nutrient concentrations of the Senegal River in West Africa. *Limnologica* **2016**, *57*, 1–13. [CrossRef]
11. Jiang, H.; Liu, W.; Li, Y.; Zhang, J.; Xu, Z. Multiple isotopes reveal a hydrology dominated control on the nitrogen cycling in the Nujiang River basin, the last undammed large river basin on the Tibetan Plateau. *Environ. Sci. Technol.* **2022**, *56*, 4610–4619. [CrossRef]
12. Li, Y.; Tu, Y.; Sun, T.; Duan, Y.; Kou, J.; Li, W.; Gao, J. Source apportionment of organic carbon and nitrogen in sediments from river and lake in the highly urbanized Changjiang Delta. *J. Hazard. Mater.* **2024**, *478*, 135590. [CrossRef]
13. Xue, D.; Botte, J.; Baets, B.D.; Accoe, F.; Nestler, A.; Taylor, P.; Boeckx, P. Present limitations and future prospects of stable isotope methods for nitrate source identification in surface and groundwater. *Water Res.* **2009**, *43*, 1159–1170. [CrossRef] [PubMed]
14. Tian, J.; Yuan, Z.; Mao, X.; Ma, T. Quantifying natural and anthropogenic impacts on riverine total nitrogen concentration and load in the Yellow River Basin. *Environ. Pollut.* **2025**, *382*, 126641. [CrossRef]
15. Xu, Y.; Liu, W.; Xu, B.; Xu, Z. Riverine sulfate sources and behaviors in arid environment, Northwest China: Constraints from sulfur and oxygen isotopes. *J. Environ. Sci.* **2024**, *137*, 716–731. [CrossRef]
16. *GB3838-2002*; Environmental Quality Standards for Surface Water. State Environment Protection Bureau of China: Beijing, China, 2002.
17. WHO (World Health Organization). *Guidelines for Drinking-Water Quality*, 4th ed.; World Health Organization: Geneva, Switzerland, 2017.
18. Popa, P.; Murariu, G.; Timofti, M.; Georgescu, L.P. Multivariate statistical analyses of water quality of danube river at galati, romania. Environ. *Eng. Manag. J.* **2018**, *17*, 1249–1266.
19. Esquivel-Hernández, G.; Sanchez-Murillo, R.; Villalobos-Córdoba, D.; Monteiro, L.R.; Villalobos-Forbes, M.; Sánchez-Gutiérrez, R.; Matiatos, I. Exploring the acid neutralizing effect in rainwater collected at a tropical urban area: Central Valley, Costa Rica. *Atmos. Pollut. Res.* **2023**, *14*, 101845. [CrossRef]
20. Ren, C.; Liu, L. Under the strong influence of human activities: The patterns and controlling factors of river water chemistry changes—A case study of the lower Yellow River. *Water* **2024**, *16*, 1886. [CrossRef]
21. Jiang, H.; Zhang, Q.; Liu, W.; Zhang, J.; Zhao, T.; Xu, Z. Climatic and anthropogenic driving forces of the nitrogen cycling in a subtropical river basin. *Environ. Res.* **2021**, *194*, 110721. [CrossRef] [PubMed]

22. Pant, R.R.; Zhang, F.A.; Rehman, F.U.; Wang, G.; Ye, M.; Zeng, C. Spatiotemporal variations of hydrogeochemistry and its controlling factors in the Gandaki River Basin, Central Himalaya Nepal. *Sci. Total Environ.* **2017**, *622–623*, 770–782. [CrossRef]
23. Niu, C.; Zhang, Q.; Xiao, L.; Wang, H. Spatiotemporal variation in groundwater quality and source apportionment along the Ye River of North China using the PMF model. *Int. J. Environ. Res. Public Health* **2022**, *19*, 1779. [CrossRef]
24. Fu, C.C.; Li, X.Q.; Ma, J.F.; Ling, X.L.; Ming, G.; Zhan, X.B. A hydrochemistry and multi-isotopic study of groundwater origin and hydrochemical evolution in the middle reaches of the Kuye River basin. *Appl. Geochem.* **2018**, *98*, 82–93. [CrossRef]
25. Xing, J.; Long, W.; Jie, Z.; Tian, L. Hydrochemical variation characteristics and driving factors of surface water in arid areas—A case study of Beichuan River in Northwest China. *Front. Environ. Sci.* **2024**, *12*, 1493390. [CrossRef]
26. Ye, H.; Han, Z.; Wu, P.; Zha, X.; Li, X.; Hou, E.; Peñuelas, J. Disentangling sources and transformation mechanisms of nitrogen, sulfate, and carbon in water of a Karst Critical Zone. *Sci. Total Environ.* **2024**, *922*, 171310. [CrossRef] [PubMed]
27. Zhang, Q.; Sun, J.; Liu, J.; Huang, G.; Lu, C.; Zhang, Y. Driving mechanism and sources of groundwater nitrate contamination in the rapidly urbanized region of South China. *J. Contam. Hydrol.* **2015**, *182*, 221–230. [CrossRef] [PubMed]
28. Liu, C.; Li, S.; Lang, Y.; Xiao, H. Using $\delta^{15}N$- and $\delta^{18}O$-values to identify nitrate sources in Karst groundwater, Guiyang, Southwest China. *Environ. Sci. Technol.* **2006**, *40*, 6928–6933. [CrossRef]
29. Yue, F.J.; Liu, C.Q.; Li, S.L.; Zhao, Z.Q.; Liu, X.L.; Ding, H. Analysis of $\delta^{15}N$ and $\delta^{18}O$ to identify nitrate sources and transformations in Songhua River, Northeast China. *J. Hydrol.* **2014**, *519*, 329–339. [CrossRef]
30. Meghdadi, A.; Javar, N. Quantification of spatial and seasonal variations in the proportional contribution of nitrate sources using a multi-isotope approach and Bayesian isotope mixing model. *Environ. Pollut.* **2018**, *235*, 207–222. [CrossRef]
31. Jiang, H.; Liu, W.; Zhang, J.; Zhou, L.; Zhou, X.; Pan, K.; Zhao, T.; Wang, Y.; Xu, Z. Spatiotemporal variations of nitrate sources and dynamics in a typical agricultural riverine system under monsoon climate. *J. Environ. Sci.* **2020**, *93*, 98–108. [CrossRef]
32. Wang, X.; Liu, Z.; Xu, Y.J.; Mao, B.; Jia, S.; Wang, C.; Ji, X.; Lv, Q. Revealing nitrate sources seasonal difference between groundwater and surface water in China's largest fresh water lake (Poyang Lake): Insights from sources proportion, dynamic evolution and driving forces. *Sci. Total Environ.* **2025**, *958*, 178134. [CrossRef] [PubMed]
33. Zhao, X.; Xu, H.; Kang, L.; Zhu, G.; Paerl, H.W.; Li, H.; Liu, M.; Zhu, M.; Zou, W.; Qin, B.; et al. Nitrate sources and transformations in a river-reservoir system: Response to extreme flooding and various land use. *J. Hydrol.* **2024**, *638*, 131491. [CrossRef]
34. Iqbal, J.; Su, C.; Abbas, H.; Jiang, J.; Han, Z.; Baloch, M.Y.J.; Xie, X. Prediction of nitrate concentration and the impact of land use types on groundwater in the Nansi Lake Basin. *J. Hazard. Mater.* **2025**, *487*, 137185. [CrossRef]
35. Qin, R.; Wu, Y.; Xu, Z.; Xie, D.; Zhang, C. Assessing the impact of natural and anthropogenic activities on groundwater quality in coastal alluvial aquifers of the lower Liaohe River Plain, NE China. *Appl. Geochem.* **2013**, *31*, 142–158. [CrossRef]
36. Wang, L.; Zhang, Q.Q.; Wang, H.W. Rapid urbanization has changed the driving factors of groundwater chemical evolution in the large groundwater depression funnel area of northern China. *Water* **2023**, *15*, 2917. [CrossRef]
37. Arıman, S.; Soydan-Oksal, N.G.; Beden, N.; Ahmadzai, H. Assessment of Groundwater Quality through Hydrochemistry Using Principal Components Analysis (PCA) and Water Quality Index (WQI) in Kızılırmak Delta, Turkey. *Water* **2024**, *16*, 1570. [CrossRef]
38. Zhu, Z.; Ding, J.; Du, R.; Zhang, Z.; Guo, J.; Li, X.; Jiang, L.; Chen, G.; Bu, Q.; Tang, N. Systematic tracking of nitrogen sources in complex river catchments: Machine learning approach based on microbial metagenomics. *Water Res.* **2024**, *253*, 121255. [CrossRef]
39. Ji, X.; Shu, L.; Li, J.; Zhao, C.; Chen, W.; Chen, Z.; Shang, X.; Dahlgren, R.A.; Yang, Y.; Zhang, M.J. Tracing nitrate sources and transformations using $\Delta^{17}O$, $\delta^{15}N$, and $\delta^{18}O$-NO_3^- in a coastal plain river network of eastern China. *J. Hydrol.* **2022**, *610*, 127829. [CrossRef]
40. Ren, X.; Yue, F.J.; Tang, J.; Li, C.; Li, S.L. Nitrate transformation and source tracking of rivers draining into the Bohai Sea using a multi-tracer approach combined with an optimized Bayesian stable isotope mixing model. *J. Hazard. Mater.* **2024**, *463*, 132901. [CrossRef] [PubMed]

water

MDPI

Article

Hydrochemical Evolution and Pollution Source Apportionment of Groundwater in Arid Regions: A Case Study of the Datong River Basin, Northwest China

Tao Liu [1,2], Jian Kang [2], Youwei Yu [1], Yue Qi [1] and Zizhao Zhang [1,*]

[1] College of Geology and Mining Engineering, Xinjiang University, Urumqi 830017, China
[2] Hydrogeological and Environmental Geological Survey Center of Xinjiang Uygur Autonomous Region Geological Bureau, Urumqi 830091, China
* Correspondence: zhangzz198511@163.com

Abstract

Understanding hydrochemical evolution and apportioning pollution sources are prerequisites for effective groundwater protection at the regional scale; nevertheless, the governing processes and anthropogenic drivers in arid regions remain poorly constrained. Here, we present a comprehensive geochemical survey of the Datong River Basin, a representative arid catchment in north-western China. Thirty-seven groundwater samples were analyzed with hydrochemical methods and Positive-Matrix Factorization (PMF) to delineate natural controls and contaminant sources. Results showed that the aquifer is dominated by HCO_3–Ca(Mg) water controlled predominantly by silicate and carbonate weathering, modified locally by evapo-concentration and human activities. Water-quality indices classify 70.3% of the samples as excellent, but spatially restricted degradation is evident. PMF resolved three independent sources: a natural end-member enriched in Mn, Na^+ and Cl^-; a mixed source reflecting domestic wastewater, agricultural fertilizers and rock weathering; and an industrial source dominated by Fe. The mixed source contributes most major ions and chemical oxygen demand (COD), whereas the industrial source accounts for 75.7% of total Fe. These findings provide a robust scientific basis for groundwater management and pollution mitigation in arid regions under similar hydrogeological settings.

Keywords: hydrochemical evolution; anthropogenic impacts; natural factors; source apportionment; positive-matrix factorization model

1. Introduction

Groundwater is the only perennial water resource capable of sustaining ecological functions, domestic supply and irrigated agriculture in arid regions [1,2]. Its evolving hydrochemistry archives an integrated signal of recharge history, water–rock interaction and anthropogenic perturbations, thereby providing a diagnostic basis for detecting water-quality degradation and for identifying the origins of contamination [3]. In arid environments, the coupled effects of intense evapoconcentration, lithological heterogeneity of aquifers and superimposed anthropogenic forcings collectively impart a high degree of complexity to groundwater hydrochemical evolution [4]. Consequently, a comprehensive understanding of hydrochemical evolution coupled with robust source apportionment is essential for designing targeted remediation strategies, implementing precise governance and achieving sustainable groundwater utilization and high-quality development in arid basins [5,6].

Extensive studies have documented that regional-scale hydrochemical evolution is governed by the interplay of natural drivers—lithology, hydrogeological conditions, water–rock interaction and precipitation—and anthropogenic forcings such as wastewater discharge and agricultural non-point-source pollution. Chen et al. (2014) [7] showed that groundwater in the Turpan Basin is controlled by calcite precipitation, evapoconcentration, cation exchange and evaporite dissolution. Zhang et al. (2017) [8] demonstrated that the evolution of pore water in the Hutuo River alluvial fan reflects inputs from domestic and industrial effluents, agricultural non-point sources and water–rock reactions. Similarly, Herojeet et al. (2016) [9] found that the chemistry of an alluvial aquifer in India's Nalagarh Valley is dominated by water–rock interaction, ion exchange, the leaching of parent material and the seepage of agricultural, industrial and municipal wastes. These findings underscore that site-specific differences in aquifer lithology, recharge regime and land-use imprint distinct hydrochemical signatures; comparative, process-oriented investigations are therefore essential for generalizing controls on groundwater quality across diverse geological settings.

Accelerated urbanization and industrialization have progressively intensified anthropogenic pressures on groundwater quality, with declining trends documented across many regions and consequent risks to safe drinking-water supplies. Gu et al. (2015) [10] identified domestic wastewater, septic-tank effluent and agrochemical fertilizers as the principal contamination sources in Qinhuangdao city, while Wang et al. (2023) [11] reported similar dominance of sewage and fertilizers in rapidly urbanized Shijiazhuang, further noting that enhanced water–rock interaction triggered by groundwater over-extraction exacerbates quality deterioration. Although these studies successfully pinpointed the main pollution origins, they remained largely qualitative and therefore provide insufficient guidance for targeted remediation. Robust, quantitative source apportionment at the regional scale is thus urgently required to underpin effective, site-specific groundwater-protection strategies.

The Datong River basin, located on the northeastern Qinghai–Tibet Plateau, is a typical alpine arid catchment that originates in the Qilian Mountains and traverses Qinghai and Gansu provinces before joining the Yellow River as a second-order tributary. Rapid economic growth has led to intensified groundwater exploitation, with measurable deterioration in water quality already reported [12]. However, the mechanisms governing groundwater hydrochemical evolution and the quantitative contribution of specific contamination sources in this plateau–arid setting remain elusive. To address this knowledge gap, this study selects the Datong River basin as the representative area and integrates hydrochemical characterization, multivariate statistical techniques and the Positive-Matrix Factorization (PMF) model to (i) accurately identify groundwater hydrochemical facies and evolutionary trends, (ii) distinguish the imprint of anthropogenic activities on water quality and (iii) quantitatively apportion the principal contamination sources together with their respective contributions. The findings will provide a scientific basis for the prevention and control of groundwater pollution in plateau–arid regions.

2. Materials and Methods

2.1. Study Area Overview

The Datong River, a second-order tributary of the Yellow River, originates from the Muli Mountain springs on the Tuo-le Range in Tianjun County, Qinghai Province. It flows through Gangcha, Qilian, Haiyan and Menyuan counties in Qinghai Province, and Tianzhu and Yongdeng counties in Gansu Province, before joining the Huangshui River in Minhe County, Qinghai Province. The mainstream extends 560.7 km and drains a basin of ~1.5×10^4 km^2. Mean annual discharge is 88.7 m^3/s, yielding an average runoff volume of 1.75×10^9 m^3. Located on the northeastern Qinghai–Tibet Plateau, the Datong

 https://doi.org/10.3390/w18010105

River basin is characterized by a plateau continental climate with pronounced altitudinal climatic gradients. The basin has a multi-year mean air temperature of 3.3–8.1 °C, mean annual precipitation of 300–600 mm (≈70% concentrated between June and September) and potential evaporation of 1000–1300 mm/yr. The dominant soil types are chestnut soils, fluvisols and meadow soils, while the main land-use categories are grassland (74.7%), forest (20.9%), cropland (2.1%), construction land (1.0%) and water bodies (1.3%) (Figure 1).

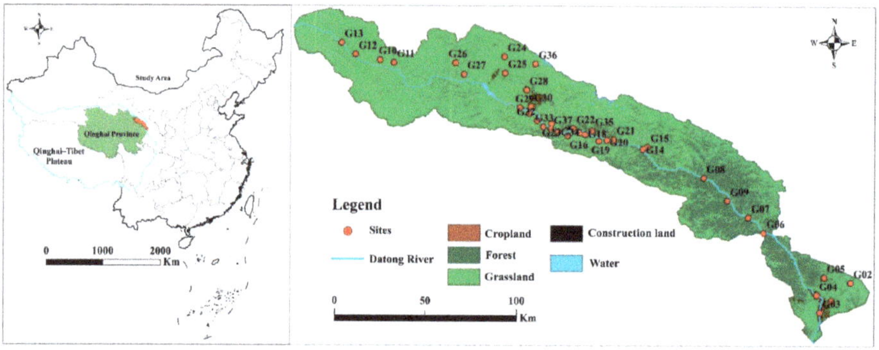

Figure 1. Study area location and sampling site distribution.

2.2. Hydrogeological Setting

Tectonically, the study area lies within the central Qilian orogenic belt. Since the Jurassic, a series of east-to-west-trending sedimentary basins—including the Xining dome, the Muli–Jiangcang–Reshui composite syncline and the Shule composite syncline—have developed along this uplift, forming the principal coal-bearing zone of Qinghai Province [13]. Groundwater in the study area flows concurrently with the river from northwest to southeast. Groundwater in the piedmont alluvial—proluvial plain is recharged predominantly by river leakage and secondarily by lateral inflow from bedrock-fissure aquifers. Within the valley corridor, the flow system is sustained chiefly by upstream subsurface inflow and river-bed infiltration, with supplementary inputs from precipitation, agricultural irrigation return-flow and canal seepage [14]. Discharge from the mountain block occurs mainly as baseflow sustaining surface runoff; in intensively exploited areas, however, artificial outlets—including domestic wells, pumping wells, mine workings and drainage galleries—constitute an important additional pathway for groundwater withdrawal for domestic, industrial and agricultural use [15].

Bedrock-fissure water occurs in the mid-mountain zone, Quaternary unconsolidated pore water in the basin and, along the surrounding low hills, Quaternary pore water in valleys, while Mesozoic–Cenozoic clastic rock pore-fissure water dominates elsewhere; alpine permafrost hosts cryotic water. Bedrock-fissure water underlies non-permafrost mountains below ~3800 m asl in Paleozoic and Lower Proterozoic metamorphic and intrusive rocks with heterogeneous productivity. Unconsolidated pore water in valleys interacts closely with surface water and, based on geomorphology and lithology, comprises two types: piedmont glaciofluvial terrace silty gravel–boulder unconfined water and valley gravel–boulder unconfined water. Clastic fracture–pore water occurs in conglomerate, sandstone and sandy-conglomerate aquifers flanking the Datong River, recharged by bedrock-fissure water and mountain runoff. Cryotic water is present above 3500–4000 m where annual precipitation exceeds 600 mm, mostly during thaw, providing ample recharge. Quaternary aquifers occupy modern glacier fronts, palaeo-cirques and trough-valley foot-

 https://doi.org/10.3390/w18010105

slopes, consisting of glacial, slope, solifluction and debris-flow deposits of clayey sandy gravelly blocks (Supplementary Figure S1), with a seasonal thaw depth of ~1 m.

2.3. Sampling and Analytical Procedures

In June 2023, 37 shallow-groundwater samples were collected from agricultural-irrigation wells and hand-pumped domestic boreholes (Figure 1). The 37 groundwater samples were collected from the same aquifer, with the water-table depth ranging from 2.3 m to 17.5 m. Wells were purged for 15–20 min until in-line pH and electrical conductivity (EC) readings stabilized. Samples were decanted into pre-cleaned polyethylene bottles (500 mL and 1 L), each rinsed three times with the target groundwater before final filling. Aliquots designated for cation and trace-metal analysis were acidified to pH < 2 with ultrapure HNO_3. All bottles were sealed with Parafilm, stored in a portable cooler at 4 °C, and transported to the laboratory for subsequent analysis.

Field pH and electrical conductivity (EC) were recorded with a portable multiparameter probe (Hach HQ40d, Loveland, CO, USA). Major cations (K^+, Na^+, Ca^{2+} and Mg^{2+}) and trace metals (Fe and Mn) were determined by inductively coupled plasma mass spectrometry (Agilent 7500ce ICP-MS, Tokyo, Japan). Anions (Cl^-, SO_4^{2-} and NO_3^-) were quantified by ion chromatography (Perkin-Elmer Lambda 35, Waltham, MA, USA); HCO_3^- was obtained by 0.05 M HCl titration. Total dissolved solids (TDS) were measured gravimetrically, total hardness (TH) by EDTA titration and chemical oxygen demand (COD) via alkaline permanganate oxidation.

2.4. Data Analysis

2.4.1. Water-Quality Index (*WQI*)

The WQI provides a quantitative, integrative measure of the suitability of groundwater for potable use. Each physicochemical parameter is weighted (W_i) according to its relative importance for human health and aesthetic acceptability are given in Supplementary Table S1. National Class III limits specified in GB/T 14848-2017 were adopted as reference values [16]; where a parameter was not regulated by the Chinese standard, guideline values given by the World Health Organization were used [17].

The stepwise calculation of *WQI* is described below.

$$RW_i = \frac{W_i}{\sum_{i=1}^{n} W_i}$$

$$Q_i = C_i / S_i \times 100$$

$$SI_i = W_i \times Q_i$$

$$WQI = \sum SI_i$$

where Q_i denotes the quality rating for the *i*-th parameter, C_i is its measured concentration (mg L^{-1}), S_i is the corresponding water-quality standard (mg L^{-1}) and SI_i is the resulting sub-index.

2.4.2. Contrast Coefficient Variance

Variance quantifies the dispersion of a random variable around its expected value. Because hydrochemical constituents span several orders of magnitude, direct comparison of their variances is meaningless. To overcome this scale effect, each observation (X_i) is normalized by the mean of its variable, $\overline{X}$, yielding the contrast coefficient $V_i = \frac{X_i}{\overline{X}}$. By definition, the mean of V is unity, so the transformed variables share a common central tendency while preserving their

 https://doi.org/10.3390/w18010105

relative variability. The variance of these contrast coefficients, denoted ${V\sigma_i}^2$, can therefore be used to compare the intrinsic variability of parameters with disparate concentrations.

The calculation is as follows:

$$ {V\sigma_i}^2 = \frac{1}{n}\sum_{i=1}^{n}\left(V_i - \overline{V_i}\right)^2 $$

where V_i is the contrast coefficient of the ith hydrochemical constituent and $\overline{V_i}$ denotes the mean contrast coefficient of that constituent across all samples.

A contrast-coefficient variance ${V\sigma_i}^2 > 1$ indicates strong anthropogenic influence on the constituent; $0.5 < {V\sigma_i}^2 < 1$ implies moderate impact; and ${V\sigma_i}^2 < 0.5$ suggests minimal human perturbation [18].

2.4.3. Positive-Matrix Factorization

Positive-Matrix Factorization (PMF, (Version 5.0)), recommended by the U.S. Environmental Protection Agency, decomposes a multivariate data matrix into factor contributions and profiles by minimizing the objective function built on the uncertainty matrix without requiring a priori source signatures. Because it needs only the measured concentrations and their analytical uncertainties, PMF is operationally simple and has become a standard tool for source apportionment in atmospheric, soil and aquatic environments [19,20].

The computational form is given by the following equation:

$$ X_{ij} = \sum_{k=1}^{p} g_{ik} f_{ki} + e_{ij} $$

where X_{ij} is the concentration of the jth species in the ith sample, g_{ik} denotes the contribution of source k to sample i, f_{kj} represents the concentration of species j in source k, and e_{ij} is the residual (i.e., the portion of X_{ij}) not explained by the model. PMF iteratively decomposes the original matrix to obtain optimal matrices G and F by minimizing the objective function Q.

3. Results

3.1. *Hydrochemical Characteristics of Groundwater in the Datong River Basin*

Table 1 summarizes the median values, coefficient of variation (CV) and exceedance rates of the main physicochemical indicators of groundwater in the Datong River Basin. The median pH is 7.59, indicating a slightly alkaline hydrochemical facies; 5.4% of the samples exhibited pH values below the Class III threshold of 6.5 specified in the Chinese Groundwater Quality Standard [14]. TDS range from 230 to 2462 mg L^{-1}, with a median of 388 mg L^{-1} and an exceedance rate of 13.5%. Among cations, the abundance follows the order $Ca^{2+} > Na^{+} > Mg^{2+} > K^{+}$; 8.1% of the samples exceed the guideline value for Na^{+} [16]. The anionic sequence is $HCO_3^{-} > SO_4^{2-} > Cl^{-} > NO_3^{-}$. Sulfate, nitrate and chloride exceed the Class III thresholds of the Chinese Groundwater Quality Standard in 10.8%, 8.1% and 2.7% of the samples, respectively, evidencing anthropogenic perturbations such as fertilizer application and sewage discharge [16]. TH represents the cumulative content of multivalent metallic cations and constitutes a key hydrochemical indicator of groundwater quality. In the Datong River Basin, TH ranges from 193 to 1052 mg L^{-1} (median: 314 mg L^{-1}), with 16.2% of the samples exceeding the Class III limit [16], implying a potential impairment of groundwater usability. Among metallic constituents, dissolved Fe and Mn median value 0.074 and 0.005 mg L^{-1}, respectively, and are above regulatory thresholds in 16.2% and 5.4% of the samples. Furthermore, The CV for Na^{+}, Cl^{-}, SO_4^{2-}, NO_3^{-}, COD and Fe all exceed 100%, evidencing pronounced spatial heterogeneity that is consistent with significant anthropogenic perturbation.

https://doi.org/10.3390/w18010105

Table 1. Characteristics of groundwater hydrochemical parameters in the Datong River Basin.

Parameters	Unit	Median	SD	CV (%)	Standard	Below Standards for All Sites (%)
pH	—	7.59	0.46	6.15	6.5–8.5	5.4
EC	μs/cm	609	683	79.8	—	—
K^+	mg/L	2.41	2.96	83.9	—	—
Na^+	mg/L	18.2	92.8	170	200	8.1
Ca^{2+}	mg/L	77.1	48.8	55.0	—	—
Mg^{2+}	mg/L	26.2	23.5	66.4	—	—
Cl^-	mg/L	12.3	107	233	250	2.7
SO_4^{2-}	mg/L	73.2	149	121	250	10.8
HCO_3^-	mg/L	295	147	44.8	—	—
NO_3^-	mg/L	10.2	26.5	128	88.6	8.1
TDS	mg/L	388	439	79.9	1000	13.5
COD	mg/L	1.07	1.53	106	3	5.4
TH	mg/L	314	194	52.7	450	16.2
Fe	mg/L	0.074	0.380	190	0.3	16.2
Mn	mg/L	0.005	0.051	89.5	0.1	5.4

Note: SD: standard deviation; CV: coefficient of variation; standard is grade III standard for groundwater quality in China (GB/T 14848-2017) [16].

3.2. Hydrochemical Facies of Groundwater in the Datong River Basin

The Piper diagram provides a concise graphical appraisal of major-ion composition, hydrochemical facies, and the dominant geochemical controls governing groundwater evolution [21]. In the Datong River Basin, the majority of samples plot within the Ca^{2+}-dominant field (zone C) for cations and the HCO_3^--dominant field (zone F) for anions (Figure 2). Accordingly, the waters were further classified using the Shchukarev nomenclature to delineate distinct hydrochemical types. Overall, the prevailing hydrochemical type is HCO_3–Ca(Mg), accounting for 48.6% of the samples, followed by $HCO_3 \cdot SO_4$–Ca·Mg (16.2%) and HCO_3–Ca (13.5%). Notably, 32.4% of the samples require SO_4^{2-} in their designation, whereas 13.5% and 5.4% incorporate Na^+ and Cl^-, respectively. This conspicuous participation of SO_4^{2-}, Na^+ and Cl^- underscores the influence of anthropogenic activities on basin-wide groundwater chemistry [22].

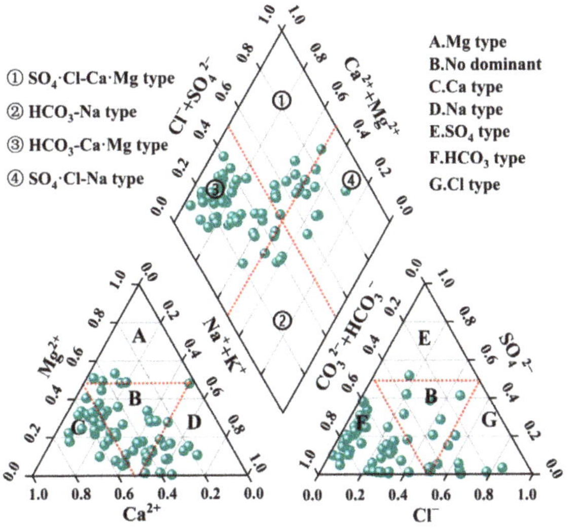

Figure 2. Hydrochemical type of groundwater in the Datong River basin [23].

 https://doi.org/10.3390/w18010105

3.3. *Water Quality Assessment of Groundwater in the Datong River Basin*

Groundwater quality in the Datong River Basin was quantitatively evaluated by means of a comprehensive water-quality index (WQI). Thirteen routinely measured parameters—pH, TDS, Na^+, Ca^{2+}, Mg^{2+}, SO_4^{2-}, Cl^-, HCO_3^-, NO_3^-, COD, TH, Fe and Mn—were incorporated into the WQI model to generate composite scores for each sampling site. Based on the established classification scheme (Table 2), 70.3% of the 37 monitoring wells were assigned to the "excellent" category, 18.9% were classified as "good", whereas 10.8% fell into the "poor" class, indicating localized degradation of groundwater quality.

Table 2. Assessment results of groundwater quality in the Datong River basin.

WQI Range	Comprehensive Water Quality Score	Number of Samples	Percentage of Samples (%)
Excellent water	<50	26	70.3
Good water	50–100	7	18.9
Poor water	100.1–200	4	10.8
Very poor water	200.1–300	0	0
Water unsuitable for drinking purposes	>300	0	0

4. Discussion

4.1. *Controlling Factors of Groundwater Hydrochemistry in the Datong River Basin*

Groundwater chemistry is primarily governed by the mineralogy of the aquifer matrix, incongruent weathering of vadose-zone soils, precipitation inputs and evapo-concentration within the basin [24]. Identifying the relative importance of these controls is essential for elucidating hydrochemical evolution and solute provenance. The Gibbs diagram provides a robust graphical framework for discriminating among lithogenic, atmospheric and evapotranspiration dominance, and has consequently become a standard tool in large-scale hydrochemical studies [25]. The Gibbs diagram resolves three hydrochemical end-members: atmospheric-precipitation dominance, rock-weathering dominance and evapo-concentration dominance. Samples controlled by atmospheric precipitation are characterized by TDS < 10 mg L^{-1} and high $[Na^+]/[Na^+ + Ca^{2+}]$ and $[Cl^-]/[Cl^- + HCO_3^-]$ ratios (0.5–1.0), plotting in the lower-right sector. Waters shaped predominantly by rock weathering fall in the central-left region, with TDS between 70 and 300 mg L^{-1} and both ionic ratios < 0.5. The upper-right field represents the evapo-concentration domain, distinguished by elevated TDS (>300 mg L^{-1}) and high $[Na^+]/[Na^+ + Ca^{2+}]$ and $[Cl^-]/[Cl^- + HCO_3^-]$ values (0.5–1.0) [26].

As shown in Figure 3, the basin-wide groundwater data cluster in the mid- to upper portion of the Gibbs template, albeit with considerable scatter. Approximately 86.5% of the samples exhibit TDS between 100 and 1000 mg L^{-1}, and 83.8% yield $[Na^+]/[Na^+ + Ca^{2+}] < 0.5$, confirming Ca^{2+} as the dominant cation. Simultaneously, $[Cl^-]/[Cl^- + HCO_3^-]$ ranges from 0.04 to 0.787. These signatures collectively indicate that rock weathering is the primary driver of hydrochemical evolution, whereas evapo-concentration exerts a secondary, spatially limited control. Atmospheric precipitation contributes negligibly to the observed ionic budgets. This result is consistent with previous findings; Zhou et al. (2021) [27] reported that groundwater hydrochemistry in the arid region of northwestern China is primarily controlled by rock weathering and evaporation-concentration processes.

https://doi.org/10.3390/w18010105

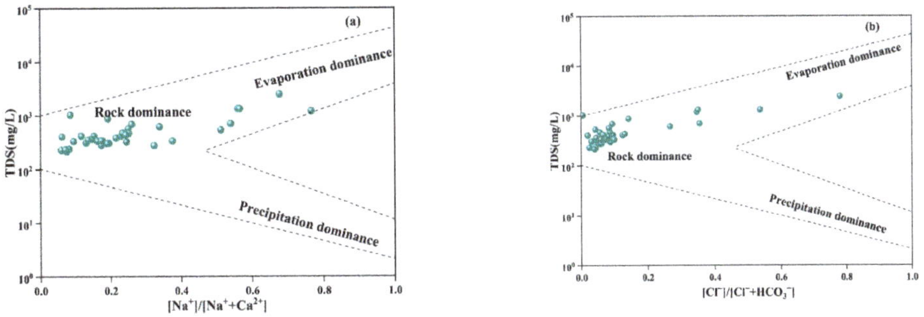

Figure 3. Identification of hydrochemical control factors for groundwater in the Datong River basin (**a**) $[Na^+]/[Na^+ + Ca^{2+}]$ VS TDS; (**b**) $[Cl^-]/[Cl^- + HCO_3^-]$ VS TDS [28].

4.2. Impact of Anthropogenic Activities on Groundwater Hydrochemistry in the Datong River Basin

4.2.1. Identifying Anthropogenic Influence Through Contrast Coefficient Variance

Groundwater chemistry is shaped by two overarching drivers: natural processes—predominantly water–rock interaction—and anthropogenic inputs [29]. Apportioning solute provenance, therefore, requires explicit discrimination between these end-members and quantification of their relative dominance. Constituents subjected to human disturbance typically exhibit markedly wider concentration ranges and elevated temporal variability relative to their geogenic counterparts [18]. Contrast-coefficient variance analysis can thus be used to determine whether the evolution of a given ion assemblage is dominated by natural processes or by anthropogenic inputs, and to rank individual solutes according to the degree of human perturbation.

Contrast-coefficient variances (σi^2) were computed for each major ion in the Datong River Basin dataset and are summarized in Table 3. Values of $\sigma^2 > 1$ for Cl^-, Na^+, SO_4^{2-} and Fe unequivocally flag these solutes as strongly impacted by anthropogenic inputs. $\sigma^2 > 0.5$ for K^+ and Mn indicates a moderate human imprint, whereas the low variances of HCO_3^- and Ca^{2+} imply that their distributions are governed predominantly by natural water–rock interaction rather than by external perturbations.

Table 3. Variance of contrast coefficients of groundwater chemical components in the Datong River Basin.

Parameters	Value
K^+	0.704
Na^+	2.899
Ca^{2+}	0.303
Mg^{2+}	0.441
SO_4^{2-}	1.473
Cl^-	5.414
HCO_3^-	0.200
NO_3^-	1.691
Fe	2.740
Mn	0.655

4.2.2. Identify Anthropogenic Impacts on Groundwater Chemistry by Using Ionic Ratios Methods

Ionic ratios are widely used as qualitative tracers of anthropogenic impacts on groundwater. Mining effluent is typically characterized by elevated $[SO_4^{2-}/Ca^{2+}]$, whereas

 https://doi.org/10.3390/w18010105

agricultural inputs yield high [NO_3^-/Ca^{2+}] [30]. Cross-plots of [NO_3^-/Ca^{2+}] versus [SO_4^{2-}/Ca^{2+}] can therefore discriminate between industrial, domestic and agricultural contamination. Figure 4 shows that the Datong River Basin samples do not cluster in any single source field, but plot in an intermediate position, indicating that groundwater chemistry is controlled by a mixture of industrial discharge, domestic sewage and agrochemical fertilizers. The monitoring wells selected for this study are all situated within 5 km of the Datong River. As evident in Figure 1, extensive farmland and rural settlements flank both sides of the riparian zone in the middle reach of the study area. Agrochemical residues, domestic sewage and industrial effluents discharged to the river readily infiltrate the shallow aquifer, constituting the principal sources of observed groundwater contamination.

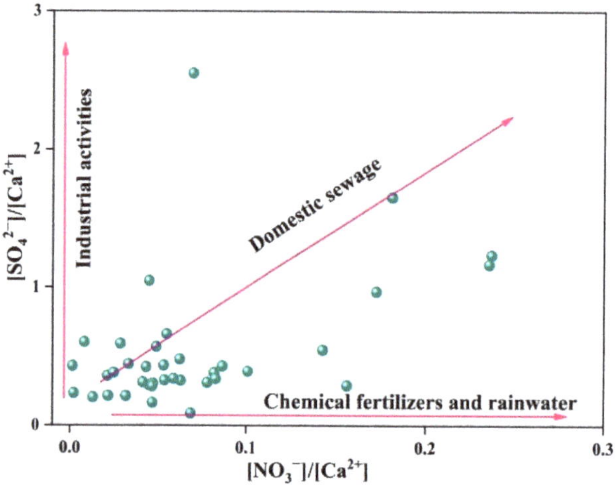

Figure 4. Relationship diagram of groundwater [NO_3^-/Ca^{2+}] VS [SO_4^{2-}/Ca^{2+}] in the Datong River Basin.

4.3. Sources of Major Ion Chemistry in the Datong River Basin Groundwater

To further elucidate the hydrochemical evolution and identify the sources of groundwater solutes, ionic-ratio coefficients were employed to quantify the governing hydrogeochemical processes and their genetic variability [31]. Molar ratios of the (Ca^{2+}/Na^+) and (Mg^{2+}/Na^+) were used to discriminate among silicate, carbonate and evaporite end-members. Published benchmark ratios indicate that silicate weathering yields $Ca^{2+}/Na^+ = 0.35 \pm 0.15$ and $Mg^{2+}/Na^+ = 0.24 \pm 0.12$, carbonate weathering produces 50 ± 20 and 20 ± 8, respectively, whereas evaporite dissolution gives 0.17 ± 0.09 and 0.02 ± 0.013 [32]. Groundwater in the Datong River Basin exhibits Ca^{2+}/Na^+ and Mg^{2+}/Na^+ ranges of 0.15–7.98 and 0.24–17.17 (Figure 5), clustering between the silicate and carbonate fields but skewed toward the silicate end-member. These signatures demonstrate that basin-wide chemistry is governed predominantly by silicate weathering, with a subsidiary contribution from carbonate dissolution.

The molar Na^+/Cl^- ratio is widely used to constrain the origin of these conservative ions in groundwater systems. When Na^+ and Cl^- are derived exclusively from halite dissolution, a stoichiometric Na^+/Cl^- ratio of unity is expected [33]. Figure 6a shows a strong linear relationship between Na^+ and Cl^- ($R^2 = 0.867$), with most samples plotting slightly above the 1:1 line, indicating that halite dissolution is the primary source of both ions. The modest excess of Na^+ relative to Cl^- further implies an additional contribution, maybe from silicate weathering.

https://doi.org/10.3390/w18010105

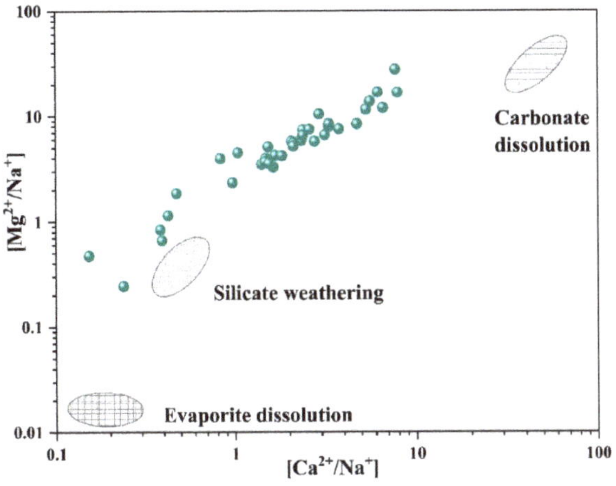

Figure 5. Relationship between the molar concentration ratios of $[Ca^{2+}/Na^{+}]$ and $[Mg^{2+}/Na^{+}]$.

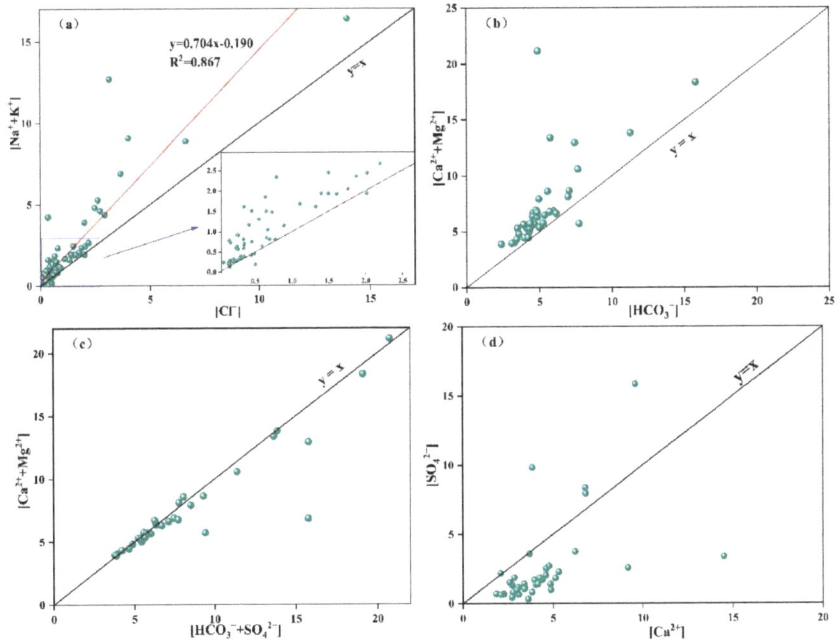

Figure 6. Relationship between the molar concentration ratios of (**a**) $[Cl^{-}]$ and $[Na^{+}]$; (**b**) $[Ca^{2+} + Mg^{2+}]$ and $[HCO_3^{-}]$; (**c**) $[Ca^{2+} + Mg^{2+}]$ and $[HCO_3^{-} + SO_4^{2-}]$; (**d**) $[Ca^{2+}]$ and $[SO_4^{2-}]$.

The bivariate plot of $[Ca^{2+} + Mg^{2+}]$ versus $[HCO_3^{-}]$ is commonly employed to discriminate the origin of alkaline-earth cations and bicarbonate [34]. A 1:1 stoichiometry (y = x) signifies exclusive carbonate dissolution, whereas simultaneous carbonate and sulfate weathering is indicated by a 1:1 relation between $[Ca^{2+} + Mg^{2+}]$ and $[HCO_3^{-} + SO_4^{2-}]$ [35]. In Figure 6b, the majority of Datong River Basin samples plot above the y = x line relative to $[Ca^{2+} + Mg^{2+}]$ versus $[HCO_3^{-}]$, demonstrating that carbonate weathering is supplemented by additional Ca^{2+} and Mg^{2+} released from silicate and/or sulfate mineral dissolution.

 https://doi.org/10.3390/w18010105

Figure 6c shows that the samples scatter symmetrically around the y = x line for [Ca^{2+} + Mg^{2+}] versus [HCO_3^- + SO_4^{2-}], corroborating that carbonate and sulfate weathering jointly supply Ca^{2+} and Mg^{2+}. In contrast, the [Ca^{2+}] versus [SO_4^{2-}] plot (Figure 6d) reveals most points below the 1:1 line with negligible correlation, ruling out significant evaporite dissolution.

4.4. Source Apportionment of Groundwater Contaminants in the Datong River Basin

Accurate source identification is a prerequisite for effective groundwater-quality management [36]. Here, a PMF model was employed to apportion contaminant sources in the Datong River Basin. Thirteen routinely monitored parameters—TDS, TH, Na^+, K^+, Ca^{2+}, Mg^{2+}, HCO_3^-, SO_4^{2-}, NO_3^-, Cl^-, COD, Fe and Mn—were incorporated into the PMF receptor model to quantify the contribution of each pollution source.

For the PMF 5.0 analysis, concentration matrices and their corresponding uncertainty estimates for the 13 variables in 37 groundwater samples were imported. Iterative refinement of the "strong"/"weak" specification and factor number was performed until the coefficient of determination (R^2) between measured and predicted concentrations exceeded 0.5 for every species, and the difference between the calculated Q value and its theoretical minimum was <10%. A three-factor solution was ultimately identified as optimal.

Factor 1 accounts for 19.2% of the total variance and is dominated by Mn, with appreciable loadings of Na^+ and Cl^- (Table 4). Cl^- in aquatic environments is chiefly derived from the natural dissolution of halide-bearing minerals, agrochemicals (e.g., potash and other chloride-based fertilizers), animal-waste leachate, septic-tank effluent and road-deicing salts [37,38]. In this study, Section 4.3 demonstrated that halite dissolution is the primary source of Na^+ and Cl^- in the basin. Mn is typically released from lithogenic weathering, domestic effluent, industrial discharge or agrochemicals [39]. Although Mn displays a high loading on Factor 1, its concentration was below the detection limit at 68% of the groundwater sites (median value is 0.005 mg L^{-1}). Given its co-variation with Na^+ and Cl^- and its overall low abundance, we attribute Factor 1 to natural origin; the PMF apportionment indicates that this end-member contributes 79.1% and 37.3% of the observed Mn and Na^+, respectively (Figure 7).

Table 4. Source profiles obtained from the PMF model.

Parameters	Factor 1	Factor 2	Factor 3
K^+	0.249	1.53	0.682
Na^+	3.08	5.18	0.000
Ca^{2+}	7.04	51.9	3.33
Mg^{2+}	2.11	14.0	1.34
SO_4^{2-}	4.91	29.5	0.000
Cl^-	1.22	3.71	0.270
HCO_3^-	34.2	186	16.7
NO_3^-	0.026	1.07	0.170
Fe	0.007	0.041	0.151
Mn	0.015	0.002	0.002
TDS	48.9	246	7.55
TH	27.4	204	13.3
COD	0.174	0.883	0.171
Possible sources	Natural sources	Agricultural non-point pollution sources, domestic sewage and water–rock interaction	Industrial wastewater
Contribution (%)	19.2	67.5	13.3

https://doi.org/10.3390/w18010105

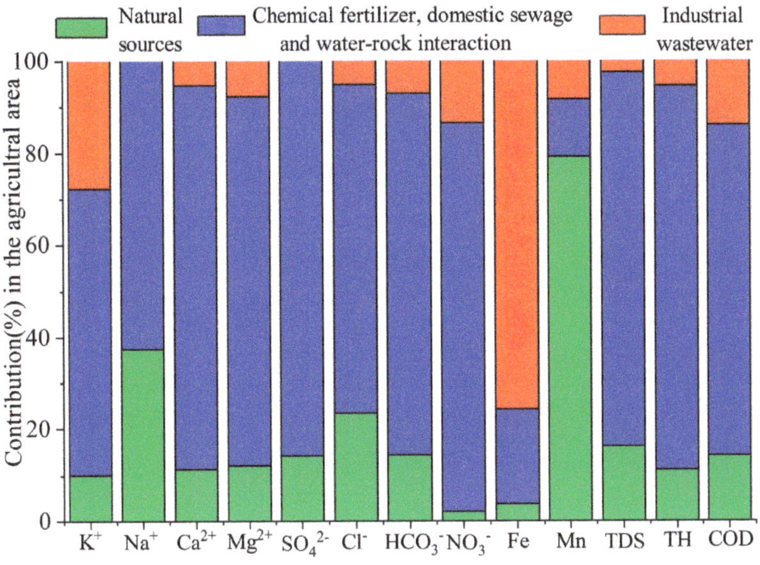

Figure 7. Cumulative contribution rate of PMF source apportionment for groundwater in the Datong River basin.

Factor 2 explains 67.5% of the total variance and exhibits high loadings for TDS, TH, Na^+, K^+, Ca^{2+}, Mg^{2+}, HCO_3^-, SO_4^{2-}, NO_3^-, Cl^- and COD (Table 4). Major cations (Na^+, K^+, Ca^{2+} and Mg^{2+}) and anions (HCO_3^- and SO_4^{2-}) are conventionally ascribed to silicate and carbonate weathering [40], a pathway corroborated for Ca^{2+}, Mg^{2+}, HCO_3^- and SO_4^{2-} in Section 4.3. Typically, NO_3^- and Cl^- in aquatic environments originate primarily from agricultural fertilizers, manure and sewage effluents and atmospheric deposition [41,42]. In this study, Section 4.2 further demonstrates that nitrate in the basin originates predominantly from chemical fertilizer, domestic sewage and industrial effluents. Consequently, Factor 2 is interpreted as a mixed source integrating domestic sewage, agrochemicals and natural weathering. According to the PMF apportionment (Figure 7), this mixed end-member contributes 62.1% of K^+, 62.7% of Na^+, 83.3% of Ca^{2+}, 80.2% of Mg^{2+}, 85.7% of SO_4^{2-}, 71.3% of Cl^-, 78.5% of HCO_3^-, 84.5% of NO_3^-, 81.3% of TDS, 83.4% of TH and 71.9% of COD.

Factor 3 accounts for 13.3% of the total variance and is dominated exclusively by Fe (Table 4). Elevated dissolved Fe in aquifers can originate from lithogenic weathering, industrial effluents, agrochemicals or urban runoff [43]. Factors 1 and 2 already capture rock weathering and the combined influence of domestic sewage/agrochemicals, respectively; the mutual orthogonality of the three factors therefore implies that Factor 3 represents an independent industrial source. Indeed, several large industrial parks—including metallurgical, chemical and hydropower facilities—are located along the middle-lower reaches of the Datong River. Fe-bearing process waters, metal scraps and atmospheric particulates generated during production, processing and transportation can infiltrate via rainfall runoff or riverbank leakage, elevating Fe concentrations in the adjacent aquifer. This interpretation is consistent with the observed 16.2% exceedance rate of the Chinese Class III standard for Fe. PMF apportionment attributes 75.7% of the measured Fe to this industrial end-member (Figure 7).

https://doi.org/10.3390/w18010105

5. Limitations and Future Perspectives of This Study

This study integrated hydrochemical characterization with PMF to quantitatively elucidate the hydrochemical signatures of an arid-region aquifer and apportion contaminant sources. The resultant framework provides a scientific basis for groundwater quality management in arid basins worldwide that share comparable hydrogeological settings. Nevertheless, three limitations warrant acknowledgement and outline future research directions:

(1) Sampling Bias Toward the Low Flow Period: All groundwater samples were collected during a single dry season campaign; hydrochemical facies and pollution vectors may differ markedly under high flow or transitional regimes typical of north-western arid catchments. Future efforts should adopt high-frequency sampling strategies (e.g., wet-season, dry-season and intermediate-flow periods) and nested temporal scales to capture the full spectrum of hydrochemical variability.

(2) Source Apportionment Uncertainty for Multi-Source Solutes: Although PMF reliably resolved major contamination patterns, quantitative attribution of redox-sensitive or biogeochemically reactive species (e.g., NO_3^-, SO_4^{2-}) remains equivocal because of their numerous sources and complex transformation pathways. Coupling PMF with dual-isotope fingerprints ($\delta^{15}N$–$\delta^{18}O$ for nitrate; $\delta^{34}S$–$\delta^{18}O$ for sulfate) will refine source partitioning and constrain reaction progress [44,45].

(3) Potential of Machine-Learning Augmentation: Advanced algorithms such as random forest regression, support vector machines and gradient boosting models can be integrated with hydrochemical datasets to disentangle non-linear controls on water quality evolution, rank predictor importance and forecast future hydrochemical trajectories under changing land-use or climate scenarios.

6. Conclusions

This study provides an integrated characterization of groundwater hydrochemistry and pollution source apportionment in the Datong River Basin, a typical arid catchment in north-western China. The aquifer is dominated by HCO_3–Ca(Mg) water types, controlled primarily by silicate and carbonate weathering, with secondary modification by evapo-concentration. Anthropogenic impacts are evident in elevated Cl^-, Na^+, SO_4^{2-}, NO_3^- and Fe concentrations. Positive-Matrix Factorization resolved three independent sources: natural weathering, a mixed source (domestic sewage + agriculture + weathering) and an industrial source, each dominantly contributing to specific chemical components. Although the overall water quality is good, localized contamination trends are emerging and require targeted control measures. Future work should adopt high-frequency sampling strategies, integrate isotopic tracers and employ machine-learning algorithms to enhance the accuracy and predictive capacity of source apportionment. Our findings offer a methodological benchmark and geochemical dataset for groundwater management and protection in analogous arid regions worldwide.

Supplementary Materials: The following supporting information can be downloaded at: https://www.mdpi.com/article/10.3390/w18010105/s1, Figure S1: Hydrogeological cross-section of the study area [46]; Table S1: Relative weight of water quality parameters and water quality standard (all units of the parameters are in mg/L except pH). References [16,17,46] are cited in Supplementary Materials.

Author Contributions: T.L., investigation, methodology and writing—original draft preparation and editing. J.K., investigation, methodology, data curation and supervision. Y.Y., investigation, methodology and data curation. Y.Q., investigation, methodology, software and data curation. Z.Z.,

https://doi.org/10.3390/w18010105

supervision, methodology and writing—review and editing. All authors have read and agreed to the published version of the manuscript.

Funding: This research received no external funding.

Data Availability Statement: The data presented in this study are available from the corresponding author upon request due to privacy restrictions.

Acknowledgments: The authors gratefully acknowledge the editor and anonymous reviewers for their valuable comments on this manuscript. The authors also appreciate the financial support from the different organizations.

Conflicts of Interest: The authors declare no conflicts of interest.

References

1. Wei, S.; Zhang, Y.; Cai, Z.; Bi, D.; Wei, H.; Zheng, X.; Man, X. Evaluation of groundwater quality and health risk assessment in Dawen River Basin, North China. *Environ. Res.* **2025**, *264*, 120292. [CrossRef]
2. Li, M.Y.; Xie, Y.Q.; Dong, Y.H.; Wang, L.H.; Zhang, Z.Y. Review: Recent progress on groundwater recharge research in arid and semiarid areas of China. *Hydrogeol. J.* **2024**, *32*, 9–30. [CrossRef]
3. Hinton, R.G.; Kalin, R.M.; Banda, L.C.; Kanjaye, M.B.; MacLeod, C.J.; Troldborg, M.; Kamtukule, S. Mixed-method analysis of anthropogenic groundwater contamination of drinking-water sources in Malawi. *Sci. Total Environ.* **2024**, *957*, 177418. [CrossRef] [PubMed]
4. Wang, L.; Yang, N.; Zhao, Y.; Zhang, Q. Research on the features and driving factors of shallow groundwater quality in arid areas, Northwest China. *Water* **2025**, *17*, 934. [CrossRef]
5. Wang, J.; Xue, B.; Wang, Y.; Wang, G.; Han, D. Identification of pollution source and prediction of water quality based on deep learning techniques. *J. Contam. Hydrol.* **2024**, *261*, 104287. [CrossRef]
6. Ruan, D.; Bian, J.; Wang, Y.; Wu, J.; Gu, Z. Identification of groundwater pollution sources and health risk assessment in the Songnen Plain based on PCA-APCS-MLR and trapezoidal fuzzy number-Monte Carlo stochastic simulation model. *J. Hydrol.* **2024**, *632*, 130897. [CrossRef]
7. Chen, L.; Wang, G.; Hu, F.; Wang, Y.; Liu, L. Groundwater hydrochemistry and isotope geochemistry in the Turpan Basin, northwestern China. *J. Arid Land* **2014**, *6*, 378–388. [CrossRef]
8. Zhang, Q.; Wang, H.; Wang, Y.; Yang, M.; Zhu, L. Groundwater quality assessment and pollution source apportionment in an intensely exploited region of northern China. *Environ. Sci. Pollut. Res.* **2017**, *24*, 16639–16650. [CrossRef] [PubMed]
9. Herojeet, R.; Rishi, M.S.; Lata, R.; Sharma, R. Application of environmetrics statistical models and water quality index for groundwater quality characterization of alluvial aquifer of Nalagarh Valley, Himachal Pradesh, India. *Sustain. Water Resour. Manag.* **2016**, *2*, 39–53. [CrossRef]
10. Gu, H.; Chi, B.; Li, H.; Jiang, J.; Qin, W.; He, W. Assessment of groundwater quality and identification of contaminant sources of Liujiang basin in Qinhuangdao, North China. *Environ. Earth Sci.* **2015**, *73*, 6477. [CrossRef]
11. Wang, L.; Zhang, Q.; Wang, H. Rapid urbanization has changed the driving factors of groundwater chemical evolution in the large groundwater depression funnel area of northern China. *Water* **2023**, *15*, 2917. [CrossRef]
12. Wang, Z.; Li, X.; Hou, X. Hydrogeochemistry of river water in the upper reaches of the Datong River Basin, China: Implications of anthropogenic inputs and chemical weathering. *Acta Geol. Sin.-Engl.* **2021**, *95*, 962–975. [CrossRef]
13. Yi, B.; Liu, J.T.; Lü, X.L.; He, W.; Zhu, L.; Zhang, Y.X.; Yang, M.N. Hydrochemical and isotopic evidence for groundwater conversion of surface water in alpine arid areas: A case study of Datong River Basin. *Environ. Sci.* **2023**, *44*, 752–760. (In Chinese)
14. Jie, F.; Fei, L.; Li, S.; Hao, K.; Liu, L.; Li, J.; Liu, N. Quantitative effects of vadose zone thickness on delayed recharge of groundwater for an irrigation district in an arid area of Northwest China. *J. Hydrol. Reg. Stud.* **2022**, *40*, 101022. [CrossRef]
15. Ma, B.; Li, D. Stages of the neotectonic movement of the menyuan basin in the middle segment of the qilian mountains. *J. Geomech.* **2008**, *14*, 201–211. (In Chinese)
16. *GB/T 14848-2017*; Standard for Groundwater Quality. Ministry of Natural Resources of the People's Republic of China: Beijing, China, 2017.
17. WHO. *Guidelines for Drinking-Water Quality*, 4th ed.; World Health Organization: Geneva, Switzerland, 2017.
18. Ren, C.; Zhang, Q. Groundwater chemical characteristics and controlling factors in a region of Northern China with intensive human activity. *Int. J. Environ. Res. Public Health* **2020**, *17*, 9126. [CrossRef]
19. Wang, Q.; Ou, X.; Wu, C.; Yao, S.; Huang, S.; Zhu, C.; Liao, Z.; Wang, K.; Zhang, S.; Chen, J. Machine learning-driven PMF modeling for accurate and objective source identification of VOCs. *Atmos. Pollut. Res.* **2025**, 102721. [CrossRef]

20. Mao, K.; Wang, X.; Yu, K.; Li, M.; Wang, Y.; Gao, G.; Geng, S.; Song, H.; Ning, W.; An, H.; et al. Source apportionment and ecological risk assessment of heavy metals in sediments of Dongting Lake based on PCA-PMF model. *Sci. Rep.* **2025**, *15*, 32026. [CrossRef] [PubMed]
21. Yang, C.; Chen, S.; Dong, J.; Zhang, Y.; Wang, Y.; Kang, W.; Zhang, X.; Liang, Y.; Fu, D.; Yan, Y.; et al. Hydrochemical characteristics, controlling factors, and high nitrate hazards of shallow groundwater in an urban area of southwestern China. *Toxics* **2025**, *13*, 516. [CrossRef] [PubMed]
22. Martín del Campo, M.A.; Esteller, M.V.; Expósito, J.L.; Hirata, R. Impacts of urbanization on groundwater hydrodynamics and hydrochemistry of the Toluca Valley aquifer, Mexico. *Environ. Monit. Assess.* **2014**, *186*, 2979–2999. [CrossRef]
23. Piper, A.M. A graphic procedure in the geochemical interpretation of water-analyses. *Eos Trans. Am. Geophys. Union* **1944**, *25*, 914–928.
24. Li, B.; Wu, D.; Yu, D.; Li, C.; Wang, J.; Xing, L.; Wang, Y. Comprehensive analysis of groundwater hydrochemistry and nitrate health risks in the Baiquan basin, Northern China. *PeerJ* **2025**, *13*, e19233. [CrossRef] [PubMed]
25. Liu, Y.; Zhou, L.; Ma, X.; Wen, D.; Li, W.; Shi, Z. The genesis and hydrochemical formation mechanism of karst springs in the central region of Shandong Province, China. *Water* **2025**, *17*, 1805. [CrossRef]
26. Chen, W.; Wu, C.; Pan, S.; Shi, L. Analysis on the spatiotemporal evolutions of groundwater hydrochemistry and water quality caused by over-extraction and seawater intrusion in eastern coastal China. *Front. Earth Sci.* **2024**, *12*, 1391235. [CrossRef]
27. Zhou, B.; Wang, H.; Zhang, Q. Assessment of the evolution of groundwater chemistry and its controlling factors in the Huangshui River basin of northwestern China, using hydrochemistry and multivariate statistical techniques. *Int. J. Environ. Res. Public Health* **2021**, *18*, 7551. [CrossRef]
28. Gibbs, R.J. Mechanisms controlling world water chemistry. *Science* **1970**, *170*, 1088–1090. [CrossRef]
29. Wu, D.; Li, B.; Li, Y.; Li, Q.; Sheng, C.; Liu, J.; Wang, M.; Min, Y.; Feng, J.; Zhang, Y.; et al. Characterization of groundwater hydrochemistry and temporal dynamics of water quality in the northern Baiquan Spring Basin. *Water* **2024**, *16*, 2519. [CrossRef]
30. Zhang, Q.; Wang, H.; Xu, Z.; Li, G.; Yang, M.; Liu, J. Quantitative identification of groundwater contamination sources by combining isotope tracer technique with PMF model in an arid area of northwestern China. *J. Environ. Manag.* **2023**, *325*, 116588. [CrossRef]
31. Zhu, L.; Lu, W.; Luo, C. A high-precision and interpretability-enhanced direct inversion framework for groundwater contaminant source identification using multiple machine learning techniques. *J. Hydrol.* **2025**, *659*, 133237. [CrossRef]
32. Jin, X.; Long, W.; Jie, Z.; Tianlun, Z. Hydrochemical variation characteristics and driving factors of surface water in arid areas—A case study of Beichuan River in Northwest China. *Front. Environ. Sci.* **2024**, *12*, 1493390. [CrossRef]
33. Pant, R.R.; Zhang, F.A.; Rehman, F.U.; Wang, G.; Ye, M.; Zeng, C.; Tang, H. Spatiotemporal variations of hydrogeochemistry and its controlling factors in the Gandaki River Basin, Central Himalaya, Nepal. *Sci. Total Environ.* **2018**, *622–623*, 770–782. [CrossRef]
34. Niu, C.; Zhang, Q.; Xiao, L.; Wang, H. Spatiotemporal variation in groundwater quality and source apportionment along the Ye River of North China using the PMF model. *Int. J. Environ. Res. Public Health* **2022**, *19*, 1779. [CrossRef]
35. Li, P.Y.; Wu, J.H.; Qian, H. Assessment of groundwater quality for irrigation purposes and identification of hydrogeochemical evolution mechanisms in Pengyang County, China. *Environ. Earth Sci.* **2013**, *69*, 2211–2225. [CrossRef]
36. Le, T.V.; Nguyen, B.T. Heavy metal pollution in surface water bodies in provincial Khanh Hoa, Vietnam: Pollution and human health risk assessment, source quantification, and implications for sustainable management and development. *Environ. Pollut.* **2024**, *343*, 123216. [CrossRef]
37. Liu, C.Q.; Li, S.L.; Lang, Y.C.; Xiao, H.Y. Using $\delta^{15}N$-and $\delta^{18}O$ values to identify nitrate sources in karst ground water, Guiyang, Southwest China. *Environ. Sci. Technol.* **2006**, *40*, 6928–6933. [CrossRef]
38. Wei, Y.; Li, Y.; Zhang, L.; Liu, C.; Meng, Q.; Yin, J.; Wang, L. A Study on the Hydrochemical Evolution Property and Pollution Source Attribution of Groundwater in Highly Urbanized Areas: A Case Study of Shenzhen City. *Water* **2025**, *17*, 2945. [CrossRef]
39. Wei, D.; Yang, S.; Zou, L.; Torres-Martínez, J.A.; Zheng, Y.; Hu, Q.; Zhang, Y. Appraisal of potential toxic elements pollution, sources apportionment, and health risks in groundwater from a coastal area of SE China. *J. Environ. Manag.* **2025**, *377*, 124691. [CrossRef]
40. Li, Y.; Zhang, Z.; Fei, Y.; Chen, H.; Qian, Y.; Yu, D. Investigation of quality and pollution characteristics of groundwater in the Hutuo River alluvial plain, North China Plain. *Environ. Earth Sci.* **2016**, *75*, 581. [CrossRef]
41. Li, L.; Li, P.; He, S.; Wang, D.; Tian, Y.; Niu, L. Groundwater nitrate pollution source apportionment under varying land use/land cover patterns. *Expo. Health* **2025**, *17*, 353–369. [CrossRef]
42. Zheng, J.; Guo, X.; Cai, Y.; Liu, J.; Wang, D.; Gao, X.; Tan, B. Source apportionment and health risks of nitrate pollution in shallow groundwater in the agricultural northern Xiaoxing'an Mountains region of China. *J. Hydrol. Reg. Stud.* **2025**, *59*, 102394. [CrossRef]
43. Bu, H.; Tan, X.; Li, S.; Zhang, Q. Temporal and spatial variations of water quality in the Jinshui River of the south Qinling Mts., China. *Ecotoxicol. Environ. Saf.* **2010**, *73*, 907–913. [CrossRef] [PubMed]

 https://doi.org/10.3390/w18010105

44. Liu, W.; Jiang, H.; Guo, X.; Li, Y.; Xu, Z. Time-series monitoring of river hydrochemistry and multiple isotope signals in the Yarlung Tsangpo River reveals a hydrological domination of fluvial nitrate fluxes in the Tibetan Plateau. *Water Res.* **2022**, *225*, 119098. [CrossRef] [PubMed]
45. Jakóbczyk-Karpierz, S.; Ślósarczyk, K. Isotopic signature of anthropogenic sources of groundwater contamination with sulfate and its application to groundwater in a heavily urbanized and industrialized area (Upper Silesia, Poland). *J. Hydrol.* **2022**, *612*, 128255. [CrossRef]
46. Yi, B. Response of Organic Matter to Surface Water-Groundwater Interactions in the Datong River Basin—A Case Study in the Datong River Basin. Master's Thesis, China University of Geosciences, Beijing, China, 2023. [CrossRef]

 https://doi.org/10.3390/w18010105

water

MDPI

Article

Seasonal and Spatial Variations in Riverine Functional Genes Related to Phosphorus Cycling and Their Responses to Environmental Factors in the Chishui River Basin

Junhai Wu [1,2], Yufeng Xiao [2], Qiusheng Wu [2,3], Qingguang Li [1,*], Yun He [2], Yu Tang [2] and Jingfu Wang [2,3,4,*]

[1] School of Resources and Environmental Engineering, Guizhou University, Guiyang 550025, China; w1849695523@outlook.com
[2] State Key Laboratory of Environmental Geochemistry, Institute of Geochemistry, Chinese Academy of Sciences, Guiyang 550081, China
[3] University of Chinese Academy of Sciences, Beijing 100049, China
[4] Guizhou Province Field Scientific Observation and Research Station of Hongfeng Reservoir Ecosystem, Guiyang 551499, China
* Correspondence: leeqg12@163.com (Q.L.); wangjingfu@vip.skleg.cn (J.W.); Tel.: +86-15885511070 (J.W.); Fax: +86-0851-84396010 (J.W.)

Abstract

In our research we analyzed a series of water quality parameters and conducted a metagenomic analysis of the microbial community in the Chishui River (in southwestern China), aiming to explore the microbial driving mechanisms of the phosphorus cycle in the river ecosystem. The research results indicated that the concentrations of total phosphorus (TP) and soluble reactive phosphorus (SRP) were higher in summer, suggesting seasonal differences in exogenous input and water body biogeochemical processes. The concentration of manganese (Mn) is higher in autumn, and it shows a significant positive correlation with Soluble reactive phosphorus (SRP). This may indicate the contribution of endogenous release from sediments to phosphorus in the water body. There were significant differences in the abundance of phosphorus cycling functional genes between summer and autumn. For example, in summer, the abundances of high-affinity phosphate transporter (*pstB*), inorganic phosphate dissolution (*pqqC*), and polyphosphate decomposition (*ppx*) were significantly higher. This might be to adapt to high productivity and the potential lack of phosphorus, or it could be that the microorganisms carrying these genes have a greater advantage during the summer. In contrast, the relative abundance of phosphonate (*phn*) and glycerophosphate (*ugpQ*) was significantly higher in autumn, indicating that the metabolic focus of the microorganisms has shifted towards the utilization of organic phosphorus, or that the microorganisms that are adept at utilizing organic phosphorus have taken the dominant ecological position in this situation. Moreover, the analysis of the microbial community showed that the Proteobacteria phylum (Pseudomonad phylum) was the main phylum, and the relative abundance of key functional bacterial genera (such as *Limnohabitans*, *Acinetobacter*) reflected seasonal changes, which was consistent with the above functional gene patterns. Spearman correlation analysis indicated that environmental physical and chemical parameters (such as iron, dissolved oxygen, dissolved organic carbon, pH value) jointly regulated the composition and distribution of phosphorus cycling functional genes. Our research results demonstrated that the microbial community plays a crucial regulatory role in the biogeochemical cycle of the river ecosystem through the transformation of functional genes and the changes in community structure. The research results emphasize that attention must be paid to the phosphorus cycling process regulated by microorganisms and its impact, in order to control water body eutrophication and maintain the stability of the ecosystem.

Keywords: phosphorus; Chishui river; phosphorus cycling functional genes; metagenomics analysis

1. Introduction

As a crucial component of global water resources, the water quality of rivers is significantly affected by human activities. Excessive nutrients from human sources such as fertilizer application in farmlands and discharge of domestic sewage often lead to deterioration of river water quality and reduction in ecosystem diversity and stability [1,2]. Among them, phosphorus is a key element that limits primary productivity in aquatic ecosystems. Excessive input of phosphorus can promote the rapid proliferation of algae, trigger harmful algal blooms, release toxic substances, and significantly reduce the diversity of microbial communities and dissolved organic matter, thereby weakening the structural and functional balance of the ecosystem [3–5]. Therefore, a thorough understanding of the phosphorus cycling mechanism in water bodies is of great significance for revealing the mechanism of eutrophication and maintaining the health of river ecosystems.

Microorganisms play a crucial role in the biogeochemical functions of freshwater systems and are the core drivers of the food web and nutrient cycling [6]. Bacteria and other microorganisms directly regulate the form and transformation of phosphorus in water bodies through a series of processes such as absorption, release, adsorption, precipitation, and dissolution. These processes are strictly regulated by microbial functional genes, which determine the functional roles of microorganisms in the ecosystem. Under conditions of limited soluble reactive phosphorus (SRP), microbial communities can enhance the synergy and coupling of phosphorus cycling functional genes to improve phosphorus utilization efficiency [7]. As the availability of nutrients in water bodies changes, the metabolic strategies of microorganisms will also be adjusted accordingly. For example, in phosphorus stress environments, some microorganisms regulate the abundance of genes related to organic phosphorus mineralization, inorganic phosphorus dissolution, phosphorus absorption and transport, and phosphorus starvation responses to adapt to low phosphorus environments. Multiple studies have shown that dissolved organic phosphorus (DOP) is an efficient phosphorus source that microorganisms can easily utilize [8], and its remineralization rate is significantly higher than that of dissolved organic carbon and nitrogen, further highlighting the core role of microorganisms in the rapid transformation of phosphorus [9]. In open ocean areas, the presence of phosphorus-rich organic matter promotes the growth of phytoplankton and can even lead to the massive proliferation of algae [10]. Moreover, human activities such as dam construction can increase the abundance and coupling of microbial carbon, nitrogen, phosphorus, and sulfur cycling functional genes and accelerate the overall nutrient cycling rate of the system [11]. These studies all indicate that the phosphorus cycle has a profound impact on the functions of water body ecosystems, and microorganisms and their functional genes are key research objects for analyzing the biogeochemical behavior of phosphorus.

The key functional genes that regulate phosphorus cycling mainly participate in processes such as organic phosphorus mineralization, inorganic phosphorus dissolution, phosphorus transport, and phosphorus starvation response [12,13]. For example, genes encoding alkaline phosphatase (such as *phoD*, *phoA*), acidic phosphatase (such as *phoN*, *olpA*),

 https://doi.org/10.3390/w18040456

inorganic pyrophosphatase (*ppa*), and C-P lyase (*phn*) give microorganisms the ability to decompose organic phosphorus compounds and dissolve inorganic phosphorus [14]. Among them, microorganisms carrying genes such as *phoC* and *phoD* are particularly sensitive to external phosphorus input [15], and by enhancing organic phosphorus mineralization and inorganic phosphorus dissolution, they promote the release of soluble phosphorus, thereby increasing the bioavailability of phosphorus [12]. The activity of phosphorus cycling and related microorganisms is regulated by various environmental factors. Studies have shown that an increase in temperature can promote the release of bioavailable phosphorus in lake sediments [16,17]; in arid ecosystem, the relative abundance of genes related to organic phosphorus mineralization (such as *ppx*, *phoD*, *phn*) is significantly negatively correlated with soil moisture content [18]; and in river environments, warming accelerates the decomposition of animal carcasses, further enhancing the complexity of the phosphorus cycling gene network [19]. Therefore, Studying the spatiotemporal changes in genes related to the phosphorus cycling process is helpful for understanding the influence of human and natural factors on the phosphorus cycle in river ecosystems from the perspective of microbial mechanisms, and for assessing its potential responses to environmental changes. The aim of this study is to clarify: (1) The transformation potential of phosphorus mediated by microorganisms at different temporal and spatial scales; (2) The regulatory mechanisms of microbial functions influenced by natural and human factors.

The Chishui River, as a first-class tributary of the Yangtze River in the upper reaches, is a crucial water source for the production of renowned liquor along its banks. It also meets the daily water needs of the residents in the basin. Therefore, this study focuses on the Maotai section of the Chishui River. Based on metagenomic sequencing technology, we systematically analyzes the impact of key environmental factors such as temperature and total phosphorus (TP) in the water bodies during summer and autumn on the functional genes related to phosphorus cycling and the microbial communities, in order to provide scientific basis for in-depth understanding of the mechanism of water body eutrophication and the maintenance of the diversity and stability of the river ecosystem.

2. Materials and Methods

2.1. Water Sample Collection and Physicochemical Parameters Measurement

During July and September 2023, surface water samples were collected from 12 sampling points in the upper, middle, lower sections of the Maotai section of the Chishui River, as well as its tributaries (Figure S1). Water samples were collected using sterile water sampling bags at each sampling point, with 10 L of water collected from each point. The collected water samples were sent back to the laboratory on the same day and were then processed and stored. At each sampling point, 1500 mL of water was taken and three replicates were made for the microbial-related measurements. The samples were filtered using 0.22 μm acetic acid fiber membranes. The filtered membranes were stored in an ultra-low temperature refrigerator at −80 °C, and then used for DNA extraction. In the field, a multi-parameter water quality analyzer (YSI, 6600V2, Yellow Springs, OH, USA) was used to measure the pH value, water temperature (WT), and dissolved oxygen (DO) content of the water samples. In the laboratory, total nitrogen (TN), total phosphorus (TP), reactive phosphorus (SRP) were determined by spectrophotometry (UV1200, Lab Tech, Shanghai, China), chemical oxygen demand (COD) was determined by the permanganate titration method, and anions and cations were analyzed by ion chromatography (IC-800, DIONEX, Osaka, Japan). Dissolved organic carbon (DOC) samples were filtered through a 0.45 μm glass fiber membrane and determined using a total organic carbon analyzer (Elementar, Langenselbold, Germany). The heavy metal content in water samples was

https://doi.org/10.3390/w18040456

determined by inductively coupled plasma mass spectrometry (IC-MS, Agilent 7700, Santa Clara, CA, USA). All chemical analyses included parallel samples and blank samples. The above analyses were all referenced to the series of documents (respectively, HJ 636-2012, GB 11893-1989, HJ 828-2017, HJ 700-2014, HJ 812-2016, HJ 84-2016) of "National Environmental Protection Standards of the People's Republic of China".

2.2. Metagenomic Sequencing and AnnoTAtion

The genomic DNA was extracted from the samples using the DNA extraction kit provided by Guangdong Magigene Biotechnology Co., Ltd. (Guangzhou, China). The integrity and purity of the DNA were assessed using 1% agarose gel [20]. The DNA concentration and purity were also measured using Qubit 4.0 (Thermo Fisher Scientific, Waltham, MA, USA) and NanoDrop One (Thermo Fisher Scientific, Waltham, MA, USA) [20]. The qualified DNA was used to construct a library using the ALFA-SEQ library construction kit and was sequenced on the Illumina or MGI platform with PE150 (Guangdong Magigene Biotechnology Co., Ltd., Guangzhou, China). The raw data after quality control were processed using the MEGAHIT (version 1.2.9) software for single-sample and mixed assembly, resulting in scaffolds longer than 500 bp [21]. Gene prediction was performed using Prodigal (version 2.6.3) [22], and a non-redundant gene catalog was constructed without redundancy using MMseqs2 (https://github.com/soedinglab/MMseqs2, accessed on 1 September 2025) [23]; gene abundance was calculated using BBMap (http://jgi.doe.gov/data-and-tools/bbtools/, accessed on 1 September 2025). Subsequently, species classification annotation was carried out based on the NCBI NR database using Diamond comparison and MEGAN's LCA algorithm (https://github.com/bbuchfink/diamond/, accessed on 1 September 2025) [24], and functional database annotations such as KEGG, eggNOG, and CAZy were completed simultaneously to obtain the abundance tables of species and phosphorus cycle-related functional genes.

2.3. Statistical Analysis

After obtaining the abundance data of phosphorus cycle-related functional genes, a difference analysis was conducted using the R software packages tidyverse (Version 2.0.0) [25] and restatix [26], and the phosphorus cycle functional genes that showed significant differences in seasons were selected. Based on this, the species abundance table of related microorganisms was screened out, and an α-diversity analysis was performed to study the species diversity within the samples. The richness, abundance-based coverage estimator (ACE), Shannon index, and Pielou index were calculated using the R software package vegan (version 2.7-2) [27]. The t-test function in R (version 4.1.2) was used to perform the t-test [28]. Principal coordinate analysis (PCoA) was conducted using the R software package ggClusterNet (version 2.0) [29], key microorganisms were screened using the R software package vegan, and redundancy analysis (RDA) was performed using the R software package vegan [27]. Spearman correlation analysis diagrams between environmental parameters and phosphorus cycle functional genes were calculated and plotted using Origin 2024 [30].

3. Results and Discussion

3.1. The Temporal and Spatial Variation Characteristics of River Environmental Parameters

Through systematic monitoring and assessment of the surface water environmental parameters of the Chishui River in July (summer) and September (autumn) of 2023, this study found that many indicators exhibited obvious temporal and spatial variation patterns (Figure S2, Table S1). For instance, total phosphorus (TP), soluble reactive phosphorus

 https://doi.org/10.3390/w18040456

(SRP), total nitrogen (TN), dissolved oxygen (DO), and pH values all changed obviously between seasons. TP decreased from 0.173 mg/L in summer to 0.117 mg/L in autumn, SRP from 0.074 mg/L to 0.041 mg/L, TN from 4.83 mg/L to 3.50 mg/L, indicating that external inputs might be more active in summer, possibly related to the increase in rainfall runoff [31]. DO increased from 6.02 mg/L to 7.70 mg/L, suggesting that microbial oxygen consumption was stronger in summer [32]; pH decreased from 8.45 to 8.00, which might be related to the greater consumption of CO_2 by photosynthesis in summer [33]. In autumn, the manganese (Mn) content obviously increased, especially in the contaminated areas. It is speculated that this might be due to the manganese ions released from the sediment-water interface under anoxic conditions, which entered the upper surface water through diffusion or re-suspension. In terms of spatial distribution, the contents of TP, SRP, and COD showed a consistent trend, indicating that these pollutants have a homogenous input characteristic [34]. Based on environmental indicators such as TP, SRP, TN, and COD, the river was divided into three regions: severely polluted area (YJ1, YJ2), characterized by high TP, SRP, and COD, especially with TP reaching 0.657 mg/L in summer; moderately polluted area (YJ3, MR1), with an unusually high TN in autumn (5.38 mg/L); and relatively clean area (the remaining sites). The trend of manganese concentration in the surface water in autumn is similar to that of dissolved reactive phosphorus (SRP) concentration. This may suggest that the internal release from the sediment is an important way to maintain high SRP concentration in the autumn tributaries of the Yanjin River (YJ1, YJ2, YJ3).

In conclusion, the chemical composition of the surface water of the Chishui River is the result of the combined action of external input, internal biological activities, and material exchange at the sediment-water interface. The seasonal variations in various environmental indicators not only reflect the temporal and spatial differences in external pollution, but also to some extent demonstrate the profound impact of the internal sources of sediments releasing through interface processes on surface water. The obvious. seasonal fluctuations of parameters such as TP, TN, and DO reveal the importance of summer as a critical period for pollution control, and the spatial division based on environmental indicators such as TP, COD, etc., clarifies that the Yanjin River tributaries and other areas should be prioritized for control.

3.2. *Spatio-Temporal Variation of Phosphorus Cycling Functional Genes*

At different levels of phosphorus cycling gene relative abundance, there were 4 genes that showed significant differences between the summer and autumn periods ($p < 0.05$), 2 genes with very significant differences ($p < 0.01$), and 12 genes with extremely significant differences ($p < 0.001$) (Figure 1, Table S2).

It can be clearly seen that most genes have significant relative abundance differences between summer and autumn (Figure 1). For example, the organic phosphorus mineralization gene shows significant seasonal differences, and the alkaline phosphatase genes *phoD* and *phoX* exhibit higher relative abundance in summer (such as *phoD*, $p < 0.05$), while multiple genes in the *phn* operon related to phosphonate degradation (such as *phnN*, *phnO*) are significantly increased in autumn. This finding indicates that microorganisms may utilize different organic phosphorus sources in different seasons. The high relative abundance of *phoD* and *phoX* in summer may be related to the potential limitation of soluble reactive phosphorus (SRP) in the water body, which enhances the demand for dissolved organic phosphorus (DOP) mineralization by microorganisms. Many studies have shown that *phoD* is a key gene for organic phosphorus mineralization in soil and water bodies, and its relative abundance is strongly induced in high biological activity and potential phosphorus limitation in summer [35]. In contrast, phosphonates are a more stubborn component

 https://doi.org/10.3390/w18040456

of the DOP pool, and the cleavage of the C-P bond requires a complex enzyme system encoded by the *phn* operon. The significant increase in the relative abundance of certain genes (such as *phnN* and *phnO*) in autumn may indicate that the microbial community is taking advantage of the seasonal window period for utilizing phosphonic organic phosphorus, or that the abundance of microorganisms possessing these genes has significantly increased. This is consistent with the finding of Tan et al. [36], that the degradation of phosphonates is more active under specific environmental conditions (such as specific carbon source availability or redox state), and these conditions may be more favorable in autumn. Between summer and autumn, the genes related to inorganic phosphorus dissolution, *pqqC* (encoding pyrroloquinoline quinone synthase), show significantly higher relative abundance in summer ($p < 0.01$). *pqqC* is a key gene for phosphate dissolution bacteria (PSB) to produce and secrete the *pqq* cofactor through the "glucose acid pathway", which is the core mechanism for microorganisms to dissolve and fixate fixed inorganic phosphorus (such as iron, aluminum, calcium-bound phosphorus) [37]. After conducting a differential analysis of the phosphorus cycling functional genes among the river sections, it was found that there were no significant differences among the other three river sections (LR, MR, UR), while the microbial communities of these three sections showed significant differences in phosphorus metabolic strategies compared to the YJ river section (Table S3). The microbial communities of the other river sections exhibited comprehensive phosphorus stress adaptation characteristics: genes encoding alkylphosphate transport and degradation such as the relative abundance of *phnCEFGH* has generally increased significantly. (such as *phnX* with an FC of 3.52, $p < 0.001$), while the relative abundance of the supposed negative regulatory gene *phnS* was significantly reduced. (FC = 0.10, $p < 0.001$), which suggested that when faced with an inorganic phosphorus deficiency, microorganisms can significantly activate the mineralization potential of organic phosphorus or that the abundance of the microbial functional groups related to organic phosphorus mineralization has significantly increased [38,39]. At the same time, genes' relative abundances involved in the classical phosphorus stress response were also simultaneously increased., including genes encoding alkaline phosphatase such as *phoX* and *phoD* ($p < 0.01$), regulators *phoR* ($p < 0.01$), and polyphosphate decomposition genes *ppx* ($p < 0.01$), which further confirmed the strong phosphorus limitation pressure in the environment. In contrast, the relative abundance of the above genes in the YJ river section's microorganisms were generally lower, suggesting that the phosphorus environment in YJ was relatively abundant. The results of this study indicate that the microbial communities of the other river sections play a more active role as "organic phosphorus activators" in the river phosphorus cycle.

Our results indicate that summer may be the most active period for PSB. This is likely a direct response to the increased phosphorus demand caused by high primary productivity in summer. Similar patterns have also been observed in lake ecosystems, where the relative abundance of *pqqC* gene is highest in summer, consistent with the increase in phosphorus release flux, highlighting the key role of microbial-mediated inorganic phosphorus dissolution in alleviating seasonal phosphorus limitation [40]. The *pstB* gene in the pst system (high-affinity phosphate-specific transporter system) shows higher relative abundance in summer ($p < 0.001$), while the regulatory gene *phoB* ($p < 0.001$) and the glycerol-3-phosphate transporter system gene *ugpQ* ($p < 0.001$) are more active in autumn. The strong induction of the *pst* system in low phosphorus environments is to efficiently capture trace phosphates in the environment [41]. The higher relative abundance of *pstB* in summer is a strong molecular evidence of seasonal phosphorus limitation in the water body, which is consistent with the phenomenon of a sharp decrease in dissolved phosphorus due to algal blooms in summer in many eutrophic water bodies [42]. *phoB*, as the regulatory gene

 https://doi.org/10.3390/w18040456

of the *PhoBR* two-component system, has a complex abundance variation pattern. Its higher relative abundance in autumn may indicate the persistence of phosphorus stress, but the form of stress or signal transduction pathway has changed seasonally. Another possibility is that *phoB*, in addition to regulating inorganic phosphorus metabolism, also participates in the regulation of the utilization of organic phosphorus substrates (such as phosphonates), which is in line with the active phase of the *phn* genes in autumn [43]. The significant increase in the relative abundance of *ugpQ* further indicates that the microorganisms in autumn may more frequently utilize organic phosphorus compounds such as glycerol-3-phosphate as a phosphorus source, demonstrating the flexibility of microorganisms in their phosphorus acquisition strategies. In summer, the relative abundance of the polyphosphate exonuclease gene *ppx* was significantly higher ($p < 0.05$). Polyphosphate (polyphosphate) serves as a reserve of phosphorus and energy in microorganisms. The activity of *ppx* is responsible for decomposing polyphosphate to release inorganic phosphorus, and its increase usually indicates that the cells are using their internal reserves to cope with the shortage of environmental phosphorus [44]. The high relative abundance of *ppx* in summer, combined with the high relative abundance of the *pst* system, constitutes a "combined strategy" for microorganisms to cope with acute phosphorus stress: on one hand, they actively absorb from the outside, and on the other hand, they decompose internal reserves. This strategy has been confirmed in planktonic bacteria and is an important mechanism for responding to pulsed nutrient supply [45]. It should be noted that these inferences are based on relative abundance data; at the same time, concurrent changes in the total biomass of microorganisms may also affect these proportional changes.

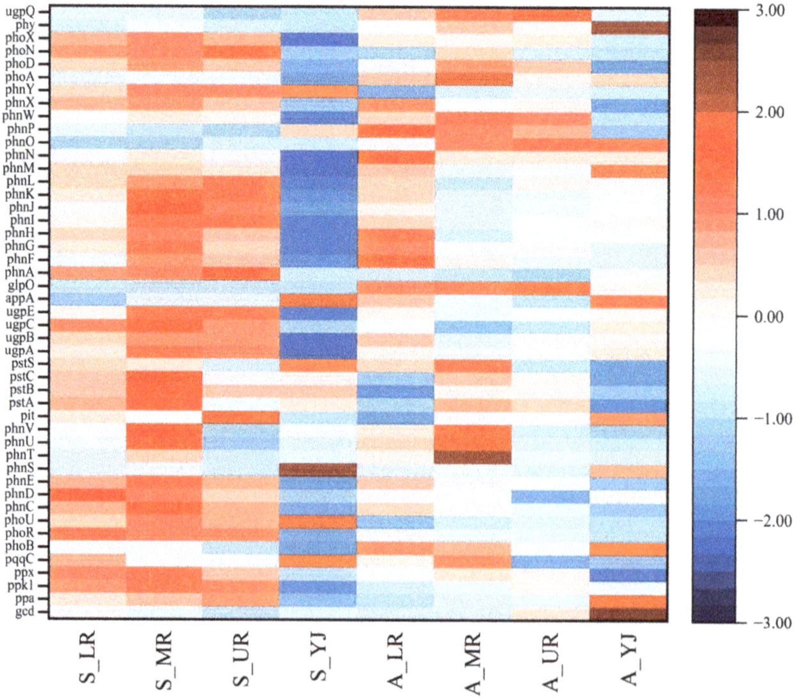

Figure 1. Heatmap of functional genes related to the phosphorus cycle. S and A, respectively, represent summer and autumn, while LR, MR, UR and YJ are the four sections of the river.

https://doi.org/10.3390/w18040456

3.3. The Key Microbial Groups, Ecological Niches and Diversity That Drive the Phosphorus Cycle

Based on the detected functional genes of the phosphorus cycle that showed significant seasonal differences, this study analyzed the composition of the associated microbial communities. It was found that the Proteobacteria (Pseudomonadota) occupied an absolute advantage in the majority of samples, with relative abundances exceeding 60% at multiple sampling points in summer and autumn (Figure 2a), which was consistent with previous research results on phosphorus cycling microorganisms in freshwater ecosystems [46].

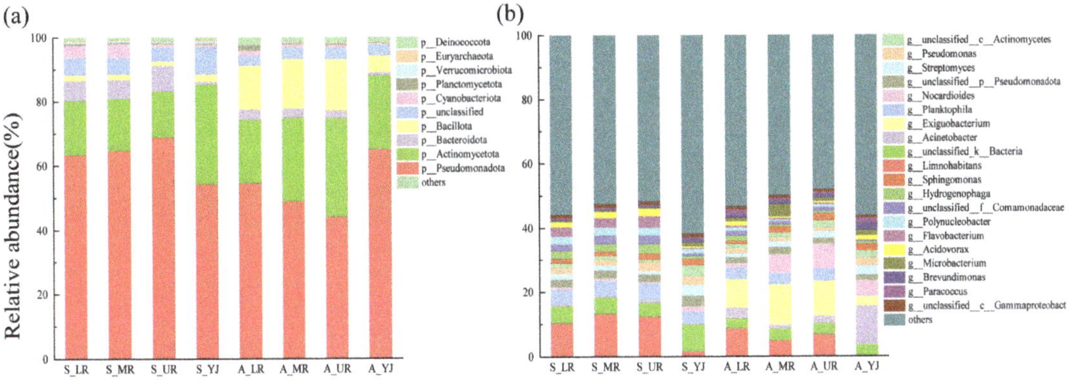

Figure 2. Composition of microbila communities related to phosphorus cycle, phylum level top 10 (**a**), genus level top 20 (**b**).

The Proteobacteria phylum contains various genera with phosphate dissolution capabilities, such as *Pseudomonas* and *Acinetobacter*, which have high abundances at multiple sites (Figure 2b), especially at MR2 and UR3 sites. These microorganisms enhance the availability of soluble phosphorus in water bodies by secreting phosphatases or producing organic acids to dissolve inorganic phosphorus [47]. It is notable that the *Hydrogenophaga* and *Acidovorax* genera within the Proteobacteria phylum were also detected at higher abundances at multiple sites, and these genus known to have various phosphorus metabolism-related genes, including the alkaline phosphatase gene *phoD* and the phosphate-specific transporter system gene *pstS*, which can effectively utilize organic and inorganic phosphorus in the environment [14]. At the genus level, typical freshwater bacteria such as *Limnohabitans* and *Planktophila* were significantly enriched at multiple sites, especially in LR1. These genera are known to participate in the mineralization process of organic phosphorus and can hydrolyze phosphate ester compounds to release bioavailable inorganic phosphorus [48]. As a typical freshwater bacterium, *Limnohabitans* has been proven to have abundant phosphatase encoding genes, capable of effectively decomposing soluble organic phosphorus [49]. *Planktophila,* as an important member of the Actinomycetota phylum, has its genome containing multiple genes encoding alkaline phosphatase, and these genes' abundance are significantly increased under phosphorus limitation conditions. Moreover, the Actinomycetota phylum (Actinomycetota) also had a relatively high proportion in some samples, with the representative genera *Streptomyces* and *Nocardioides* having strong phosphatase activity, especially under phosphorus limitation conditions. These genus not only produce various phosphatases but also secrete iron carriers to promote the dissolution of insoluble phosphates, playing an important role in the phosphorus biogeochemical cycle [50]. It is worth noting that in some sites, the Cyanobacteriota phylum showed a lower abundance in summer but increased in autumn (Figure 3a), which may be related to

https://doi.org/10.3390/w18040456

seasonal changes in phosphorus concentration. Cyanobacteria usually dominate during water body eutrophication, but their phosphorus competition ability is weak, and they may rely on interactions with other phosphorus cycling microorganisms [51]. Studies have shown that there are complex interactions between cyanobacteria and phosphorus cycling microorganisms, such as certain cyanobacterial strains being able to release organic phosphorus compounds, which can be further decomposed and utilized by the associated Proteobacteria phylum bacteria [52]. Additionally, in MR1 and UR1, *Exiguobacterium* from the Bacillota phylum significantly increased in the autumn, and this genus has potential functions in alkaline phosphatase activity, which may be activated under phosphorus starvation conditions [53]. Previous studies have found that *Exiguobacterium* not only has alkaline phosphatase genes but also contains multiple genes encoding phosphate transport proteins, enabling it to effectively obtain phosphorus sources in low-phosphorus environments [54]. From the perspective of ecological niche differentiation, different microbial groups may play different roles in the phosphorus cycle. For instance, during autumn, the genera *Brevundimonas* and *Paracoccus* in the Proteobacteria phylum showed higher abundances in UR1 and YJ3. These genus-level groups are known to possess multiple phosphorus metabolic pathways, including phosphate absorption, storage, and reuse [55]. Additionally, during summer, the higher abundance of the *Flavobacterium* genus in LR3 and UR1 is also noteworthy. This genus of bacteria can produce various extracellular enzymes, including phosphatases, participating in the decomposition of organic phosphorus [45]. This functional diversity enables the microbial community to respond more effectively to changes in the availability of phosphorus in the environment and maintain the balance of the phosphorus cycle in the ecosystem. From summer to autumn, the relative abundance of *Polynucleobacter* decreased at most sampling sites, while the relative abundances of *Acinetobacter* and *Exiguobacterium* increased. This community structure variation may be closely related to seasonal changes in water body temperature, dissolved oxygen, and nutrient salt concentrations [40].

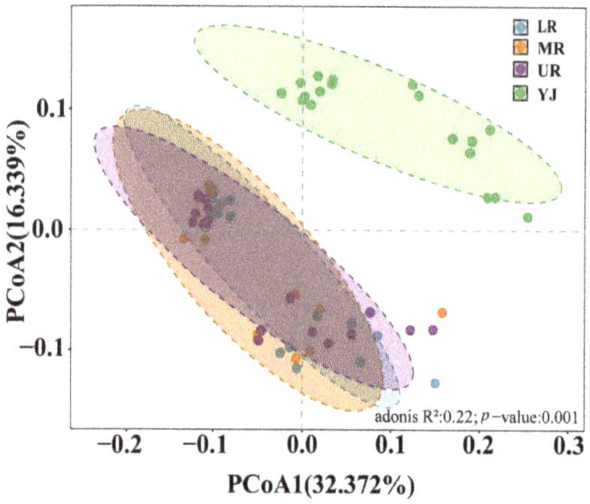

(a)

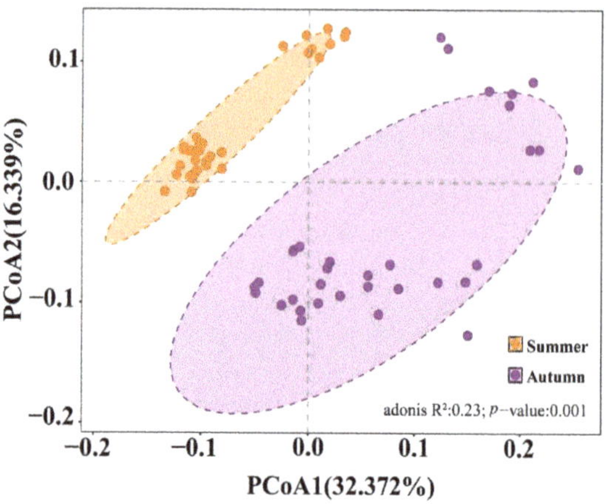

(b)

Figure 3. Principal coordinate analysis (PCoA) of microbial communities in different reaches (**a**) and seasons (**b**) of the Chishui River.

The α-diversity indices in summer and autumn showed differences in some indicators (Table S4), and the Shannon index exhibited significant seasonal differences ($p < 0.05$),

 https://doi.org/10.3390/w18040456

indicating that the species diversity of the microbial community changes significantly with the seasons, which may affect the structure of the functional groups of the phosphorus cycle; the Pielou evenness also showed a marginal significant trend ($p = 0.056$). Other indices such as Simpson, Richness, Chao, and ACE-Sobs did not show significant differences between seasons. The PCoA analysis (Figure 3a,b) showed that the microbial community structure at the genus level differed significantly among different samples (adonis $R^2 = 0.19$, $p = 0.001$), and the main coordinates explained 45.31% of the variation (PCoA1: 31.572%, PCoA2: 13.738%), indicating that seasonal or environmental changes have a significant impact on the overall microbial composition. The RDA (Figure S3a,b) further revealed the driving effect of environmental factors on the microbial community. At the seasonal scale, RDA1 and RDA2 together explained 60.5% of the variation, where TP (total phosphorus) was closely related to seasonal changes, indicating that phosphorus content is a key factor regulating the seasonal differentiation of the microbial community. At the river section scale, the community distribution of different river sections (LR, MR, UR, YJ) was also closely associated with phosphorus-related environmental variables such as TP, reflecting the importance of the phosphorus cycle in regulating the spatial pattern of water body microorganisms.

3.4. The Regulatory Mechanism of Environmental Factors on the Composition of Phosphorus Cycling Functional Genes

Finally, this study revealed a complex interaction network between environmental factors and phosphorus cycling functional genes through correlation analysis (Figure 4). TP were significantly positively correlated with key functional genes, such as *phnD* encoding phosphonate transporter ($p < 0.05$) and *phnA*, a key component of carbon–phosphorus cleavage enzyme (TP: $p < 0.05$), indicating that the abundance of phosphorus in the environment directly regulates the acquisition and degradation ability of microorganisms for organic phosphorus (especially phosphonates) [12]. It is noteworthy that there is a significant correlation between iron (Fe) and SRP as well as multiple functional genes (such as *phnA*: $p < 0.01$), suggesting that iron oxides may indirectly regulate the phosphorus availability in the surrounding microenvironment through adsorption–desorption processes, thereby driving the abundance and distribution of microbial phosphorus metabolic genes [56,57]. Dissolved oxygen (DO) was significantly negatively correlated with multiple genes, such as *phnA* ($p < 0.01$) and *phnY* ($p < 0.01$), indicating that the anaerobic environment with low oxygen may preferentially activate alternative phosphorus acquisition strategies such as phosphonate utilization, and at the same time, low oxygen conditions are more conducive to the reductive release of iron-bound phosphorus in sediments, providing phosphorus sources for microorganisms [58]. The pH value shows a certain selective influence. Among them, the alkaline phosphatase gene *phnY* has an extremely significant correlation with pH ($p < 0.01$), while the gene regulated by *phoB* has a significantly negative correlation with pH ($p < 0.01$). This may indicate that different microbial populations have different adaptability to pH. The phosphatases secreted by them can only exert maximum activity at the optimal pH value, or that the biological actions of algae and other organisms have changed the pH value, and the regulation of genes by pH is mapped onto the pH value [59]. There was a significant positive correlation between dissolved organic carbon (DOC) and the *phy* gene (phosphoric acidase gene) ($p < 0.05$), indicating that microorganisms capable of producing phosphoric acid enzymes have a close relationship with the activity and abundance of organic matter. Phytic acid is itself an organic carbon–phosphorus compound, and a high-DOC environment may mean abundant organic matter (such as plant residues), providing sufficient substrates and energy for these microorganisms. At the same time, a

https://doi.org/10.3390/w18040456

significant positive correlation between Ca and DOC and phy ($p < 0.05$), as well as a strong correlation between Ca and Mn ($p < 0.01$), suggested that metal–organic complexes may affect the biological availability of phosphorus. The activity of many phosphoric acidases requires metal ions as cofactors to exert maximum efficacy.

These ions may directly participate in the catalytic process of the enzymes. Additionally, the process of phosphoric acidase hydrolyzing phytic acid itself releases these metal ions that are chelated by phytic acid, thus forming a symbiotic relationship in the microenvironment. These findings collectively reveal that the physical and chemical conditions of the water body are the key conditions shaping the microbial phosphorus cycling functional community. The synergistic relationship between functional genes further reveals the microbial regulatory mechanism of phosphorus cycling. The significant positive correlation between the high-affinity phosphate transporter system gene *pstB* and the regulatory protein gene *phoU* ($p < 0.01$) verified their core synergistic regulatory role in the response to phosphorus starvation [41]. While the strong positive correlation between the gene encoding glycerophosphate diphosphate phosphatase *glpO* and the gene possibly involved in the synthesis of phosphonate precursors *phnO* ($p < 0.01$) suggested that microorganisms may have metabolic pathways crossing and synergy when degrading different forms of organic phosphorus (such as phosphonates and phosphonates), this coupling may enhance the overall efficiency of organic phosphorus mineralization. *Phy* is highly significantly positively correlated with *phoB* ($p < 0.01$). *PhoB* is the core response regulatory factor of the bacterial *Pho* regulatory system, mainly responsible for sensing the inorganic phosphorus starvation stress in the environment. The positive correlation between *phy* and *phoB* indicates that the gene abundance of phosphatase is likely to be regulated by the *Pho* system. When easily absorbable inorganic phosphorus (such as SRP) is scarce in the environment, bacteria will activate the *Pho* system and increase the abundance of various phosphorus acquisition genes, including phosphatase, to obtain phosphorus resources from complex organic phosphorus (such as phytic acid). *phnC* is highly significantly positively correlated with *phnD* ($p < 0.01$). These two genes belong to the phosphonate operon of microorganisms. The proteins encoded by them are two indispensable components of the ABC transporter system, jointly responsible for capturing environmental phosphonate molecules and transporting them into the cell. This indicates that phosphonate organic phosphorus in the environment may account for a large proportion, thus exerting strong natural selection pressure on microorganisms carrying this gene cluster. The abundance of *phnX* and *phoX* genes was significantly negatively correlated with soluble reactive phosphorus (SRP). This suggests that in low-phosphorus environments, their expression levels would increase to enhance the phosphorus uptake capacity of microorganisms, or that microorganisms containing these genes would have a greater advantage in such environments (Figure S4). This is consistent with the induced expression mechanism of the alkaline phosphatase gene *phoX* under phosphorus restriction conditions [60]. *phoR* is negatively correlated with dissolved oxygen (DO), indicating that it may be involved in phosphorus signal transduction and metabolic regulation in a low-oxygen deposition environment. In addition, *phnC* shows an extremely significant positive correlation with most phn genes, *ugpB* and *phoX*, indicating that it may serve as the core hub of the phosphorus metabolism regulatory network, coordinating the collaboration of multiple functional genes in response to environmental phosphorus changes. The significant association of this gene module with environmental factors such as DOC and salinity further reveals that in low-organic-carbon, low-salinity freshwater habitats, microorganisms adapt to the ecological strategy of co-limiting phosphorus and carbon by synergistically regulating phosphorus acquisition and organic phosphorus mineralization processes [61].

https://doi.org/10.3390/w18040456

Through the analysis of the correlation between important environmental parameters and phosphorus cycling functional genes, this study reveals the microbial regulatory mechanism of phosphorus cycling in rivers. The form of phosphorus and its bioavailability may control the expression of corresponding phosphorus acquisition genes in microorganisms or the changes in microbial communities. Moreover, the interactions among microbial genes also reflect the evolution of their strategies for efficiently obtaining phosphorus under environmental phosphorus limitation conditions, as well as the cooperation among different functional microorganisms. The interactions among these genes collectively constitute the complex phosphorus cycling microbial metabolic network in the river ecosystem, and their changes directly affect the geochemical cycling process of phosphorus in water bodies.

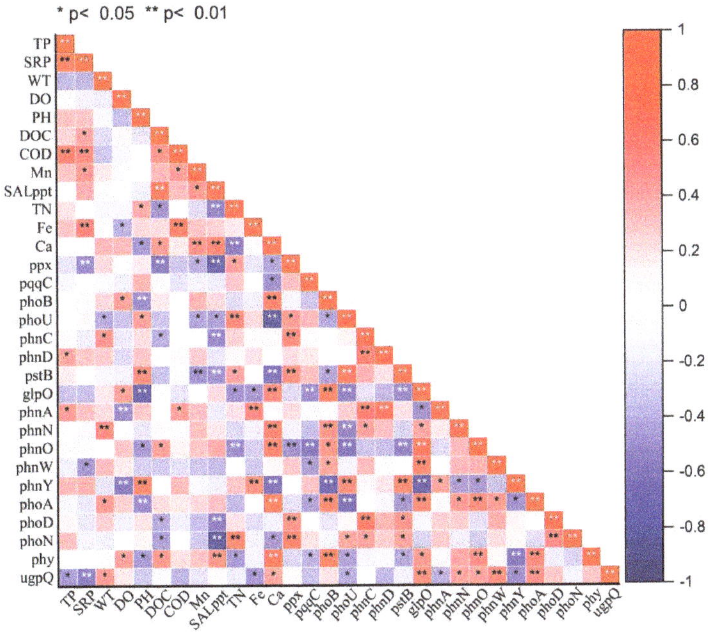

Figure 4. Spearman correlation analysis between phosphorus genes and environmental factors in the Maotai reach of the Chishui River (Genes with significant differences in phosphorus cycling functions across seasons). * indicates $0.01 < p < 0.05$, ** indicate $p < 0.01$, $-0.4 < R < 0.4$ and $p < 0.05$ represents significant correlation.

4. Conclusions

This study systematically revealed the temporal and spatial dynamics of the phosphorus cycle in the Chishui River Basin and the microbial driving mechanisms. The results shows that the environmental parameters of the Chishui River water body exhibited significant temporal and spatial variations. During the summer, affected by rainfall runoff, external inputs led to a significant increase in nutrient salt concentrations such as total phosphorus and total nitrogen; while in the autumn, DO and Mn concentrations significantly increased, and combined with their positive correlation with SRP, it may reflect the contribution of the release of substances within the sediment to the phosphorus content in the water body. At the microbial regulation level, the abundance of phosphorus cycling functional genes shows significant seasonal variations. In summer, due to po-

https://doi.org/10.3390/w18040456

tential phosphorus limitation, the abundance of genes such as high-affinity phosphate transporter system (*pst*), enhancing inorganic phosphate dissolution (*pqqC*), and decomposing internal polyphosphate (*ppx*) increases significantly; while in autumn, phosphonate (*phn* gene cluster) and other stubborn organic phosphate reservoir-related genes increase significantly, possibly reflecting their flexible metabolic strategies in response to changes in phosphorus source forms. The microbial community analysis indicates that these functions are jointly undertaken by various types of microorganisms (such as proteobacteria and actinobacteria), and the seasonal changes in community structure are coupled with the patterns of these functional genes. Furthermore, the relevant analysis revealed a complex environmental regulatory network, in which factors such as iron, dissolved oxygen, pH value, and soluble organic carbon may jointly shape the distribution pattern of phosphorus cycling functional genes by influencing the biological availability of phosphorus, acting as cofactors for enzymes, or regulating the growth of algae. Additionally, the crucial interrelationships between genes highlight the efficient metabolic networks that microorganisms have evolved under phosphorus limitation stress, as well as the collaborative relationships among different microbial functional groups. In summary, the phosphorus cycling in the Qingshui River is regulated by external inputs, endogenous releases, and microbial metabolic processes. The microbial community, as the core "transformer", the dynamic changes in its functional genes are crucial for regulating the phosphorus biogeochemical cycle in the water body.

Supplementary Materials: The following supporting information can be downloaded at: https://www.mdpi.com/article/10.3390/w18040456/s1, Table S1. The changes in water quality parameters of the Maotai section of the Chishui River. Table S2. The analysis results of the differences in phosphorus cycling functional genes between summer and autumn (only listing the genes with significant differences). Table S3. Analysis results of the differences in phosphorus cycling functional genes between other river reaches and YJ (only listing the genes with significant differences). Table S4. T-test for alpha diversity index between Summer and Autumn. Figure S1. Distribution map of sampling points in the study area. Figure S2. Characteristics of changes in different physical and chemical indicators in the Maotai reach of the Chishui River. Figure S3. Redundancy analysis (RDA) of microbial community structure and environmental factors in different seasons (a) and reaches (b) of the Chishui River. Figure S4. Spearman correlation analysis between phosphorus genes and environmental factors in the Maotai reach of the Chishui River (Genes with significant differences in phosphorus cycling functions across seasons and reaches).

Author Contributions: Conceptualization: Q.L. and J.W. (Jingfu Wang); Methodology: J.W. (Jingfu Wang); Formal analysis and investigation: J.W. (Junhai Wu), Y.X., Q.W., Y.H. and Y.T.; Writing—original draft preparation: J.W. (Junhai Wu); Writing—review and editing: J.W. (Junhai Wu), Q.L., J.W. (Jingfu Wang) and Q.W.; Funding acquisition: Q.L. and J.W. (Jingfu Wang); Supervision: Q.L. and Q.W. All authors have read and agreed to the published version of the manuscript.

Funding: This study was sponsored jointly by the Science and technology Project of Kweichow Moutai Liquor Co., Ltd. (No. MTGF2022011, GFJS20231990), the Guizhou Provincial Major Scientific and Technological Program (No. [2024]006, [2024]009, [2024]013), the National Key R&D Plan of China (No. 2023YFF0806000), the Chinese NSF Project (No. 42277253, 42573057), the Guizhou Provincial Science and Technology Program (No. Qiankehe Platform Talents-YQK [2023]034, Qiankehe Platform-YWZ[2023]006), the Innovation and Development Fund of Science and Technology of Institute of Geochemistry, CAS (2025-02), and the Outstanding member of Youth Innovation Promotion Association CAS (No. Y2023105). The funder was not involved in the study design, collection, analysis, interpretation of data, the writing of this article, or the decision to submit it for publication.

Data Availability Statement: The raw data supporting the conclusions of this article will be made available by the authors on request.

Conflicts of Interest: The authors declare that they have no known competing financial interests or personal relationships that could have appeared to influence the work reported in this paper. The funder was not involved in the study design, collection, analysis, interpretation of data, the writing of this article, or the decision to submit it for publication.

References

1. Wu, Z.; Li, J.; Sun, Y.; Peñuelas, J.; Huang, J.; Sardans, J.; Jiang, Q.; Finlay, C.J.; Britten, L.G.; Follows, J.M.; et al. Imbalance of global nutrient cycles exacerbated by the greater retention of phosphorus over nitrogen in lakes. *Nat. Geosci.* **2022**, *15*, 464–468. [CrossRef]
2. Carpenter, S.R. Phosphorus control is critical to mitigating eutrophication. *Proc. Natl. Acad. Sci. USA* **2008**, *105*, 11039–11040. [CrossRef] [PubMed]
3. Geng, M.; Zhang, W.; Hu, T.; Wang, R.; Cheng, X.; Wang, J. Eutrophication causes microbial community homogenization via modulating generalist species. *Water Res.* **2022**, *210*, 118003. [CrossRef] [PubMed]
4. Chen, G.L.; Qian, C.; Du, M.; Tong, M.J.; Chen, J.J.; Yu, H.Q. Quantifying assembly processes of dissolved organic matter pools in eutrophication using high-resolution mass spectrometry and ecological models. *Water Res.* **2025**, *282*, 123781. [CrossRef] [PubMed]
5. Vincent, S.G.T.; Jennerjahn, T.C.; Ramasamy, K. *Microbial Communities in Coastal Sediments: Structure and Functions*; Elsevier: Amsterdam, The Netherlands, 2021.
6. Lin, L.; Xiong, J.; Yue, T.C.; Xu, W.; Liu, L.; Wang, F.; Yang, S.; Cao, W. Phosphorus starvation response genes and function coupling: A mechanism to regulate phosphorus availability in a subtropical estuary. *Sci. Total Environ.* **2024**, *928*, 172575. [CrossRef]
7. Letscher, R.T.; Moore, J.K.; Teng, Y.C.; Primeau, F. Variable C: N: P stoichiometry of dissolved organic matter cycling in the Community Earth System Model. *Biogeosciences* **2015**, *12*, 209–221. [CrossRef]
8. Letscher, R.T.; Moore, J.K. Preferential remineralization of dissolved organic phosphorus and non-Redfield DOM dynamics in the global ocean: Impacts on marine productivity, nitrogen fixation, and carbon export. *Glob. Biogeochem. Cycles* **2015**, *29*, 325–340. [CrossRef]
9. Tu, C.; Jin, Z.; Che, F.; Cao, X.; Song, X.; Lu, C.; Huang, W. Characterization of phosphorus sorption and microbial community in lake sediments during overwinter and recruitment periods of cyanobacteria. *Chemosphere* **2022**, *307*, 135777. [CrossRef]
10. Yoshimura, T.; Nishioka, J.; Ogawa, H.; Kuma, K.; Saito, H.; Tsuda, A. Dissolved organic phosphorus production and decomposition during open ocean diatom blooms in the subarctic Pacific. *Mar. Chem.* **2014**, *165*, 46–54. [CrossRef]
11. Wang, P.; Xing, X.; Wang, X.; Hu, B. Spatial distribution and coupling of carbon, nitrogen, phosphorus, and sulfur functional genes in a highly dam-regulated international river. *Environ. Res.* **2025**, *287*, 122988. [CrossRef]
12. Bergkemper, F.; Schöler, A.; Engel, M.; Lang, F.; Krüger, J.; Schloter, M.; Schulz, S. Phosphorus depletion in forest soils shapes bacterial communities towards phosphorus recycling systems. *Environ. Microbiol.* **2016**, *18*, 1988–2000. [CrossRef] [PubMed]
13. Liang, J.L.; Liu, J.; Jia, P.; Yang, T.T.; Zeng, Q.W.; Zhang, S.C.; Liao, B.; Shu, W.S.; Li, J.T. Novel phosphate-solubilizing bacteria enhance soil phosphorus cycling following ecological restoration of land degraded by mining. *ISME J.* **2020**, *14*, 1600–1613. [CrossRef] [PubMed]
14. Siles, J.A.; Starke, R.; Martinovic, T.; Fernandes, M.L.P.; Orgiazzi, A.; Bastida, F. Distribution of phosphorus cycling genes across land uses and microbial taxonomic groups based on metagenome and genome mining. *Soil Biol. Biochem.* **2022**, *174*, 108826. [CrossRef]
15. Ragot, S.A.; Kertesz, M.A.; Meszaros, E.; Frossard, E.; Bünemann, K.E. Soil phoD and phoX alkaline phosphatase gene diversity responds to multiple environmental factors. *FEMS Microbiol. Ecol.* **2017**, *93*, fiw212. [CrossRef]
16. Wang, J.; Chen, J.; Ding, S.; Luo, J.; Xu, Y. Effects of temperature on phosphorus release in sediments of Hongfeng Lake, southwest China: An experimental study using diffusive gradients in thin-films (DGT) technique. *Environ. Earth Sci.* **2015**, *74*, 5885–5894. [CrossRef]
17. Feng, W.; Yang, F.; Zhang, C.; Liu, J.; Song, F.; Chen, F.; Zhu, Y.; Liu, S.; Giesy, P.J. Composition characterization and biotransformation of dissolved, particulate and algae organic phosphorus in eutrophic lakes. *Environ. Pollut.* **2020**, *265*, 114838. [CrossRef]
18. Li, W.; Wang, J.; Jiang, L.; Lv, G.; Hu, D.; Wu, D.; Yang, X. Rhizosphere effect and water constraint jointly determined the roles of microorganism in soil phosphorus cycling in arid desert regions. *Catena* **2023**, *222*, 106809. [CrossRef]

 https://doi.org/10.3390/w18040456

19. Han, Q.; Mu, X.; Gan, X.; Wang, S.; Yu, Q.; Li, H. Temperature rise alters phosphorus pool in corpse polluted water by inhibiting organic phosphorus mineralization and phosphorus transports. *J. Clean. Prod.* **2024**, *434*, 140161. [CrossRef]
20. Bolger, A.M.; Lohse, M.; Usadel, B. Trimmomatic: A flexible trimmer for Illumina sequence data. *Bioinformatics* **2014**, *30*, 2114–2120. [CrossRef]
21. Li, D.; Liu, C.M.; Luo, R.; Sadakane, K.; Lam, T.W. MEGAHIT: An ultra-fast single-node solution for large and complex metagenomics assembly via succinct de Bruijn graph. *Bioinformatics* **2015**, *31*, 1674–1676. [CrossRef]
22. Hyatt, D.; Chen, G.L.; LoCascio, P.F.; Land, M.L.; Larimer, F.W.; Hauser, L.J. Prodigal: Prokaryotic gene recognition and translation initiation site identification. *BMC Bioinform.* **2010**, *11*, 119. [CrossRef]
23. Steinegger, M.; Söding, J. MMseqs2 enables sensitive protein sequence searching for the analysis of massive data sets. *Nat. Biotechnol.* **2017**, *35*, 1026–1028. [CrossRef] [PubMed]
24. Buchfink, B.; Xie, C.; Huson, D.H. Fast and sensitive protein alignment using DIAMOND. *Nat. Methods* **2015**, *12*, 59–60. [CrossRef] [PubMed]
25. Wickham, H.; Averick, M.; Bryan, J.; Chang, W. Welcome to the Tidyverse. *J. Open Source Softw.* **2019**, *4*, 1686. [CrossRef]
26. Kassambara, A. rstatix: Pipe-Friendly Framework for Basic Statistical Tests. 2019. Available online: https://cran.r-project.org/web/packages/rstatix/index.html (accessed on 20 September 2025).
27. Oksanen, J.; Blanchet, F.G.; Kindt, R.; Legendre, P. vegan: Community Ecology Package. 2013. Available online: https://cran.r-project.org/web/packages/vegan/index.html (accessed on 21 September 2025).
28. R Core Team. *R: A Language and Environment for Statistical Computing*; R Foundation for Statistical Computing: Vienna, Austria, 2021.
29. Wen, T.; Xie, P.; Yang, S.; Niu, G.; Liu, X.; Ding, Z.; Xue, C.; Liu, Y.X.; Sheng, Q.; Yuan, J. ggClusterNet: An R package for microbiome network analysis and modularity-based multiple network layouts. *Imeta* **2022**, *1*, e32. [CrossRef]
30. Smith, T.J.; Stevenson, K.J. Origin 7.5 OriginLab Corporation, One Roundhouse Plaza, Northampton, MA 01060. 1-800-969-7720. www.OriginLab.com. Suggested price $699.00 (retail, single user), $489.00 (educational, single user). Contact company for other pricing options. *J. Am. Chem. Soc.* **2004**, *126*, 6834. [CrossRef]
31. Wang, Y.; Xu, H.; Zhao, X.; Kang, L.; Qiu, Y.; Paerl, H.; Zhu, G.; Li, H.; Zhu, M.; Qin, B.; et al. Rainfall impacts on nonpoint nitrogen and phosphorus dynamics in an agricultural river in subtropical montane reservoir region of southeast China. *J. Environ. Sci.* **2025**, *149*, 551–563. [CrossRef]
32. Diaz, R.J.; Rosenberg, R. Spreading dead zones and consequences for marine ecosystems. *Science* **2008**, *321*, 926–929. [CrossRef]
33. Anderson, L.G.; Björk, G.; Jutterström, S.; Pipko, I.; Shakhova, N.; Semiletov, I.; Wåhlström, I. East Siberian Sea, an Arctic region of very high biogeochemical activity. *Biogeosciences* **2011**, *8*, 1745–1754. [CrossRef]
34. Carpenter, S.R.; Caraco, N.F.; Correll, D.L.; Howarth, R.W.; Sharpley, A.N.; Smith, V.H. Nonpoint pollution of surface waters with phosphorus and nitrogen. *Ecol. Appl.* **1998**, *8*, 559–568. [CrossRef]
35. Wang, S.; Yan, J.; Heal, K.V.; Li, H.; Yu, Y.; Zhou, C. Rhizosphere microbial roles in phosphorus cycling during successive plantings of Chinese fir plantations. *For. Ecol. Manag.* **2024**, *570*, 122227. [CrossRef]
36. Tan, H.; Barret, M.; Mooij, M.J.; Rice, O.; Morrissey, J.P.; Dobson, A.; Griffiths, B.; O'Gara, F. Long-term phosphorus fertilisation increased the diversity of the total bacterial community and the phoD phosphorus mineraliser group in pasture soils. *Biol. Fertil. Soils* **2013**, *49*, 661–672. [CrossRef]
37. Bi, Q.F.; Li, K.J.; Zheng, B.X.; Liu, X.P.; Li, H.Z.; Jin, B.J.; Ding, K.; Yang, X.R.; Lin, X.Y.; Zhu, Y.G. Partial replacement of inorganic phosphorus (P) by organic manure reshapes phosphate mobilizing bacterial community and promotes P bioavailability in a paddy soil. *Sci. Total Environ.* **2020**, *703*, 134977. [CrossRef] [PubMed]
38. Metcalf, W.W.; Wanner, B.L. Involvement of the Escherichia coli phn (psiD) gene cluster in assimilation of phosphorus in the form of phosphonates, phosphite, Pi esters, and Pi. *J. Bacteriol.* **1991**, *173*, 587–600. [CrossRef]
39. Hu, Y.; Xia, Y.; Sun, Q.I.; Liu, K.; Chen, X.; Ge, T.; Zhu, B.; Zhu, Z.; Zhang, Z.; Su, Y. Effects of long-term fertilization on phoD-harboring bacterial community in Karst soils. *Sci. Total Environ.* **2018**, *628*, 53–63. [CrossRef]
40. Sun, T.; Zhu, L.; Huang, T.; Tao, P.; Bao, Y.; Wang, B.; Sun, Q.; Chen, K. Seasonal distribution patterns of P-cycling-related microbes and its association with internal phosphorus release in the eutrophic Lake Chaohu, China. *J. Environ. Sci.* **2025**, *154*, 226–237. [CrossRef]
41. Muñoz-Martín, M.Á.; Mateo, P.; Leganés, F.; Piñas, F.F. Novel cyanobacterial bioreporters of phosphorus bioavailability based on alkaline phosphatase and phosphate transporter genes of *Anabaena* sp. PCC 7120. *Anal. Bioanal. Chem.* **2011**, *400*, 3573–3584. [CrossRef]
42. Paerl, H.W.; Otten, T.G. Harmful cyanobacterial blooms: Causes, consequences, and controls. *Microb. Ecol.* **2013**, *65*, 995–1010. [CrossRef]

 https://doi.org/10.3390/w18040456

43. VanBogelen, R.A.; Olson, E.R.; Wanner, B.L.; Neidhardt, F.C. Global analysis of proteins synthesized during phosphorus restriction in Escherichia coli. *J. Bacteriol.* **1996**, *178*, 4344–4366. [CrossRef]
44. Rao, N.N.; Gómez-García, M.R.; Kornberg, A. Inorganic polyphosphate: Essential for growth and survival. *Annu. Rev. Biochem.* **2009**, *78*, 605–647. [CrossRef]
45. Dyhrman, S.N.T.; Ammerman, J.W.; Van Mooy, B.A.S. Microbes and the marine phosphorus cycle. *Oceanography* **2007**, *20*, 110–116. [CrossRef]
46. Ren, J.; Wang, J.; Dong, Y.; Xiao, L.; Wang, L.; Ji, J.; Liu, Y. Microbial community dynamics and its relationship with biogeochemical processes under geochemical perturbations. *Water Res.* **2025**, *289*, 124889. [CrossRef]
47. Lidbury, I.D.E.A.; Scanlan, D.J.; Murphy, A.R.J.; Oleza, J.A.C.; Ferretjans, M.M.A.; Hitchcock, A.; Daniell, T.J. A widely distributed phosphate-insensitive phosphatase presents a route for rapid organophosphorus remineralization in the biosphere. *Proc. Natl. Acad. Sci. USA* **2022**, *119*, e2118122119. [CrossRef] [PubMed]
48. Ivashchenko, K.; Sushko, S.; Selezneva, A.; Ananyeva, N.; Zhuravleva, A.; Kudeyarov, V.; Makarov, M.; Blagodatsky, S. Soil microbial activity along an altitudinal gradient: Vegetation as a main driver beyond topographic and edaphic factors. *Appl. Soil Ecol.* **2021**, *168*, 104197. [CrossRef]
49. Šimek, K.; Kasalický, V.; Jezbera, J.; Jezberová, J.; Hejzlar, J.; Hahn, M.W. Broad habitat range of the phylogenetically narrow R-BT065 cluster, representing a core group of the betaproteobacterial genus *Limnohabitans*. *Appl. Environ. Microbiol.* **2010**, *76*, 631–639. [CrossRef]
50. Mitra, D.; Mondal, R.; Khoshru, B.; Senapati, A.; Radha, T.K.; Mahakua, B.; Uniyal, N.; Myo, E.M.; Boutaj, H.; Sierra, B.E.G. Actinobacteria-enhanced plant growth, nutrient acquisition, and crop protection: Advances in soil, plant, and microbial multifactorial interactions. *Pedosphere* **2022**, *32*, 149–170. [CrossRef]
51. Tanvir, R.U.; Hu, Z.; Zhang, Y.; Lu, J. Cyanobacterial community succession and associated cyanotoxin production in hypereutrophic and eutrophic freshwaters. *Environ. Pollut.* **2021**, *290*, 118056. [CrossRef]
52. Liang, J.; Yan, M.; Zhu, Z.; Lu, L.; Ding, J.; Zhou, Q.; Gao, X.; Tang, N.; Li, S.; Li, X.; et al. The role of microorganisms in phosphorus cycling at river-lake confluences: Insights from a study on microbial community dynamics. *Water Res.* **2025**, *268*, 122556. [CrossRef]
53. Cui, S.; Wang, Z.; Kong, L.; Zhang, M.; Xu, H.; Xiao, P.; Tao, Y. Dynamics of microbial communities and functional genes under the interactions between phosphorus inactivation agents and phosphorus-solubilizing bacteria in sediments. *J. Water Process Eng.* **2025**, *77*, 108395. [CrossRef]
54. Srivastava, N.; Shiburaj, S.; Khare, S.K. Pan-genomic comparison of a potential solvent-tolerant alkaline protease-producing *Exiguobacterium* sp. TBG-PICH-001 isolated from a marine habitat. *3 Biotech* **2023**, *13*, 371. [CrossRef]
55. Yang, W.; Ma, J.; Zhen, Y.; Li, W.; Yao, Z.; Feng, W. Community characteristics and functional gene response analysis of phosphorus-metabolizing bacteria in plateau saline lake sediments. *Front. Environ. Sci.* **2022**, *10*, 994104. [CrossRef]
56. Dittrich, M.; Chesnyuk, A.; Gudimov, A.; McCulloch, J.; Quazi, S.; Young, J.; Winter, J.; Stainsby, E.; Arhonditsis, G. Phosphorus retention in a mesotrophic lake under transient loading conditions: Insights from a sediment phosphorus binding form study. *Water Res.* **2013**, *47*, 1433–1447. [CrossRef] [PubMed]
57. Martins, G.; Peixoto, L.; Brito, A.G.; Nogueira, R. Phosphorus–iron interaction in sediments: Can an electrode minimize phosphorus release from sediments? *Rev. Environ. Sci. Bio/Technol.* **2014**, *13*, 265–275. [CrossRef]
58. Wu, X.; Peng, J.; Liu, P.; Bei, Q.; Rensing, C.; Li, Y.; Yuan, H.; Liesack, W.; Zhang, F.; Cui, Z. Metagenomic insights into nitrogen and phosphorus cycling at the soil aggregate scale driven by organic material amendments. *Sci. Total Environ.* **2021**, *785*, 147329. [CrossRef]
59. Ragot, S.A.; Kertesz, M.A.; Bünemann, E.K. phoD alkaline phosphatase gene diversity in soil. *Appl. Environ. Microbiol.* **2015**, *81*, 7281–7289. [CrossRef]
60. Dai, J.; Chen, D.; Gao, G.; Tang, X.; Wu, S.; Wu, X.; Zhou, J. Recovery of novel alkaline phosphatase-encoding genes (phoX) from eutrophic Lake Taihu. *Can. J. Microbiol.* **2014**, *60*, 167–171. [CrossRef]
61. Shi, J.; Ou-Yang, T.; Yang, S.; Zhao, L.; Ji, L.; Wu, Z. Transcriptomic responses to phosphorus in an invasive cyanobacterium, *Raphidiopsis raciborskii*: Implications for nutrient management. *Harmful Algae* **2022**, *111*, 102150. [CrossRef]

 https://doi.org/10.3390/w18040456

MDPI AG
Grosspeteranlage 5
4052 Basel
Switzerland
Tel.: +41 61 683 77 34

Water Editorial Office
E-mail: water@mdpi.com
www.mdpi.com/journal/water